ADVANCED SERIES ON THEORETICAL PHYSICAL SCIENCE
A Collaboration between World Scientific and Institute of
Theoretical Physics

Series Editors: Dai Yuan-Ben, Hao Bai-Lin, Su Zhao-Bin
                 (Institute of Theoretical Physics Academia Sinica)

Vol. 1: Yang-Baxter Equation and Quantum Enveloping Algebras
        (Zhong-Qi Ma)

Vol. 6: Differential Geometry for Physicists
        (Bo-Yu Hou & Bo-Yuan Hou)

Forthcoming:

Vol. 2: Geometric Methods in the Elastic Theory of Membrane in Liquid
        Crystal Phases *(Ouyang Zhong-Can, Xie Yu-Zhang and Liu Ji-Xing)*

Vol. 3: Geometric Methods in Physics *(Wu Yong-Shi)*

Vol. 5: Group Theory for Condensed Matter Physics *(Tao Rui-Bao)*

SPECIAL RELATIVITY AND ITS EXPERIMENTAL FOUNDATIONS

Advanced Series on Theoretical Physical Science | Volume **4**

SPECIAL RELATIVITY AND ITS EXPERIMENTAL FOUNDATIONS

Yuan Zhong Zhang

Institute of Theoretical Physics,
Academia Sinica, China

World Scientific
Singapore • New Jersey • London • Hong Kong

Published by

World Scientific Publishing Co. Pte. Ltd.

P O Box 128, Farrer Road, Singapore 912805

USA office: Suite 1B, 1060 Main Street, River Edge, NJ 07661

UK office: 57 Shelton Street, Covent Garden, London WC2H 9HE

Library of Congress Cataloging-in-Publication Data

Chang, Yüan-chung, 1940–
    Special relativity and its experimental foundations / Yuan Zhong Zhang.
      p.   cm. -- (Advanced series on theoretical physical science ; v. 4)
    Includes bibliographical references and index.
    ISBN 9810227493
    1. Special relativity (Physics)   2. Special relativity (Physics) -- Experiments.
    I. Title.   II. Series.
    QC173.65.C465   1997
    530.11--dc21   97-36351
                                               CIP

British Library Cataloguing-in-Publication Data
A catalogue record for this book is available from the British Library.

This book is printed on acid-free paper.

Printed in Singapore by Uto-Print

PREFACE

Einstein's theory of special relativity as the first flat space–time theory has altered our concepts of space and time, and been supported directly by a variety of experiments and its successful applications to modern physics. Einstein's theory and quantum mechanics have already been the main bases of modern physics.

The key point in Einstein's theory is the postulate concerning the constancy of the (one-way) velocity of light, which contradicts the classical (nonrelativistic) addition law of velocities. This postulate is needed only for constructing well-defined inertial frames of reference or, in other words, only for synchronizing clocks (i.e., defining simultaneity). It is not possible to test the one-way velocity of light because another independent method of clock synchronization has not yet been found.

However, some different flat space–time theories, such as Edwards', Robertson's, and Mansouri–Sexl's (MS') test theories, can be found in the literature although corresponding dynamical theories have not yet been constructed. Only a few authors paid their attention to Edwards' and Robertson's theories while many authors incorrectly claimed that comparison between MS' theory and an experiment would give a test of the one-way speed of light. For this reason, many pages of this book are used to discuss different definitions of simultaneity and their implications.

The book shows that the main differences between these test theories and Einstein's special relativity are nothing but different definitions of simultaneity. In other words, they involve different postulates concerning the velocity of light. Relationships among these flat space–time theories can be summarized as follows: (a) Edwards' and MS' theories are physically, respectively, equivalent to Einstein's and Robertson's theories (see Secs. 6.5 and 7.4 for kinematics, and see Secs. 6.6 and 7.4.6 for dynamics). This means that the one-way velocity of light is not observable because any independent method of clock synchronization other than that by use of a light signal has not yet been developed; (b) Robertson's theory is physically not equivalent to Einstein's theory, so that it may be really regarded as a test theory of special relativity; (c) MS' theory is a generalization of Robertson's theory just as Edwards' theory is a generalization of Einstein's theory. We see, therefore, that MS' theory is a trivial theory in both physics and mathematics because we have known Edwards' and Robertson's theories.

In the remainder of the book we introduce the performed experiments which are regarded as the tests of special relativity. For simplicity, most of them are compared with Einstein's theory, while only a few of them are used to compare with the test theories. Of course, one could use the experiments to yield limits on the parameters in Robertson's transformations but not on the directional parameter q in Edwards' and MS' theories.

As mentioned above, corresponding dynamical theories have not yet been constructed. However, if we assume that for all physical events the four-dimensional space–time intervals defined in Secs. 6.4, 7.2.3, and 7.4.5 is invariant, the above relation (a) would be valid too for dynamic phenomena. We can say, therefore, that only Robertson's dynamics is needed to be constructed.

Yuan Zhong ZHANG
December 1996

CONTENTS

Preface   v

Part 1.   Einstein's Theory of Special Relativity

Chapter 1.   Foundations of Space-Time Theories   **3**
1.1.   Introduction   3
1.2.   Definition of Inertial Reference Frame   3
    1.2.1.   Definition of Space Coordinates   3
    1.2.2.   Definitions of Time Coordinate   7
    1.2.3.   Definition of "Inertial" Frames   7
1.3.   Simultaneity and Clock Synchronization   8
    1.3.1.   Newtonian Absolute Simultaneity   9
    1.3.2.   Einstein's Definition of Simultaneity   10
    1.3.3.   Edwards' Definition of Simultaneity   11
    1.3.4.   Coordinate Transformation Between Einstein
        and Edwards' Frames   11
    1.3.5.   Robertson's Definition of Simultaneity   14
    1.3.6.   Mansouri-Sexl's Definition of Simultaneity   15
    1.3.7.   Coordinate Transformation Between Robertson
        and MS Frames   16
1.4.   Principle of Relativity   16
1.5.   Velocity and Simultaneity   17

Chapter 2.   Relativistic Kinematics   **21**
2.1.   Galilean Transformation   21
2.2.   Lorentz Transformation   22
2.3.   Four-Dimensional Minkowski Space-Time   27
2.4.   Einstein's Law of the Addition of Velocities   32
2.5.   Transformation of Accelerations   35
2.6.   Infinitesimal Lorentz Transformation   36
2.7.   Simultaneity and Causality   37
2.8.   Contraction of a Moving Body   39
2.9.   Time Dilation of a Moving Clock   40
2.10. Aberration and Doppler Effect   41
2.11. The Thomas Precession   45

Chapter 3.   Relativistic Mechanics   **51**
3.1.   Mass, Momentum, Force, Work and Energy   51
3.2.   Transformations of Mass, Momentum, Energy and Force   55

Chapter 4.   Electrodynamics in Media   **57**
4.1.   The Fundamental Equations   57
4.2.   Relativistic Transformations of Electromagnetic Quantities   58
4.3.   Propagation of Electromagnetic Waves in a Medium   60
4.4.   Reflection and Refraction of Electromagnetic Waves   62

Chapter 5.   The VPROCA Vector Field   **67**
5.1.   Covariant Form of Maxwell's Field Equations   67
5.2.   Proca's Vector Field Equations   70
5.3.   Dispersion in Vacuum   71

Part 2.   Test Theories of Special Relativity

Chapter 6.   Edwards' Theory   **75**
6.1.   Introduction   75
6.2.   One-Way and Two-Way Velocities of Light   76
6.3.   Edwards Transformation   78
6.4.   Anisotropic Four-Dimensional Space-Time   82
6.5.   Comparison Among Edwards' Theory and Experiments   88
       6.5.1.   Edwards' law of the Addition of Velocities   88
       6.5.2.   On Reciprocity of Relative Velocities   89
       6.5.3.   Time Dilation of a Moving Clock   90
       6.5.4.   The Römer Experiment   91
       6.5.5.   Aberration and Doppler Effect   94
       6.5.6.   Contraction of a Moving Body   98
       6.5.7.   Conclusion   99
6.6.   On Dynamics of Edwards' Theory   100

Chapter 7.   The General Test Theories   **103**
7.1.   Introduction   103
7.2.   Robertson's Test Theory   103
       7.2.1.   Robertson Transformation   103
       7.2.2.   Comparison among Robertson and Lorentz
                Transformations   105

7.2.3.	Four-Dimensional Space-Time	106
7.2.4.	Kinematical Effects	109
7.2.5.	Comparison between Robertson's Theory and Experiments	111
7.3.	Mansouri-Sexl (MS) Transformation	113
7.4.	Comparison between the MS Transformations and Experiments	117
7.4.1.	MS' Law of the Addition of Velocities	118
7.4.2.	Time Dilation	119
7.4.3.	The Römer experiment and Slow Transport of Clocks	120
7.4.4.	Transverse Doppler Effect and Mössbauer Rotor Experiment	122
7.4.5.	Anisotropic Four-Dimensional Space-Time	123
7.4.6.	Dynamics	126
7.5.	Relationships among Lorentz and Generalized Transformations	126
7.6.	Comparison of Different Conventions	128

Part 3.   Experimental Tests of Special Relativity

Chapter 8.   The Tests of Einstein's Two Postulates   **135**

8.1.	Introduction	135
8.2.	Tests of Directionality	136
8.2.1.	The Closed-Path Experiments	136
8.2.2.	The One-Way Experiments	145
8.3.	The Experiments with Moving Sources of Light	150
8.3.1.	The Astronomical Verification	151
8.3.2.	The Laboratory Evidences	154
8.4.	Summary of the Tests of the Constancy of the Velocity of Light	171
8.5.	Tests of the Principle of Special Relativity	174

Chapter 9.   The Tests of Time Dilation   **175**

9.1.	The Problem of Clock Paradox	175
9.2.	Around-the-World Atomic Clocks	180
9.3.	Doppler Effect	183
9.3.1.	The Hydrogen Canal Ray Experiment	183
9.3.2.	Doppler Effect of γ-Rays from Capture Reactions	188

9.3.3. The Doppler Effect Experiments Based on
        Mössbaue Effect   190
9.3.4. Other Measurements of the Second-Order Doppler
        Effect   195
9.4. Lifetime Dilation of Moving Mesons   195
9.4.1. The μ-Mesons in the Cosmic Rays   196
9.4.2. Measurements of the Lifetime of the μ-Mesons in
        the Cosmic Rays   197
9.4.3. Measurements of the Lifetime of the Mesons
        Produced by Accelerators   197

Chapter 10.   The Electromagnetism Experiments   201
10.1. Introduction   201
10.2. Electromagnetic Induction of Moving Bodies   202
10.2.1. Unipolar Induction   202
10.2.2. Magnetic Effects of Moving Bodies   204
10.2.3. The Wilson-Wilson Experiment   206
10.3. The Fresnel Drag Effect   207
10.3.1. Fizeau's and Zeeman's Experiments   209
10.3.2. The Transverse "Drag" Experiment   214
10.3.3. Other "Drag" Experiments   218
10.4. Reflection at Moving Mirrors   222

Chapter 11.   The Tests of Relativistic Mechanics   225
11.1. The Test of Variation of Mass with Velocity   225
11.1.1. Deflection of Charged Particles in Electric and
         Magnetic Fields   225
11.1.2. High Energy Synchro-Cyclotron   231
11.1.3. Other Measurements   232
11.1.4. The Measurements of Flight-Time for Moving Particles 233
11.1.5. Elastic Collisions   234
11.1.6. The Fine Structure of Atomic Spectra   238
11.2. Relation of Mass and Energy   239

Chapter 12.   The Upper Bounds on Photon Mass   245
12.1. Dispersion Effect of Velocity of Light in Vacuum   245
12.1.1. Measurements of the Velocity of Light   246
12.1.2. Arrival Time of Light Ray from Stars   246

12.2. The Tests of Coulomb's Law 248
12.3. The Magnetostatic Effect of Photon Mass 253
    12.3.1. Schrödinger's (External Field) Method 253
    12.3.2. Altitude–Dependence of Geomagnetic Field 256
    12.3.3. Eccentric Dipole or "Vertical Current" Effect 257
12.4. The Magnetohydrodynamic Effects 258
    12.4.1. Hydromagnetic Waves 258
    12.4.2. Dissipation of the Interplanetary Magnetic Field 261
    12.4.3. Instability of Interstellar Plasma 264
12.5. Other Methods 265
12.6. Summary 266

Chapter 13. The Tests of Thomas Precession **269**
13.1. The Fine Structure of Atomic Spectra 269
13.2. Measurements for $(g - 2)$ Factor of Leptons 270

References **275**

Index **285**

Part I

Einstein's Theory of Special Relativity

CHAPTER 1

FOUNDATIONS OF SPACE–TIME THEORIES

1.1. Introduction

The first step for constructing a flat space–time theory is to define an inertial reference frame (an inertial system of reference) in vacuum, which will be denoted by, e.g., $F(xyzt)$ with (x, y, z) and t being space and time coordinates, respectively. To define an inertial reference frame, the following three problems must be solved:

(i) What are the space coordinates (x, y, z)?

(ii) What is the time coordinate t?

(iii) What is "inertial"?

Responding the first and third problems will be simple and unique in spite of the definition of the vacuum. However the answer to the second problem will be complex and variant. The main difference among Newtonian mechanics, Einstein's theory and test theories of special relativity is just the different answer to the second problem. In this chapter, we shall mainly clarify some definitions of the time coordinate, i.e., some definitions of simultaneity. In Einstein's theory of special relativity, the second postulate, i.e., the so-called principle of the constancy of the velocity of light, establishes the definition of Einstein simultaneity. As shown in chapter 8, it is the two-way speed but not the one-way speed of light has been measured.

The second step is to assume the principle of relativity. In Einstein's theory of special relativity the so-called Einstein's principle of relativity is just a simple generalization of the Galilean principle of relativity in classical mechanics. This principle makes us possible to establish transformation equations of any physical quantities in the well-defined inertial frames.

1.2. Definition of Inertial Reference Frame

Let $F(xyzt)$ denote an inertial reference frame, where x, y, z are called the space coordinates which are given in a Cartesian coordinate system $S(xyz)$ in a three dimensional vacuum space and t is the time coordinate representing a "common" time within the whole system S. In this section we shall give the definition of the inertial reference frame.

1.2.1. Definition of Space Coordinates

We assume that the vacuum, a three-dimensional Euclidean space, exists. Imagine that a framework of three orthogonal rods (i.e., x-, y-, and z-axis) which extends into the space can be rigidly attached to a reference body. This conceptual framework is regarded as a rectangular Cartesian coordinate system, $S(xyz)$. Using this Cartesian system, we characterize any space point, e.g., P_1 by three numbers, the coordinates x_1, y_1, z_1 of that space point. We use $P_1(x_1, y_1, z_1)$ to denote that point. Let $P_2(x_2, y_2, z_2)$ be another point. The coordinate differences between these two points are denoted by $\Delta x, \Delta y, \Delta z$. The distance Δs between the two points is given by

$$(\Delta s)^2 = (\Delta x)^2 + (\Delta y)^2 + (\Delta z)^2, \tag{1.2.1}$$

or

$$(\Delta s)^2 = \delta_{ij} \Delta x_i \Delta x_j, \quad i, j = 1, 2, 3, \tag{1.2.2}$$

where

$$\Delta x_1 \equiv \Delta x = x_2 - x_1,$$
$$\Delta x_2 \equiv \Delta y = y_2 - y_1,$$
$$\Delta x_3 \equiv \Delta z = z_2 - z_1,$$

and δ_{ij} is the so-called Kronecker symbol defined by the equations

$$\delta_{ij} = 1, \quad i = j,$$
$$\delta_{ij} = 0, \quad i \neq j. \tag{1.2.3}$$

The convention is used: Each index that occurs once in a product (i.e., an open index) assumes all its values; Each index which occurs twice in a product (i.e., a contracted index) is a summation index, where the summation is to be carried over all possible values.

Consider another Cartesian system $S'(x'y'z')$. Let the origin point of S' coincide with that of S. The differences between the corresponding coordinates (x_1', y_1', z_1'), and (x_2', y_2', z_2') of the two points P_1 and P_2 as measured in the system S' are denoted by $\Delta x_1' \equiv x_2' - x_1'$, $\Delta x_2' \equiv y_2' - y_1'$, and $\Delta x_3' \equiv z_2' - z_1'$. The coordinate transformations between $\Delta x_i'$ and Δx_i are given by

$$\Delta x_i' = D_{ij} \Delta x_j, \tag{1.2.4}$$

where D_{ij} are constants in x_i and t. These equations keep the three-dimensional space interval (1.2.1) invariant:

$$(\Delta s)^2 = (\Delta s')^2, \tag{1.2.5}$$

with

$$(\Delta s')^2 = (\Delta x')^2 + (\Delta y')^2 + (\Delta z')^2. \tag{1.2.6}$$

The transformations (1.2.4) are the orthogonal transformations where the transformation matrix (D_{ij}) satisfies the condition:

$$D_{ki} D_{li} = D_{ik} D_{il} = \delta_{kl}. \tag{1.2.7}$$

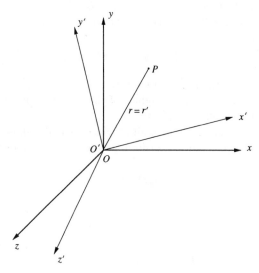

Fig. 1. Special rotation of Cartesian coordinate systems.

The inverse matrix $(D_{ij})^{-1}$ gives the inverse transformations of Eq. (1.2.4). Equations (1.2.4) together with Eqs. (1.2.7) define the group of orthogonal transformation, the $O(3)$ group.

It is easy to prove that the determinant of the matrix D_{ij} is equal to ± 1:

$$det\,(D_{ij}) \equiv |\,(D_{ij})\,| = \pm 1. \tag{1.2.8}$$

The value $+1$ of the determinant belongs to the proper rotations, the $SO(3)$ group, while the value -1 belongs to orthogonal transformations involving a reflection.

The transformations (1.2.4) give the general transformation law of vectors with respect to orthogonal transformations. In other words, a vector $\mathbf{A}$ is defined as a set of three quantities A_i which transform like coordinate differences:

$$A'_j = D_{ji}A_i. \tag{1.2.9}$$

Therefore, when the vector components are given with respect to any one Cartesian coordinate system, they can be calculated with respect to every other Cartesian coordinate system.

The norm of the vector is defined as the sum of the squared vector components: $A = |\mathbf{A}| = (A_i A_i)^{1/2}$. The norm is an invariant with respect to the orthogonal transformations,

$$A'_i A'_i = A_i A_i. \tag{1.2.10}$$

The scalar product of two vectors $\mathbf{A} = (A_x, A_y, A_z) = (A_i, i = 1, 2, 3)$ and $\mathbf{B} = (B_x, B_y, B_z) = (B_i, i = 1, 2, 3)$ is defined as the sum of the products of corresponding vector components,

$$\mathbf{A} \cdot \mathbf{B} = A_i B_i = AB \cos\theta, \tag{1.2.11}$$

where $B = |\mathbf{B}|$, and θ is the angle between the directions of $\mathbf{A}$ and $\mathbf{B}$. We see that the norm of a vector is the scalar product of the vector by itself. Thus the scalar product of two vectors is an invariant too.

In general, a tensor has N indices, all of which take all values 1 to 3. Thus, the tensor has 3^N components, which transform according to the transformation law

$$T'_{ijk\ldots} = D_{il} D_{jm} D_{kn} \ldots T_{lmn\ldots}. \tag{1.2.12}$$

The number of index, N, is referred to its rank. Scalars may be called tensors of rank 0. Vectors are tensors of rank 1. Quantities with $N \geq 2$ are called tensors of rank N.

The values of the Kronecker symbol (the three-dimensional Euclidean metric) keep invariant under the orthogonal transformations (1.2.4):

$$\delta'_{ij} = D_{ik} D_{jl} \delta_{kl} = D_{ik} D_{jk} = \delta_{ij}. \tag{1.2.13}$$

One can define other kinds of quantities, the tensor densities, which transform like the tensors, except that they are also multiplied by the transformation determinant (1.2.8). As long as this determinant equals $+1$, that is, when the transformation is a proper orthogonal transformation without reflection, there is no difference between a tensor and a tensor density. But a density undergoes a change of sign (compared with a tensor) when a reflection of the coordinate frame is carried out. A typical tensor density is the constant tensor density of rank 3, the Levi–Civita tensor density, defined as follows. ϵ_{ijk} is skewsymmetric in its three indices; therefore, all those components which have at least two indices equal vanish. The values of the non-vanishing components are ± 1, the sign depending on whether (i, j, k) is an even or an odd permutation of $(1, 2, 3)$. The contraction of two Levi–Civita tensor densities is related to Kronecker symbols:

$$\epsilon_{ijk} \epsilon_{ilm} = \delta_{jl} \delta_{km} - \delta_{jm} \delta_{kl}. \tag{1.2.14}$$

The vector product of $\mathbf{A}$ and $\mathbf{B}$ can be expressed in terms of the Levi–Civita tensor density as follows:

$$(\mathbf{A} \times \mathbf{B})_i = \epsilon_{ijk} A_j B_k. \tag{1.2.15}$$

1.2.2. Definitions of Time Coordinate

We have already defined a (space) frame of reference $S(xyz)$, and hence the location of a body can be designated by its three coordinates. However, in order to describe the equation of motion for the body, we must define a "common" time within the whole system. For this reason, assume that at each one of space points in the Cartesian frame $S(xyz)$ there is a standard clock at rest. A reading of the clock at each point defines a "local time" at that point. The name "local time" means that comparison among the readings of different clocks at different space points will have no physical meaning because those clocks have not yet been synchronized one another. The synchronization among the clocks at different space points is called the definition of simultaneity.

Consider two physical events $E_1\,(x_1, y_1, z_1, t_1)$ and $E_2\,(x_2, y_2, z_2, t_2)$ which occur at two different points $P_1\,(x_1, y_1, z_1)$ and $P_2\,(x_2, y_2, z_2)$, respectively. Note that t_1 and t_2 are given by the two clocks C_{P_1} and C_{P_2} at the two points P_1 and P_2, respectively. In other words, t_1 and t_2 represent only the so-called "local times" before C_{P_1} and C_{P_2} are synchronized each other. Thus, to compare the times of the two events, we must firstly synchronize the two clocks at those two points. A reading of a synchronized clock represents the "common" time of the frame, which is called the time coordinate t.

After completed the above steps, we can say that we have already constructed a general frame of reference $F(xyzt)$ which consists of a Cartesian frame $S(xyz)$ involving a well-defined "common" time t. Of course, definition of simultaneity, i.e., clock synchronization, is a complex problem to be discussed in section 1.3.

1.2.3. Definition of "Inertial" Frames

Definition of an *"inertial"* frame is based on the law of inertia. The law of inertia can be stated as follows: *Bodies when removed from interaction with other bodies will continue in their states of rest or straight-line uniform motion.*

A frame of reference $F(xyzt)$ is regarded as an *inertial* frame of reference if the law of inertia is valid within it. In other words, the motion of a free particle in the inertial frame of reference F is the inertial motion: The particle is unaccelerated,

$$\frac{d^2x}{dt^2} = 0, \quad \frac{d^2y}{dt^2} = 0, \quad \frac{d^2z}{dt^2} = 0,$$

or equivalently of the constant velocity,

$$\frac{dx}{dt} = \text{const}, \quad \frac{dy}{dt} = \text{const}, \quad \frac{dz}{dt} = \text{const}.$$

Note that for given any particle we can always introduce a frame of reference with respect to which it is at rest or of constant velocity and, therefore, unaccelerated. So the key point is that the particle is "free", i.e., removed from any interactions with other bodies.

We know from the law of inertia that if F is an inertial frame of reference, then another frame F' moving at a constant velocity relative to F is an inertial frame of reference too.

1.3. Simultaneity and Clock Synchronization

We emphasize again that the key point for constructing an inertial frame is the clock synchronization, i.e., the definition of simultaneity. In the present section we shall discuss in detail clock synchronization.

Let us first imagine that we could find an idea signal, the velocity of which in a given frame had a certain value not depending upon the states of motion of its emitter. Therefore, one could use such a signal in clock synchronizing. We assume that the one-way velocity of that signal in the direction $\mathbf{r}/r$ is c_r as seen in the frame F. Let C_O and C_P denote two standard clocks at rest at the origin $O(0,0,0)$ and the point $P(x, y, z)$, respectively. Consider that the signal is emitted from the origin O at the reading t_O of the clock C_O, and then reaches the point P at the reading t_P of the clock C_P. The clocks C_P and C_O are regarded as synchronization if and only if

$$t_P = t_O + \frac{r}{c_r}, \tag{1.3.1}$$

where $r = \sqrt{x^2 + y^2 + z^2}$ is the distance between the two points O and P. Of course, one can synchronize the two clocks by means of a signal propagating from P at the time t_P to O at the time t'_O. Similarly, C_O and C_P are regarded as synchronization if and only if

$$t'_O = t_P + \frac{x}{c_{-r}}, \tag{1.3.2}$$

where c_{-r} is the one-way velocity of the signal in the direction $-\mathbf{r}/r$. The time given by the synchronized clock is just the common time, the time coordinate of the frame.

In another point of view, the clock synchronization equations, (1.3.1) and (1.3.2), are equivalent to the definitions of the one-way velocities,

$$c_r = \frac{r}{t_P - t_O}, \qquad c_{-r} = \frac{r}{t'_O - t_P}, \tag{1.3.3}$$

where the differences $(t_P - t_O)$ and $(t'_O - t_P)$ are called *the coordinate time intervals* which are given by the two clocks at the two different points and, hence, dependent on the definition of simultaneity. Therefore we can explain the clock synchronization equations (1.3.1) and (1.3.2) as follows: *The clocks C_O and C_P are synchronized in such a way that the one-way velocities of the signal as measured by use of the two clocks are c_r and c_{-r}.*

We now introduce the two-way velocity $\bar{c}_r$ of the signal traveling from the point O to the point P and, by reflection, back to O. By definition we have

$$\bar{c}_r \equiv \frac{2r}{(t'_O - t_O)}, \tag{1.3.4}$$

where $(t'_O - t_O)$ is the difference between the two readings t'_O and t_O of the same clock C_O and, therefore, called *the proper time interval* not relevant to the definition of simultaneity. By using Eqs. (1.3.3), equation (1.3.4) becomes

$$\bar{c}_r = \frac{2r}{(t'_O - t_P) + (t_P - t_O)} = \frac{2}{(1/c_{-r}) + (1/c_r)} = \frac{2c_r c_{-r}}{c_r + c_{-r}}, \qquad (1.3.5a)$$

or

$$\frac{1}{c_r} + \frac{1}{c_{-r}} = \frac{2}{\bar{c}_r}. \qquad (1.3.5b)$$

This gives the relationship between the one-way and the two-way velocities of the signal. In contrast to the one-way speed, the two-way speed $\bar{c}_r$ is a measurable quantity because it is related to the proper time interval $(t'_O - t_O)$.

The one-way velocity may be parameterized by introducing a directional parameter $\mathbf{q}$ as follows:

$$c_r = \frac{\bar{c}_r}{1 - q_r}, \qquad c_{-r} = \frac{\bar{c}_r}{1 + q_r}, \qquad (1.3.6)$$

where $q_r = \mathbf{q} \cdot \mathbf{r}/r$.

Equation (1.3.5) implies that the choices of c_r and c_{-r} are restricted in such a way that the sense of cause is preserved. In other words, the signal starting at O cannot reach P before it leaves O. Since $(t_P - t_O)$ and $(t'_O - t_P)$ must be positive, so must c_r and c_{-r} be positive, we see from (1.3.5b) that neither can be smaller than $\bar{c}_r/2$. This leads to the restriction

$$\frac{\bar{c}_r}{2} \leq c_r(c_{-r}) \leq \infty. \qquad (1.3.7)$$

Using (1.3.6) in (1.3.7), we get the limit on the directional parameter

$$-1 \leq q_r \leq +1. \qquad (1.3.8)$$

This is to say that we have the following conclusion: If the one-way velocity of the signal were known in advance, then one could use it in clock synchronizing. In the above example, when the observer at the point $P(x, y, z)$ receives the signal, he should adjust the reading t_P of his clock C_P to the value that satisfies the clock synchronization equation (1.3.1); Conversely, if the clocks C_O and C_P were synchronized in advance by means of some given method, then they could be used for the measurement of the one-way velocity according to equation (1.3.3).

1.3.1. Newtonian Absolute Simultaneity

Newtonian absolute simultaneity is equivalent to postulating the existence of an instantaneously propagating signal with an infinite velocity, i.e., $c_r = c_{-r} = \bar{c}_r = \infty$. This, according to Eq. (1.3.1), implies that when the signal reaches the point P the reading t_P of the clock C_P should be adjusted to be the same as its emitting time t_O,

$$t_P = t_O. \qquad (1.3.9)$$

Similarly, when that signal reflected from P returns to O, we obtain from (1.3.2)

$$t'_O = t_P. \tag{1.3.10}$$

It is well known that in classical mechanics the time coordinate within an inertial frame of reference is defined by *Newtonian absolute simultaneity*. Such a frame denoted by $N(xyzt)$ will be called a Newtonian inertial frame of reference or a *Newtonian frame* for short, so as to distinguish it from an Einstein frame $F(xyzt)$.

We have known that there is no any *instantaneous signal* in nature and, therefore, the absolute simultaneity cannot be realized in any laboratory.

1.3.2. Einstein's Definition of Simultaneity

According to Einstein's second postulate [1], the (one-way) velocity of light in every inertial reference frame is the same constant c independent of the motion of the light source. In other words, Einstein chose a vanishing directional parameter, $\mathbf{q} = 0$, in each one of inertial frames. Thus, we have

$$c \equiv c_r = c_{-r} = \bar{c}_r. \tag{1.3.11}$$

This is the definition of *Einstein simultaneity*. Therefore, Einstein's synchronization between the clocks C_O and C_P at, respectively, the points $O(0,0,0)$ and $P(x,y,z)$ is obtained from Eq. (1.3.1)

$$t = t_O + \frac{r}{c}, \tag{1.3.12a}$$

or from Eq. (1.3.2)

$$t'_O = t + \frac{r}{c}, \tag{1.3.12b}$$

where the subscript "P" of t_P is omitted. Such a frame $F(xyzt)$, in where the time coordinate t is the reading of the clock C_P synchronized with the clock C_O, will be called an Einstein's inertial frame of reference, or an *Einstein frame* for short.

The clock synchronization equations (1.3.12) can be also explained as follows: *The clocks C_O and C_P are synchronized in such a way that the one-way velocity of light is the constant c as measured by the two clocks.*

It is well-known that one always use a light signal for the clock synchronization in a laboratory. Therefore Einstein's simultaneity can be directly realized in experiments. What we want to stress here is that only the two-way speed, but not the one-way speed, of light has been already measured in the experimental measurements, and hence the isotropy of the one-way velocity of light is just a postulate.

We shall see from Chap. 6 that a more general postulate, a choice of the anisotropy of the one-way velocity of light, together with the principle of relativity, would give the same physical predictions as Einstein's theory of special relativity.

1.3.3. Edwards' Definition of Simultaneity

A more general definition of simultaneity is the postulate made by Edwards (1963) [2]:
The two-way speed of light in a vacuum as measured in two coordinate systems moving with constant relative velocity is the same constant regardless of any assumptions concerning the one-way speed.

Therefore, in the general inertial frame the directional parameter $\mathbf{q}$ is chosen to be an arbitrary constant vector and the two-way velocity $\bar{c}$ is a universal constant which is assumed by Edwards to be the same as c in Einstein's theory. Thus, we have from (1.3.6)

$$c_r = \frac{c}{1 - q_r}, \quad c_{-r} = \frac{c}{1 + q_r}, \quad \bar{c}_r = c. \tag{1.3.13}$$

This is the definition of *Edwards simultaneity.*

Edwards simultaneity may be explained as follows. Let $\tilde{C}_O$ and $\tilde{C}_P$ denote two standard clocks at rest at the origin $O(0,0,0)$ and the point $P(x,y,z)$, respectively, in the three-dimensional Cartesian system S. A light signal starting from the origin O at the reading $\tilde{t}_O$ of the clock $\tilde{C}_O$ reaches the point P at the reading $\tilde{t}_P$ of the clock $\tilde{C}_P$, and then return to P at the reading $\tilde{t}'_O$ of $\tilde{C}_O$. The clocks $\tilde{C}_P$ and $\tilde{C}_O$ are regarded as synchronization if and only if

$$\tilde{t} = \tilde{t}_O + \frac{r}{c_r}, \tag{1.3.14a}$$

or

$$\tilde{t}'_O = \tilde{t} + \frac{r}{c_{-r}}, \tag{1.3.14b}$$

where $\tilde{t} \equiv \tilde{t}_P$, and $r = \sqrt{x^2 + y^2 + z^2}$ is the distance between the two points O and P. The general inertial frame of reference with the Edwards (time) coordinate $\tilde{t}$ is called *Edwards frame $\tilde{F}(xyz\tilde{t})$.*

It will be clear in section 6.5 that Edwards simultaneity is physically equivalent to Einstein simultaneity.

1.3.4. Coordinate Transformation Between Einstein and Edwards Frames

The relation between Einstein frame $F(xyzt)$ and Edwards frame $\tilde{F}(\tilde{x}\tilde{y}\tilde{z}\tilde{t})$ with $\tilde{x}_i = x_i$ can be obtained from their definitions. Let $S(xyz)$ be a Cartesian system of reference in a three-dimensional (vacuum) space. There are two standard clocks at the location of the origin $O(0,0,0)$ of the frame, which are denoted by C_O and $\tilde{C}_O$. It is similar that other two standard clocks, C_P and $\tilde{C}_P$, are resided at another point $P(x,y,z)$. Note that all the clocks are at rest within this Cartesian frame. Consider a light signal propagating from O to P. Let the readings of C_O and $\tilde{C}_O$ be, respectively, t_O and $\tilde{t}_O$ when the signal leaves O, and the readings of C_P and $\tilde{C}_P$ be, respectively, t and $\tilde{t}$ when the signal reaches P (see Fig. 2).

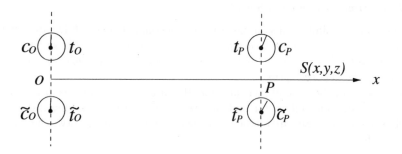

Fig. 2. Einstein clocks and Edwards clocks.

The clocks C_O and C_P are synchronized according to Einstein's definition of simultaneity, i.e., Eq. (1.3.12a),

$$t = t_O + \frac{r}{c}. \qquad (1.3.15)$$

These clocks are named *Einstein clocks*. *An Einstein frame $F(xyzt)$ is just the system $S(xyz)$ involving an Einstein clock, which defines the time coordinate t, at each one of space points.*

On the other hand, the clocks $\tilde{C}_O$ and $\tilde{C}_P$ are synchronized according to Edwards' definition of simultaneity, i.e., Eq. (1.3.14a),

$$\tilde{t} = \tilde{t}_O + \frac{r}{c_r}. \qquad (1.3.16)$$

We refer the clocks synchronized by use of (1.3.16) to *Edwards clocks*. *An Edwards frame $\tilde{F}(\tilde{x}\tilde{y}\tilde{z}\tilde{t})$ with $\tilde{x}_i = x$ is just a three-dimensional Cartesian system $S(xyz)$ involving an Edwards clock, which give the time coordinate $\tilde{t}$, at every point.*

We now discuss the necessary condition under which different definitions of simultaneity can be compared one another. Let us now consider the light signal being reflected from the point P back to the origin O. When the signal comes back to the point O, the readings of C_O and $\tilde{C}_O$ are

$$t'_O = t + \frac{r}{c} \qquad (1.3.17)$$

and

$$\tilde{t}_O = \tilde{t} + \frac{r}{c_{-r}}, \qquad (1.3.18)$$

respectively.

It is emphasized here that the difference between the readings of the two clocks C_O and $\tilde{C}_O$ must always be a constant because they are all the standard clocks and at rest at the same point O. For instance, we can choose the initial condition:

$$t_O - \tilde{t}_O = f, \tag{1.3.19}$$

where f is a constant. This is to say that the reading of C_O differs from the reading of $\tilde{C}_O$ by the value f when the light signal leaves O. Then, the difference between the readings of C_O and $\tilde{C}_O$ must be the same forever. This means that when the signal comes back to O we must have

$$t'_O - \tilde{t}'_O = f. \tag{1.3.20}$$

Then Eqs. (1.3.19) and (1.3.20) lead to

$$t_O - \tilde{t}_O = t'_O - \tilde{t}'_O = f$$

or

$$t'_O - t_O = \tilde{t}'_O - \tilde{t}_O. \tag{1.3.21}$$

In other words, the difference between the two successive readings of the clock C_O must be equal to that between the corresponding readings of the clock $\tilde{C}_O$. This implies that the two-way velocities of light in the two definitions of simultaneity must be the same. Thus, *the necessary condition under which different definitions of simultaneity can be compared each other is as follows: The two-way velocities of light obtained from the one-way velocities must have the same value.*

It is obvious that Eqs. (1.3.13)–(1.3.15) satisfy the condition (1.3.21), because the two-way speeds of light in Einstein's and Edwards' postulates are the same constant c. Therefore we can consider the difference between the Einstein and Edwards time coordinates t and $\tilde{t}$. To this end, we obtain from Eqs. (1.3.15) and (1.3.16)

$$t = \tilde{t} + r\left(\frac{1}{c} - \frac{1}{c_r}\right), \tag{1.3.22}$$

where we choose the initial condition $t_O - \tilde{t}_O = f = 0$. Using Eq. (1.3.13), we can express (1.3.22) in terms of the directional parameter as

$$t = \tilde{t} + q_r\frac{r}{c}. \tag{1.3.23a}$$

Of course, the space coordinates are the same:

$$x = \tilde{x}, \quad y = \tilde{y}, \quad z = \tilde{z}. \tag{1.3.23b}$$

Equations (1.3.23) are just the coordinate transformations between Einstein frame F and Edwards frame $\tilde{F}$. We shall see in section 6.5 that the difference between t and $\tilde{t}$ given by Eq. (1.3.23a) is unobservable in experiments.

It is useful for understanding the definitions of simultaneity to note the following comparing: The transformation (1.3.23) between Einstein and Edwards frames is similar to the transformation (1.2.4) between any two three-dimensional Cartesian frames in the sense that the relative velocities of these frames are all zero.

1.3.5. *Robertson's Definition of Simultaneity*

In Einstein's and Edwards' postulates, the two-way velocity of light as measured in any one of inertial frames is the same constant c. A more general postulate relaxes this limit: The two-way velocity $\bar{c}_r$ of light in a general inertial frame of reference, which is called the *Robertson frame* $\bar{F}(\bar{x}\bar{y}\bar{z}\bar{t})$, is equal to the one-way velocity of light, and is anisotropic and independent of the motion of the light source, i.e.,

$$\bar{c}_r = c_r = c_{-r}, \tag{1.3.24a}$$

where c_r and c_{-r} are the one-way velocities of light in the directions $\pm \bar{\mathbf{r}}/\bar{r}$ with $\bar{\mathbf{r}} = (\bar{z}, \bar{y}, \bar{z})$ and $\bar{r} = \sqrt{\bar{x}^2 + \bar{y}^2 + \bar{z}^2}$. Furthermore we assume that the two-way velocity can be expressed as

$$\bar{c}_r = \frac{\bar{c}_\| \bar{c}_\perp}{\sqrt{\bar{c}_\|^2 + (\bar{c}_\perp^2 - \bar{c}_\|^2)\cos^2\alpha}}, \tag{1.3.24b}$$

where $\bar{c}_\| \equiv \bar{c}_x$ is the two-way velocity of light along the path of the $\bar{x}$-axis, $\bar{c}_\perp$ is the two-way velocity in the direction perpendicular to the $\bar{x}$-axis, and α is the angle between the directions of $\bar{\mathbf{r}}/\bar{r}$ and the $\bar{x}$-axis, and in general $\bar{c}_\| \neq \bar{c}_\perp$.

The *Robertson frame* $\bar{F}(\bar{x}\bar{y}\bar{z}\bar{t})$ is just such a three-dimensional Cartesian inertial frame of reference $S(\bar{x}\bar{y}\bar{z})$ involving an *Robertson clock*, which will be defined below, at each point ($\bar{x}, \bar{y}, \bar{z}$, and $\bar{t}$ are called Robertson coordinates). Let a light signal starting from the origin point $O(0, 0, 0)$ at the reading $\bar{t}_O$ of the clock $\bar{C}_O$ at the origin reach the point $P(\bar{x}, \bar{y}, \bar{z})$ at the reading $\bar{t}$ of the clock $\bar{C}_P$ at that point, and then return to O at the reading $\bar{t}'_O$ of $\bar{C}_O$. The clocks $\bar{C}_P$ and $\bar{C}_O$ are regarded as synchronization if and only if

$$\bar{t} = \bar{t}_O + \frac{\bar{r}}{\bar{c}_r}, \tag{1.3.25a}$$

or

$$\bar{t}'_O = \bar{t} + \frac{\bar{r}}{\bar{c}_r}, \tag{1.3.25b}$$

where $\bar{c}_r$ is given by Eq. (1.3.24b) in terms of the two parameters $\bar{c}_\|$ and $\bar{c}_\perp$. These clocks synchronized according to Eq. (1.3.25) are called *Robertson clocks*, which define the time coordinate $\bar{t}$ of the frame $\bar{F}$. In other words, *Robertson simultaneity* [3] in the frame $\bar{F}$ is defined in such a way that the velocities of light are given by Eq. (1.4.24).

It is obvious that we can distinguish the Robertson simultaneity from the Einstein simultaneity by experiments, because any possible anisotropy of the two-way speeds of light could be tested.

1.3.6. Mansouri–Sexl's Definition of Simultaneity

$F^*(x^*y^*z^*t^*)$ is called a Mansouri–Sexl inertial system of reference or a *MS frame* in short, where (x^*, y^*, z^*) are the coordinates of a Cartesian frame of reference $S(x^*y^*z^*)$ and t^* is the time coordinate defined by the Mansouri–Sexl (MS) simultaneity [4]. The MS simultaneity is a generalization of the Robertson simultaneity, just as the Edwards simultaneity is a generalization of the Einstein simultaneity.

Therefore, in the MS frame F^* the velocity of light is given by

$$c_r = \frac{\bar{c}_r}{1 - q_r}, \qquad c_{-r} = \frac{\bar{c}_r}{1 + q_r}, \tag{1.3.26a}$$

$$\bar{c}_r = \frac{\bar{c}_\parallel \bar{c}_\perp}{\sqrt{\bar{c}_\parallel^2 + (\bar{c}_\perp^2 - \bar{c}_\parallel^2)\cos^2\alpha}}, \tag{1.3.26b}$$

$$-1 \le q_r \le +1, \tag{1.3.26c}$$

where c_r and c_{-r} are the one-way velocities of light in the directions $\pm \mathbf{r}^*/r^*$ respectively, $\mathbf{r}^* = (x^*, y^*, z^*)$, $r^* = \sqrt{x^{*2} + y^{*2} + z^{*2}}$, $\bar{c}_r$ is the corresponding two-way speed which is expressed as a function of the two parameters $\bar{c}_\parallel$ and $\bar{c}_\perp$, and $q_r = \mathbf{q} \cdot \mathbf{r}/r$ is a directional parameter. In case of $q_r = 0$, equation (1.3.26a) reduces to Eq. (1.3.24a), i.e., the MS simultaneity reduces to the Robertson simultaneity. So we can compare the difference between the two definitions of simultaneity.

The MS simultaneity or MS clock synchronization may be stated as follows: As it does in the previous sections, let a light signal starting from the point $O(0, 0, 0)$ at the reading t_O^* of the clock C_O^* at that point reach the point $P(x^*, y^*, z^*)$ at the reading t^* of the clock C_P^* at that point, and then return to O at the reading $t_O^{*\prime}$ of C_O^*. The clocks C_P^* and C_O^* are regarded as synchronization if and only if

$$t^* = t_O^* + \frac{r^*}{c_r}, \tag{1.3.27a}$$

or

$$t^{*\prime}_O = t^* + \frac{r^*}{c_{-r}}, \tag{1.3.27b}$$

where c_r and c_{-r} are given by Eq. (1.2.26a). These equations are called the *MS clock synchronization* equations or the definition of the *MS simultaneity*, and C_O^* and C_P^* are called the *MS clocks*.

Therefore, the definition of the MS frame $F^*(x^*y^*z^*t^*)$ can be stated as follows: *The MS frame F^* is just such a Cartesian inertial frame of reference $S(x^*y^*z^*)$ involving the MS clocks which define the Mansouri–Sexl time coordinate t^*.*

In other words, the space coordinates are given by the usual Cartesian frame $S(x^*y^*z^*)$, and the time coordinate t^* is defined in such a way that the one-way velocity of light should be expressed by Eqs. (1.3.26).

1.3.7. Coordinate Transformation Between the Robertson and MS Frames

It is known that the two-way velocity of light in the Robertson frame is the same as that in the MS frame. Thus, we can consider the difference between these two kinds of frames.

As done in section 1.3.4, let the space coordinates of the Robertson frame $\tilde{F}(\bar{x}, \bar{y}, \bar{z}, \bar{t})$ coincide with those of the MS frame $F^*(x^*y^*z^*t^*)$, i.e.,

$$\bar{x} = x^*, \quad \bar{y} = y^*, \quad \bar{z} = z^*. \tag{1.3.28a}$$

The time coordinates $\bar{t}$ and t^* are defined by (1.3.25) and (1.3.27) respectively, and thus the transformation between them is

$$\bar{t} = t^* + \bar{r}\left(\frac{1}{\bar{c}_r} - \frac{1}{c_r}\right) = t^* + q_r \frac{r^*}{\bar{c}_r}, \tag{1.3.28b}$$

where $r^* = \bar{r}$, and the initial condition $\bar{t}_O = t^*_O$ is chosen for simplicity. Equations (1.3.28) are called the coordinate transformations between the Robertson and MS frames. Comparing the transformation (1.3.28) with the transformation (1.3.23), we can see that the only difference between them is the different two-way velocities of light, i.e., $\bar{c}_r \neq c$. So that the MS simultaneity is a generalization of Robertson simultaneity, just as Edwards simultaneity is a generalization of Einstein simultaneity.

1.4. Principle of Relativity

Let us first recall Galilean principle of relativity. It is well known that the choice of a frame of reference determines the form of a law of nature. Using the inertial frame of reference defined in the previous sections, the law of inertia takes its simple form: In the absence of forces, the space coordinates of a mass point are linear functions of the time coordinate; Inversely, of course, that the space coordinates of a free particle takes linear form of time coordinate is just the definition of an inertial frame of reference.

Galilean principle of relativity says that all of the laws of mechanics take the same form when stated in terms of any one of the Newtonian frames.

In Einstein's theory of special relativity, the first postulate, the principle of special relativity, is simply a generalization of the Galilean principle of relativity. Of course, the definition of the inertial frames of reference in special relativity differs from that in the Newton–Galileo space–time theory. Einstein's principle of relativity says that all of the laws of physics take the same form when stated in terms of any one of Einstein frames.

1.5. Velocity and Simultaneity

In the previous sections, we have given the definitions of the inertial frames. Thus, we can now describe the motion of any body, and define any physical quantities relevant to the body. An important quantity is the velocity of the body, which is denoted by $\mathbf{u} = (u_x, u_y, u_z)$. By definitions, we have

$$u_x = \frac{dx}{dt}, \quad u_y = \frac{dy}{dt}, \quad u_z = \frac{dz}{dt}. \tag{1.5.1}$$

We want to stress that dt is a *coordinate time interval* that depends upon the definition of simultaneity. This implies that the velocity of any given body would have different values corresponding to different definitions of simultaneity. Consider the following example. Let a particle move from the point $O(0,0,0)$ to another point $P(dx,0,0)$, and the distance between O and P be dx. Let two clocks C_O and $\tilde{C}_O$ be at rest at O, while two other clocks C_P and $\tilde{C}_P$ be at rest at P. The clocks C_O and C_P are two Einstein clocks, which have been synchronized by use of a light signal according to Eq. (1.3.12a),

$$t_P = t_O + \frac{dx}{c}. \tag{1.5.2a}$$

On the other hand, the clocks $\tilde{C}_O$ and $\tilde{C}_P$ are two Edwards clocks, which have been synchronized by use of a light signal according to equation (1.3.14a),

$$\tilde{t}_P = \tilde{t}_O + \frac{dx}{c_x}. \tag{1.5.2b}$$

For simplicity, we choose the initial condition $\tilde{t}_O = t_O$. This means that the readings of the two clocks C_O and $\tilde{C}_O$ are equal to, and furthermore will be forever equal to, each other. However, the readings of the clocks C_P and $\tilde{C}_P$ are different, the difference between them being the value from (1.5.2),

$$t_P - \tilde{t}_P = dx \left(\frac{1}{c} - \frac{1}{c_x} \right) = q_x \frac{dx}{c}, \tag{1.5.3}$$

where Eq. (1.3.13) is used. Note that after the two clocks C_P and $\tilde{C}_P$ are synchronized according to Eqs. (1.5.2a) and (1.5.2b) respectively, the difference between them will forever take the same value given by Eq. (1.5.3). Let us now establish the time coordinates of a moving body by means of these synchronized clocks. Consider the body moving from the point O to the point P. Let the time $t_O = \tilde{t}_O = 0$ when the body leaves O, and the readings of C_P and $\tilde{C}_P$ be dt and $d\tilde{t}$, respectively, when the body reaches P. Then, the difference between $d\tilde{t}$ and dt should be equal to the difference $(t_P - \tilde{t}_P)$ given by Eq. (1.5.3), because, as mentioned above, the two synchronized clocks will preserve the relative difference between their readings. Thus, we have

$$dt - d\tilde{t} = q_x \frac{dx}{c}. \tag{1.5.4}$$

By definitions, the velocities of the body are given by

$$u_x \equiv \frac{dx}{dt}, \quad \tilde{u}_x \equiv \frac{dx}{d\tilde{t}}, \tag{1.5.5}$$

where u_x and $\tilde{u}_x$ are , respectively, called *Einstein velocity* and *Edwards velocity*, which are measured by use of Einstein clocks and Edwards clocks, respectively. Using (1.5.4) in (1.5.5), we obtain the relationship between u_x and $\tilde{u}_x$,

$$\tilde{u}_x = \frac{u_x}{1 - q_x u_x/c}, \tag{1.5.6a}$$

or inversely,

$$u_x = \frac{\tilde{u}_x}{1 + q_x \tilde{u}_x/c}. \tag{1.5.6b}$$

Similarly, if the body moves from $P(dx, 0, 0)$ to $O(0, 0, 0)$, then using Eqs. (1.3.12b) and (1.3.14b) we get

$$u_{-x} = \frac{\tilde{u}_{-x}}{1 - q_x \tilde{u}_{-x}/c}, \tag{1.5.6c}$$

where $u_{-x} = |\mathbf{u}_{-x}|$ and $\tilde{u}_{-x} = |\tilde{\mathbf{u}}_{-x}|$ are, respectively, Einstein velocity and Edwards velocity.

This is to say that the velocity for a given moving body would take different values according to different definitions of simultaneity. A similar conclusion is also valid for any quantity related to the definition of simultaneity. For instance, the validity of the reciprocity of relative velocities depends upon the definitions of simultaneity.

The reciprocity of relative velocities can be stated as follows: For any two inertial frames, e.g., $\tilde{F}$ and $\tilde{F}'$, the relative velocity of $\tilde{F}'$ to $\tilde{F}$ is equal to v if and only if the relative velocity of $\tilde{F}$ to $\tilde{F}'$ is equal to $-v$.

We now discuss the *condition of reciprocity of relative velocities*. At first we know that for any two Einstein frames, F and F', the reciprocity of relative velocities is valid (see Lorentz transformation in section 2.2):

$$\mathbf{v}' = -\mathbf{v}, \tag{1.5.7}$$

where $\mathbf{v}$ is the relative velocity of F' to F as seen from F, and $\mathbf{v}'$ is the relative velocity of F to F' as seen from F'. In terms of the norms of $\mathbf{v}'$ and $\mathbf{v}$, equation (1.5.7) becomes

$$v' = v. \tag{1.5.8}$$

Using the relations between the Einstein velocities (v and v') and the Edwards velocities ($\tilde{v}$ and $\tilde{v}'$) given by Eqs. (1.5.6b,c), equation (1.5.8) becomes

$$\frac{\tilde{v}'}{1 - q_x' \tilde{v}'/c} = \frac{\tilde{v}}{1 + q_x \tilde{v}/c}, \tag{1.5.9}$$

where q_x and q'_x are the directional parameters within $\tilde{F}$ and $\tilde{F}'$, respectively. The reciprocity of the relative velocities implies

$$\tilde{v}' = \tilde{v}. \tag{1.5.10}$$

Substituting (1.5.10) into (1.5.9), we get the condition of reciprocity of relative velocities,

$$q'_x = -q_x. \tag{1.5.11}$$

This result can be also obtained from Edwards transformations (6.3.4). This shows that the reciprocity of the relative velocities must not require the vanishing directional parameters; In other words, the constancy of the velocity of light (i.e., $\mathbf{q}' = \mathbf{q} = 0$) is simply a special case that the reciprocity of the relative velocities is valid. (John A. Winnie (1970)[5] assumed $q' = q$, so that he omitted the above result).

We have considered above the relations between Edwards' quantities and Einstein's quantities, while relations between Mansouri–Sexl's (MS') and Robertson's quantities can be found in Sec. 7.3.

CHAPTER 2

RELATIVISTIC KINEMATICS

2.1. Galilean Transformation

Before discuss Lorentz transformation, let us recall Galilean transformation in the Newtonian absolute space–time theory. Consider two arbitrary Newtonian frames $N(xyzt)$ and $N'(x'y'z't')$ defined in subsection 1.3.1. Let N' move in the direction of x-axis with a constant velocity v as seen in the frame N. For simplicity, we consider the initial condition: At the initial instant $t = t' = 0$, the frame N coincides with the frame N'. Let E denote a physical event, its space–time coordinates in the frames N and N' being, respectively, (x',y,z,t) and (x',y',z',t').

In order to construct transformation equations between the two sets of space–time coordinates for the event, we need a principle that is the so-called Galilean principle of relativity: The mechanical laws are valid in any one of Newtonian inertial frames of reference. By use of the Newtonian simultaneity and Galilean principle of relativity, the relations between the two sets of coordinates for the same event, the Galilean transformations, are easily obtained

$$x' = x - vt,$$

$$y' = y,$$

$$z' = z,$$

$$t' = t. \tag{2.1.1}$$

The linear forms of the Galilean transformations guarantee the validity of the law of inertia: A moving, free particle should be in state of straight-line and uniform motion relative to *both* frames.

Galilean transformations (2.1.1) give us the kinematical and dynamic effects as follows.

Newtonian law of the addition of velocities:

$$\mathbf{u}' = \mathbf{u} - \mathbf{v}, \tag{2.1.2}$$

where $\mathbf{v}$ is the relative velocity of N' to N, and $\mathbf{u} = d\mathbf{r}/dt$ and $\mathbf{u}' = d\mathbf{r}'/dt'$ are the velocities of a moving particle as measured in the frames $N(xyzt)$ and $N'(x'y'z't')$, respectively. This shows that there is no upper limit on the velocity of a moving body as seen from any one of Newtonian frames.

Absolute simultaneity: Two events occurring at two separate points at the same time as seen in the frame N will emerge simultaneously too from another frame N'.

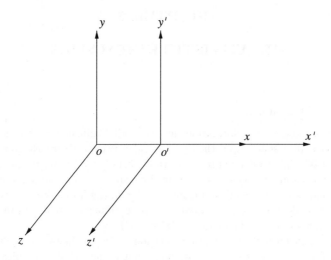

Fig. 3. Two coordinate systems without rotation.

Absolute space and time: Both the length of a rigid rod and the rate of a standard clock are constants not relevant to the Newtonian frames.

Constant inertial mass: Newtonian second law of a body acted by a force in the frame N has the same form as that in the frame N', i.e.,

$$\mathbf{F} = m\mathbf{a}, \qquad \mathbf{F}' = m\mathbf{a}', \qquad (2.1.3)$$

where the inertial mass m of the body is a constant not relevant to the reference frames , $\mathbf{a}$ and $\mathbf{a}'$ are the accelerations, and $\mathbf{F}$ and $\mathbf{F}'$ are the forces acting on the body as seen in N and N', respectively.

2.2. Lorentz Transformation

It was known in the end of 19th century and the beginning of 20th century that Newtonian mechanics could not explain some kinds of experimental results. For instance, the first one was about the electromagnetism of moving bodies. The experiments showed that the effects of electromagnetic interaction between a conductor and a magnet appear a symmetry about the relative motion of the conductor and magnet. However, Maxwell's electrodynamic equations could not be covariant under Galilean transformations (2.1.1). In other words, the Galilean principle of relativity is no longer valid for electromagnetic interaction; The second was that physicists failed in all their attempts to search for "light ether". The typical experiments were performed in 1881 and 1887 by A. A. Michelson and E. W. Morley [6,7], which showed that light rays are the independent entities but not vibrations of the so-called "light ether"; The third was about the observations of electronic mass, which exposed the dependence of the inertial masses of electrons upon their velocities. This is contrary to the constancy of inertial mass predicted in Newtonian mechanics. To solve these problems, Einstein generalized the Galilean principle of relativity and abandoned Newtonian simultaneity, and then developed his theory of special relativity in 1905.

Einstein's theory of special relativity is based on two basic postulates. The first postulate is the principle of special relativity (the word "special" means "with respect to inertial frames" but not to "more general frames"): All of the physical laws take the same form when stated in terms of any one of Einstein frames. The second postulate is the principle of constancy of velocity of light: Light signals propagate in vacuum with the constant velocity c independent of the motion of its source.

Einstein's inertial frames of reference have been defined in subsection 1.3.2 by making use of Einstein's second postulate. Next we shall use Einstein's two postulates to derive coordinate transformations between any two of Einstein frames.

Consider any two of Einstein frames, e.g. F and F', defined in subsection 1.3.2 and a physical event E, coordinates of which are (x, y, z, t) and (x', y', z', t') as seen in F and F', respectively. As it does in the previous section, general transformations between the two sets of coordinates are chosen as linear forms

$$x' = a^1{}_1 x + a^1{}_2 y + a^1{}_3 z + a^1{}_0 t + b^1, \qquad (2.2.1a)$$

$$y' = a^2{}_1 x + a^2{}_2 y + a^2{}_3 z + a^2{}_0 t + b^2, \qquad (2.2.1b)$$

$$z' = a^3{}_1 x + a^3{}_2 y + a^3{}_3 z + a^3{}_0 t + b^3, \qquad (2.2.1c)$$

$$t' = a^0{}_1 x + a^0{}_2 y + a^0{}_3 z + a^0{}_0 t + b^0, \qquad (2.2.1d)$$

where $a^\mu{}_\nu$ and b^μ ($\mu, \nu = 0, 1, 2, 3$) are arbitrary constants. For simplicity, we choose the initial conditions [1]: The two frames coincide with each other at the initial time $t = t' = 0$ as well as the frame F' moves with a constant velocity v in the direction of x-axis in F (see Fig. 3). Using the initial conditions in Eqs. (2.2.1), we obtain $b^\mu = a^1{}_2 = a^1{}_3 = 0$ and $a^1{}_0 = -a^1{}_1 v$. We see that x' is a function of x and t but not relevant to y and z. The principle of special relativity requires that the coordinate transformation equations must contain nothing which could give one of the two frames a preferred position as compared with the other frame. The inverse transformations of (2.2.1), thus, must take the same forms as (2.2.1). In particular, x must be a function of x' and t' but not relevant to y' and z'. This requires $a^0{}_2 = a^0{}_3 = 0$. Therefore the transformations (2.2.1) reduce to

$$x' = a^1{}_1 (x - vt), \qquad (2.2.2a)$$

---

[1]This initial condition leads to the constants b^μ vanishing. Non-zero values of b^μ represent the Poincaré displacements (see Sec. 3.1).

$$y' = y, \tag{2.2.2b}$$

$$z' = z, \tag{2.2.2c}$$

$$t' = a^0{}_1 x + a^0{}_0 t. \tag{2.2.2d}$$

In order to establish the constants $a^1{}_1, a^0{}_0$ and $a^0{}_1$, we consider the fact that the time coordinates are defined in such a way that the one-way velocities of light as measured in the frames are the same constant c. This implies

$$(x')^2 + (y')^2 + (z')^2 - (ct')^2 = x^2 + y^2 + z^2 - (ct)^2 = 0. \tag{2.2.3}$$

Substituting Eqs. (2.2.2) into (2.2.3), then we obtain

$$a^1{}_1 = a^0{}_1 = \frac{1}{\sqrt{1 - v^2/c^2}},$$

$$a^0{}_0 = -\frac{v/c^2}{\sqrt{1 - v^2/c^2}}.$$

So equations (2.2.2) become

$$x' = \frac{1}{\sqrt{1 - v^2/c^2}} (x - vt), \tag{2.2.4a}$$

$$y' = y, \tag{2.2.4b}$$

$$z' = z, \tag{2.2.4c}$$

$$t' = \frac{1}{\sqrt{1 - v^2/c^2}} \left(t - \frac{v}{c^2} x \right). \tag{2.2.4d}$$

This is the Lorentz transformation without rotation.

The inverse transformations of Eqs. (2.2.4) are

$$x = \frac{1}{\sqrt{1 - v^2/c^2}} (x' + vt'), \tag{2.2.5a}$$

$$y = y', \tag{2.2.5b}$$

$$z = z', \tag{2.2.5c}$$

$$t = \frac{1}{\sqrt{1 - v^2/c^2}} \left(t' + \frac{v}{c^2} x' \right), \tag{2.2.5d}$$

where the factor $\sqrt{1 - v^2/c^2}$ is called *Lorentz contraction factor*, and its reciprocal is called the *expansion factor*. Comparing the Lorentz transformations for two events $E_i(x_i, y_i, z_i, t_i)$ with $i = 1, 2$, one has

$$\Delta x' = \frac{1}{\sqrt{1 - v^2/c^2}} (\Delta x - v\Delta t), \tag{2.2.6a}$$

$$\Delta y' = \Delta y, \qquad (2.2.6b)$$

$$\Delta z' = \Delta z, \qquad (2.2.6c)$$

$$\Delta t' = \frac{1}{\sqrt{1 - v^2/c^2}} \left(\Delta t - \frac{v}{c^2} \Delta x \right), \qquad (2.2.6d)$$

where $\Delta x \equiv x_2 - x_1$, and so on. These coordinate transformations are valid for finite as well as infinitesimal intervals due to their linear forms.

The inverse of Eqs. (2.2.6) are

$$\Delta x = \frac{1}{\sqrt{1 - v^2/c^2}} (\Delta x' + v \Delta t'), \qquad (2.2.7a)$$

$$\Delta y = \Delta y', \qquad (2.2.7b)$$

$$\Delta z = \Delta z', \qquad (2.2.7c)$$

$$\Delta t = \frac{1}{\sqrt{1 - v^2/c^2}} \left(\Delta t' + \frac{v}{c^2} \Delta x' \right). \qquad (2.2.7d)$$

To derive more general Lorentz transformations where the relative velocity $\mathbf{v}$ has an arbitrary direction, let us rewrite Eqs. (2.2.4) in a three-dimensional vector (or 3-vector for simplicity) form. To do this, we decompose the vector $\mathbf{r}$ into two parts:

$$\mathbf{r} = (x, y, z) = \mathbf{r}_\| + \mathbf{r}_\perp, \qquad (2.2.8a)$$

where $\mathbf{r}_\|$ and $\mathbf{r}_\perp$ are, respectively, the projection vectors of $\mathbf{r}$ parallel with and perpendicular to the velocity $\mathbf{v}$, which are defined by

$$\mathbf{r}_\| = \frac{(\mathbf{r} \cdot \mathbf{v}) \mathbf{v}}{v^2}, \qquad \mathbf{r}_\perp = \mathbf{r} - \frac{(\mathbf{r} \cdot \mathbf{v}) \mathbf{v}}{v^2}. \qquad (2.2.8b)$$

We have known that the direction of $\mathbf{v}$ in Eqs. (2.2.4) is in the direction of x-axis, i.e.,

$$\mathbf{r}_\| = (x, 0, 0), \qquad \mathbf{r}_\perp = (0, y, z). \qquad (2.2.9)$$

By considering Eq. (2.2.9), we can rewrite the Lorentz transformations (2.2.4) in the 3-vector form:

$$\mathbf{r}'_\| = \gamma(\mathbf{v}) \left(\mathbf{r}_\| - \mathbf{v} t \right), \qquad (2.2.10a)$$

$$\mathbf{r}'_\perp = \mathbf{r}_\perp, \qquad (2.2.10b)$$

$$ct' = \gamma(\mathbf{v}) \left(ct - \frac{\mathbf{v} \cdot \mathbf{r}_\|}{c} \right), \qquad (2.2.10c)$$

with

$$\gamma(\mathbf{v}) \equiv \frac{1}{\sqrt{1 - \mathbf{v}^2/c^2}}. \qquad (2.2.10d)$$

Similarly, by definition we have

$$\mathbf{r}' = \mathbf{r}'_{\parallel} + \mathbf{r}'_{\perp}. \tag{2.2.11}$$

Substituting Eqs. (2.2.10a,b) into (2.2.11), and using the definition (2.2.8a), we obtain the Lorentz transformations in the 3-vector form:

$$\mathbf{r}' = \gamma(\mathbf{v})(\mathbf{r}^{\dagger} - \mathbf{v}t), \tag{2.2.12a}$$

$$ct' = \gamma(\mathbf{v}) \left[ct - \frac{\mathbf{v} \cdot \mathbf{r}}{c} \right], \tag{2.2.12b}$$

where

$$\mathbf{r}^{\dagger} \equiv \frac{1}{\gamma(\mathbf{v})} \left[\mathbf{r} - (1 - \gamma(\mathbf{v})) \frac{(\mathbf{v} \cdot \mathbf{r})\mathbf{v}}{v^2} \right]. \tag{2.2.12c}$$

Since the three axes in F are parallel to the corresponding axes in F', so that the transformation equation (2.2.12) represents a general Lorentz transformation without rotation.

Let us now consider the case where the Cartesian axes in F and F' do not have the same orientation. In this case while the transformation for the time, Eq. (2.2.10c), remains valid without change, the transformations for the spatial coordinates should be changed to

$$D\mathbf{r}' = \mathbf{r} + \mathbf{v} \left[(\gamma - 1) \frac{\mathbf{r} \cdot \mathbf{v}}{v^2} - \gamma t \right], \tag{2.2.13}$$

where D is the *rotation operator* which transforms the vector $\mathbf{r}'$ into the vector $D\mathbf{r}'$ corresponding to a Lorentz transformation without rotation. Therefore the inverse operator D^{-1} represents the rotation of the Cartesian axes in F' that would give these axes the same orientation with the axes in F. Thus, the relative velocities $\mathbf{v}$ and $\mathbf{v}'$ between F and F' have the relation

$$D\mathbf{v}' = -\mathbf{v}. \tag{2.2.14}$$

Multiplying (2.2.13) by D^{-1} and applying (2.2.14), the *most general Lorentz transformation with rotation* may be written in the form

$$\mathbf{r}' = D^{-1}\mathbf{r} - \mathbf{v}' \left[(\gamma - 1) \frac{\mathbf{r} \cdot \mathbf{v}}{v^2} - \gamma t \right], \tag{2.2.15a}$$

$$t' = \gamma \left(t - \frac{\mathbf{v} \cdot \mathbf{r}}{c^2} \right). \tag{2.2.15b}$$

Since D can also be interpreted as the rotation that has to be applied to the axes in F in order to obtain the same orientation of the axes in F and F', then the inverse transformations are

$$\mathbf{r} = D\mathbf{r}' - \mathbf{v} \left[(\gamma - 1) \frac{\mathbf{r}' \cdot \mathbf{v}'}{v^2} - \gamma t' \right], \tag{2.2.16a}$$

$$t = \gamma \left(t' - \frac{\mathbf{v}' \cdot \mathbf{r}'}{c^2} \right), \qquad (2.2.16b)$$

where

$$v' = |\mathbf{v}'| = v = |\mathbf{v}|. \qquad (2.2.16c)$$

2.3. Four-Dimensional Minkowski Space–Time

The most general homogeneous Lorentz transformations for any physical event can be written in the form

$$x'^{\mu} = \Lambda^{\mu}_{\ \nu} x^{\nu}, \qquad \mu, \nu = 0, 1, 2, 3, \qquad (2.3.1)$$

where

$$(x'^{\mu}) = \left(x'^0, x'^1, x'^2, x'^3 \right) \equiv (ct', x', y', z'),$$
$$(x^{\mu}) = \left(x^0, x^1, x^2, x^3 \right) \equiv (ct, x, y, z), \qquad (2.3.2)$$

and the constant coefficient matrix $(\Lambda^{\mu}_{\ \nu})$ is called the transformation matrix. The determinant of the matrix equals to ± 1:

$$det \left(\Lambda^{\mu}_{\ \nu} \right) \equiv |\Lambda^{\mu}_{\ \nu}| = \pm 1. \qquad (2.3.3)$$

The transformation matrix $(\Lambda^{\mu}_{\ \nu})$ is an element of the group of linear transformations, the *Lorentz group*. The value $+1$ of the determinant belongs to the *proper Lorentz group*, while the value -1 belongs to the *full Lorentz group* involving a reflection. The coefficient matrix of Eqs. (2.2.4) or (2.2.12) belongs to the proper Lorentz group. The following quadratic form is invariant under the Lorentz transformations (2.3.1):

$$ds^2 = (cdt)^2 - dx^2 - dy^2 - dz^2 = (cdt)^2 - (d\mathbf{r})^2, \qquad (2.3.4)$$

where $\mathbf{r} = (x^i) = (x, y, z)$. This equation is called the line element, the metric, or the interval in the four-dimensional Minkowski space.

The Lorentz transformations (2.3.1) suggest the uniform treatment of the time and space coordinates, ct, x, y, and z. This treatment was first done by Poincaré in 1906 [8] and H. Minkowski in 1909 [9]. They considered the ordinary, three-dimensional space plus the time as a space–time continuum or simply the four-dimensional Minkowski space denoted by M_4, the *"world"*. A *"world point"* is an event, the ordinary space point at a certain time, its four coordinates ct, x, y, and z. The "interval" of the four-dimensional Minkowski space is given by Eq. (2.3.4). By introducing a "metric tensor" $\eta_{\mu\nu}$, the *Minkowski metric tensor*, with the components

$$\eta_{\mu\nu} = \mathrm{diag}(1, -1, -1, -1) = \begin{pmatrix} 1 & 0 & 0 & 0 \\ 0 & -1 & 0 & 0 \\ 0 & 0 & -1 & 0 \\ 0 & 0 & 0 & -1 \end{pmatrix}, \qquad (2.3.5)$$

we can rewrite Eq. (2.3.4) in the form

$$ds^2 = \eta_{\mu\nu}dx^\mu dx^\nu. \tag{2.3.6}$$

The Lorentz transformations are those linear coordinate transformations which carry the metric tensor $\eta_{\mu\nu}$ over into itself.

The Lorentz transformations keep the 4-interval (2.3.6) invariant:

$$\eta_{\mu\nu}dx'^\mu dx'^\nu = \eta_{\alpha\beta}dx^\alpha dx^\beta. \tag{2.3.7}$$

By substituting (2.3.1) into (2.3.7), we have

$$\eta_{\mu\nu}\Lambda^\mu{}_\alpha\Lambda^\nu{}_\beta dx^\alpha dx^\beta = \eta_{\alpha\beta}dx^\alpha dx^\beta, \tag{2.3.8}$$

and, because the dx^α are arbitrary, then

$$\eta_{\mu\nu}\Lambda^\mu{}_\alpha\Lambda^\nu{}_\beta = \eta_{\alpha\beta}. \tag{2.3.9}$$

The inverse of the Minkowski metric $\eta_{\mu\nu}$ is defined by

$$\eta^{\nu\lambda}\eta_{\mu\lambda} = \delta^\nu_\mu. \tag{2.3.10a}$$

Solutions to Eq. (2.3.10a) are

$$\eta^{\nu\lambda} = \mathrm{diag}(1,-1,-1,-1). \tag{2.3.10b}$$

It is obvious that the four-dimensional Minkowski space M_4 is analogous to the ordinary three-Euclidean space: The "interval" (2.3.6) and Minkowski metric (2.3.5) are analogous, respectively, to the ordinary space interval (1.2.2) and Euclidean metric (1.2.3); The Lorentz transformations correspond to the three-dimensional orthogonal transformations. The transformation coefficients are also subject to the conditions (2.3.9) which is similar to Eq. (1.2.7).

The time and space coordinates, dx^μ, are four components of a *four-dimensional contravariant vector* in the four-dimensional Minkowski space M_4. Any four quantities, e.g. V^μ, transforming like the coordinates form the components of a four-dimensional contravariant vector. Define the covariant vector V_μ that transforms like the derivative operators $\partial/\partial x^\mu$, i.e.,

$$V'_\mu = \Lambda_\mu{}^\nu V_\nu, \tag{2.3.11a}$$

where the coefficients $\Lambda_\mu{}^\nu$ are defined by

$$\frac{\partial}{\partial x'^\mu} = \Lambda_\mu{}^\nu \frac{\partial}{\partial x^\nu}. \tag{2.3.11b}$$

By making operation of (2.3.11b) on x^μ, we have

$$\Lambda_\mu{}^\nu = \frac{\partial x^\nu}{\partial x'^\mu}. \tag{2.3.11c}$$

This shows that $\Lambda_\mu{}^\nu$ are the coefficients of the inverse of Eq. (2.3.1). In fact, by use of the contravariant transformations (2.3.1), equations (2.3.11b) can be expressed as

$$\frac{\partial}{\partial x'^\mu} = \Lambda_\mu{}^\nu \frac{\partial}{\partial x^\nu} = \Lambda_\mu{}^\nu \frac{\partial x'^\lambda}{\partial x^\nu} \frac{\partial}{\partial x'^\lambda} = \Lambda_\mu{}^\nu \Lambda^\lambda{}_\nu \frac{\partial}{\partial x'^\lambda}. \tag{2.3.11d}$$

This indicates that the coefficients $\Lambda_\mu{}^\nu$ of the covariant transformations should be the solutions to the equations

$$\Lambda^\lambda{}_\mu \Lambda_\lambda{}^\nu = \delta^\nu_\mu. \tag{2.3.12}$$

A four-dimensional contravariant tensor has N superscripts, all of which take all values 0 to 3. The tensor has, thus, 4^N components. The 4^N components transform according to the transformation law:

$$T'^{\mu\nu\cdots} = \Lambda^\mu{}_\alpha \Lambda^\nu{}_\beta T^{\alpha\beta\cdots}, \tag{2.3.13}$$

where the coefficients $\Lambda^\mu{}_\alpha$ are defined in Eqs. (2.3.1). The number of superscripts, N, is called its rank.

Similarly, a *covariant 4-tensor* is such a quantity, the transformation law of which takes the form,

$$T'_{\mu\nu\cdots} = \Lambda_\mu{}^\alpha \Lambda_\nu{}^\beta T_{\alpha\beta\cdots}. \tag{2.3.14}$$

Another type tensor is called *mixed tensor* that has N superscripts and M subscripts, which transform according to Eqs. (2.3.13) and (2.3.14):

$$T'^{\mu\nu\cdots}_{\alpha\beta\cdots} = \Lambda^\mu{}_{\mu'} \Lambda^\nu{}_{\nu'} \Lambda_\alpha{}^{\alpha'} \Lambda_\beta{}^{\beta'} T^{\mu'\nu'\cdots}_{\alpha'\beta'\cdots}. \tag{2.3.15}$$

A 4-tensor density is defined as follows: A tensor density transforms like a tensor, except that it is also multiplied by the transformation determinant (2.3.3). As long as this determinant equals to $+1$, that is, when the transformation is a proper Lorentz transformation without reflection, there is no difference between a tensor and a tensor density. But a tensor density undergoes a change of sign (compared with a tensor) when a reflection of the Einstein frame is carried out. Therefore, the transformation law of a 4-tensor density can be given by

$$\mathcal{T}'^{\mu\nu\cdots} = \Lambda^\mu{}_{\mu'} \Lambda^\nu{}_{\nu'} |\Lambda^\alpha{}_\beta| \mathcal{T}^{\mu'\nu'\cdots}. \tag{2.3.16}$$

The laws of tensor (or tensor density) algebra and calculus are: The sum or difference of two tensors (or tensor densities) of equal rank is again a tensor (or tensor density). The product of a tensor and a tensor density is a tensor density. The product of two tensor densities is a tensor. The contraction of the Minkowski metric and a tensor (density) yields a new tensor (density) of lower rank. The derivatives of the components of a tensor (density) are the components of a new tensor (density), the rank of which is greater by 1 than the rank of the original tensor (density).

Tensors (or tensor densities) may have symmetry properties, e.g., the symmetry or the antisymmetry with respect to their indices. A tensor (density) is called to be symmetric, if it is not changed when two or more indices are exchanged. For example, $T^{\mu\nu}$ is a symmetric tensor, if

$$T^{\mu\nu} = T^{\nu\mu}.$$

When a tensor (density) remains the same or changes the sign of every component upon the permutation of certain indices, the sign depending on whether it is an even or an odd permutation, we say that the tensor is antisymmetric (also skewsymmetric or alternating) with respect to indices. Instances are

$$T^{\mu\nu} = -T^{\nu\mu},$$

$$T^{\mu\nu\alpha\beta} = T^{\nu\alpha\mu\beta} = T^{\alpha\mu\nu\beta} = -T^{\mu\alpha\nu\beta} = -T^{\nu\mu\alpha\beta} = -T^{\alpha\nu\mu\beta}.$$

All such symmetry properties of a tensor (density) are invariant under the Lorentz transformations.

The 4-interval ds^2 is a tensor of rank 0, a scalar; (dx^μ) is a contravariant tensor of rank 1, a contravariant 4-vector; Thus, the Minkowski metric $\eta^{\mu\nu}$ is a contravariant tensor of rank 2, while its inverse $\eta_{\mu\nu}$ is a covariant tensor of rank 2.

From the transformation laws (2.3.13) and (2.3.14) and the conditions (2.3.12) we can see that the product of the metric $\eta_{\mu\nu}$ and the contravariant 4-vector V^ν, $V_\mu \equiv \eta_{\mu\nu}V^\nu$, is a covariant 4-vector. It is similar that the quantity $V^\mu \equiv \eta^{\mu\nu}V_\nu$ with V_ν being a covariant 4-vector is a contravariant 4-vector. Therefore the Minkowski metric $\eta_{\mu\nu}$ or its inverse $\eta^{\mu\nu}$ can be used to raise or to lower the indices of tensors.

We can denote a 4-vector, e.g. (V^μ), by $V \equiv (V^\mu, \mu = 0-3) = (V^0, \mathbf{V})$ with V^0 being the time component and $\mathbf{V} = (V^i, i = 1, 2, 3)$ being the spatial component. The square of norm of the 4-vector V is defined as

$$V^2 = \eta_{\mu\nu}V^\mu V^\nu = \left(V^0\right)^2 - \left(V^i\right)^2 = \left(V^0\right)^2 - \mathbf{V}^2, \qquad (2.3.17)$$

which is invariant under the Lorentz transformations, just as the interval ds^2 does.

The scalar product of two 4-vectors, e.g. V and U, is defined as

$$V \cdot U = \eta_{\mu\nu}V^\mu U^\nu = V_\nu U^\nu = V^\mu U_\mu, \qquad (2.3.18)$$

where $V_\nu \equiv \eta_{\mu\nu}V^\mu$, and $U_\mu \equiv \eta_{\mu\nu}U^\nu$.

The 4-interval of $ds^2 = 0$ is called *light-like*. The two events separating by the light-like interval are called the light-like events: The two events can be just connected by a light signal which leaves the site of one event at the time it occurs and arrives at the site of the other event at the time it takes place. The 4-interval of $ds^2 > 0$ is called *time-like*. The two events separating by the time-like interval are called the time-like events: The sequence of the two events is such that a light signal starting from either event reaches to the site of the other only before it will occur. The *world lines* of all massive particles are time-like, because their velocities can never be faster than light. The 4-interval of $ds^2 < 0$ is called *space-like*. The two events separating by the space-like interval are called the space-like events: The sequence of the two events is such that a light signal starting from either event reaches to the site of the other only after it has occurred. The space-like events are those of non-causality, because there are not any signals propagating at faster-than-light.

We now apply the concept of the space-time interval to the motion of a massive particle and to the space-time points along its path. The path of a massive particle is a *time-like world line*, because its velocity is always smaller than the velocity of light. If the motion of the particle is not straight-line and uniform, we can define a parameter τ along its path by the differential equation

$$d\tau \equiv \frac{ds}{c} = \sqrt{dt^2 - \frac{d\mathbf{r}^2}{c^2}} = dt\sqrt{1 - \frac{\mathbf{u}^2}{c^2}}, \qquad (2.3.19)$$

where $\mathbf{u} \equiv d\mathbf{r}/dt$ is the velocity of the particle, $d\tau$ is the time interval shown by a clock rigidly connected with the moving particle, really its *"proper time"* (*its own time*) *interval*, and dt is the corresponding *coordinate time interval*. Therefore, Eq. (2.3.19) gives the relation between coordinate time and proper time (see Sec. 2.9),

$$\frac{d\tau}{dt} = \sqrt{1 - \frac{\mathbf{u}^2}{c^2}}. \qquad (2.3.20)$$

This relation is valid for accelerated as well as unaccelerated particles.

Both $d\tau$ and τ, which is defined by the integral

$$\tau = \int dt\sqrt{1 - \frac{\mathbf{u}^2}{c^2}}, \qquad (2.3.21)$$

are invariant under the Lorentz transformations.

The 4-velocity u^μ of the particle is defined as

$$u^\mu = \frac{dx^\mu}{d\tau} = \left(c\frac{dt}{d\tau}, \frac{d\mathbf{r}}{d\tau}\right), \qquad (2.3.22)$$

which is a 4-vector, because dx^μ is a 4-vector and $d\tau$ is a scalar.

The 4-acceleration a^μ of the particle is given by

$$a^\mu = \frac{du^\mu}{d\tau} = \frac{d^2x^\mu}{d\tau^2}, \qquad (2.3.23)$$

which is also a 4-vector.

2.4. Einstein's Law of the Addition of Velocities

Let space–time coordinates of a particle as seen in the frame F be (x, y, z, t), and the corresponding coordinates in the frame F' be denoted by (x', y', z', t'). The velocities of the particle as seen in F and F' are, respectively, given by

$$\mathbf{u} = (u_x, u_y, u_z) = \left(\frac{dx^i}{dt}, \quad i = 1, 2, 3 \right) \tag{2.4.1a}$$

and

$$\mathbf{u}' = \left(u'_x, u'_y, u'_z \right) = \left(\frac{dx'^i}{dt'}, \quad i = 1, 2, 3 \right). \tag{2.4.1b}$$

The relationship between $\mathbf{u}$ and $\mathbf{u}'$ can be found from the Lorentz transformations (2.2.6). To this end, changing Δ into the derivative d and then dividing (2.2.6a,b,c) by (2.2.6d), one arrives at

$$u'_x = \frac{u_x - v}{1 - vu_x/c^2}, \tag{2.4.2a}$$

$$u'_y = u_y \frac{\sqrt{1 - v^2/c^2}}{1 - vu_x/c^2}, \tag{2.4.2b}$$

$$u'_z = u_z \frac{\sqrt{1 - v^2/c^2}}{1 - vu_x/c^2}. \tag{2.4.2c}$$

This is Einstein's law of the addition of velocities. The inverse transformations of Eq. (2.4.2) are

$$u_x = \frac{u'_x + v}{1 + vu'_x/c^2}, \tag{2.4.3a}$$

$$u_y = u'_y \frac{\sqrt{1 - v^2/c^2}}{1 + vu'_x/c^2}, \tag{2.4.3b}$$

$$u_z = u'_z \frac{\sqrt{1 - v^2/c^2}}{1 + vu'_x/c^2}. \tag{2.4.3c}$$

The relation between the norms of $\mathbf{u}$ and $\mathbf{u}'$ is obtained from Eq. (2.4.2)

$$u'^2 = c^2 \frac{u^2 - 2vu \cos \alpha + v^2 \left[1 - (u^2/c^2) \cos^2 \alpha \right]}{c^2 - 2vu \cos \alpha + (v^2 u^2/c^2) \cos^2 \alpha}, \tag{2.4.4a}$$

where

$$u'^2 = u'^2_x + u'^2_y + u'^2_z, \quad u^2 = u^2_x + u^2_y + u^2_z, \tag{2.4.4b}$$

$$u_x = u \cos \alpha, \quad u_y = u \cos \beta, \quad u_z = u \cos \gamma, \tag{2.4.4c}$$

and $\cos^2 \alpha + \cos^2 \beta + \cos^2 \gamma = 1$ are used. Equation (2.4.4a) can be expressed in terms of the 3-vector form:

$$\mathbf{u}'^2 = \frac{1}{(1 - \mathbf{u} \cdot \mathbf{v}/c^2)^2} \left[\mathbf{u}^2 - 2\mathbf{u} \cdot \mathbf{v} + \mathbf{v}^2 - (\mathbf{u} \times \mathbf{v})^2/c^2 \right], \qquad (2.4.5)$$

where $\mathbf{u} \cdot \mathbf{v} = u^i v^i$ is a scalar product, and $(\mathbf{u} \times \mathbf{v})^i = \epsilon^{ijk} u^j v^k$ is a vector product.

The addition law of velocities can be expressed in the 3-vector form, just as the Lorentz transformations do in the section 2.2. To do this, let us start from the 4-velocity u^μ of a body, which is defined by Eq. (2.3.22),

$$u^\mu = \left(c\frac{dt}{d\tau}, \frac{d\mathbf{r}}{d\tau} \right). \qquad (2.4.6)$$

The relation between the proper time interval $d\tau$ and coordinate time interval dt is given by Eq. (2.3.20),

$$\frac{dt}{d\tau} = \frac{1}{\sqrt{1 - \mathbf{u}^2/c^2}}, \qquad (2.4.7a)$$

and hence

$$\frac{d}{d\tau} = \frac{dt}{d\tau}\frac{d}{dt} = \gamma(\mathbf{u})\frac{d}{dt}, \qquad (2.4.7b)$$

where

$$\gamma(\mathbf{u}) \equiv \frac{1}{\sqrt{1 - \mathbf{u}^2/c^2}}, \qquad (2.4.7c)$$

and $\mathbf{u} \equiv d\mathbf{r}/dt$ is the velocity of the body. Then, we can write

$$\frac{d\mathbf{r}}{d\tau} = \frac{dt}{d\tau}\frac{d\mathbf{r}}{dt} = \gamma(\mathbf{u})\mathbf{u}, \qquad (2.4.8)$$

and equation (2.4.6) becomes

$$\{\gamma(\mathbf{u})c, \ \gamma(\mathbf{u})\mathbf{u}\}. \qquad (2.4.9)$$

The components $\gamma(\mathbf{u})c$ and $\gamma(\mathbf{u})\mathbf{u}$ transform like, respectively, the coordinates ct and $\mathbf{r}$. Thus, following Eqs. (2.2.12) we have the Lorentz transformations of the 4-velocity,

$$\gamma(\mathbf{u}')\mathbf{u}' = \gamma(\mathbf{v})\gamma(\mathbf{u})(\mathbf{u}^\dagger - \mathbf{v}), \qquad (2.4.10a)$$

$$\gamma(\mathbf{u}') = \gamma(\mathbf{v})\gamma(\mathbf{u})\left(1 - \frac{\mathbf{u} \cdot \mathbf{v}}{c^2} \right), \qquad (2.4.10b)$$

where

$$\mathbf{u}^\dagger \equiv \frac{1}{\gamma(\mathbf{v})} \left[\mathbf{u} - (1 - \gamma(\mathbf{v}))\frac{(\mathbf{v} \cdot \mathbf{u})\mathbf{v}}{v^2} \right], \qquad (2.4.10c)$$

and $\gamma(\mathbf{v})$ is given by (2.2.10d). Substituting (2.4.10b) for $\gamma(\mathbf{u}')$ in (2.4.10a), we get Einstein's addition law of velocities in the 3-vector form,

$$\mathbf{u}' = \frac{\mathbf{u}^\dagger - \mathbf{v}}{1 - \mathbf{u} \cdot \mathbf{v}/c^2}. \tag{2.4.11}$$

In the special case of $\mathbf{v} = (v, 0, 0)$, equations (2.4.11) and (2.4.10c) reduce to Eqs. (2.4.2).

In the above, we have obtained the addition law of velocities in the 3-vector form. Let us now discuss some results of the addition law in some special cases.

In particular, by substituting the velocity of light for u in Eq. (2.4.4a) we have $u' = c$, the constancy of one-way velocity of light. Of course, this is a trivial result because the constancy of the one-way velocity of light is one of the starting points that we have used in derivation of Lorentz transformations. The constant velocity of light does not satisfy the Newtonian law of the addition of velocities, but is consistent with Einstein's law of addition of velocities. Therefore, the constancy of the one-way velocity of light is compatible with the principle of special relativity although it contradicts the Galilean principle of relativity. This is easy to understand if we recall that, as argued in section 1.5, velocities depend upon the definition of simultaneity, the definition of the inertial frames of reference. It is the same case that the description of the principle of relativity is also dependent on the definition of the inertial frames of reference. So we see that in this sense the principle of special relativity is not only a generalization of the Galilean principle of relativity, but also based on the different definition of simultaneity. In other words, the principle of special relativity is closely related to the principle of constancy of the velocity of light.

Lorentz transformations (2.2.4) involve the Lorentz contraction factor, $\sqrt{1 - v^2/c^2}$, which goes to zero when $v \to c$. Thus, we would be faced the three cases, $v < c, v = c$, and $v > c$:

(i) In the case of $v > c$, the value of the contraction factor $\sqrt{1 - v^2/c^2}$ becomes a purely imaginary number. It is known from the relationships among energy, momentum, mass and velocity that both energy and momentum depend on the mass factor $m_0/\sqrt{1 - v^2/c^2}$ (see section 3.1). Thus, if simultaneously $v^2/c^2 > 1$ and m_0 is replaced with the purely imaginary proper mass im^*, then the conserved quantities could remain real numbers. This is to say that a *faster-than-light* particle, which is called a *tachyon*, should possess a purely imaginary proper mass [10–12]. Many attempts have been done to search for tachyons. However, all of these failed. Therefore, to remove the tachyon becomes a necessary condition for constructing a physical field theory.

(ii)The mass–energy relationship, $E = m_0 c^2/\sqrt{1 - v^2/c^2}$ (see Sec. 3.1), shows that $v = c$ leads to $E = \infty$. The reference bodies, therefore, could not move at the velocity of light; Particles moving at the speed of light must have the vanishing proper mass, i.e., $m_0 = 0$.

(iii) From the Newtonian law of the addition of velocities, $\mathbf{u} = \mathbf{u}' + \mathbf{v}$, we know that speeds of reference bodies can become faster than light. However, Einstein's law of the addition of velocities (2.4.2) gives a different result. To see this, let us consider the case of $u'_y = u'_z = 0$. From (2.4.2) we have $u_y = u_z = 0$ and

$$u' = \frac{u - v}{1 - uv/c^2},\tag{2.4.12}$$

where $u' = u'_x$ and $u = u_x$. If $u' < c$, then Eq. (2.4.12) becomes

$$c(u - v) < c^2 - uv, \quad or \quad u(c + v) < c(c + v).\tag{2.4.13}$$

This leads to $u < c$. Inversely, if $u < c$, then (2.4.3) leads to $u' < c$. Therefore, in Einstein's theory of special relativity, one cannot transform a subluminal speed into a superluminal speed by use of Lorentz transformations. Thus, the tachyons and the ordinary particles cannot be able to transfer each other.

2.5. Transformation of Accelerations

To derive the transformation law of accelerations, let us recall the 4-acceleration (2.3.23),

$$a^\mu = \frac{du^\mu}{d\tau} = \left(c\frac{d^2t}{d\tau^2}, \frac{d^2\mathbf{r}}{d\tau^2} \right).\tag{2.5.1}$$

By use of Eq. (2.4.7a), the components of a^μ become

$$c\frac{d^2t}{d\tau^2} = c\frac{d}{d\tau}\frac{dt}{d\tau} = c\frac{d\gamma(\mathbf{u})}{d\tau} = c\gamma(\mathbf{u})\frac{d\gamma(\mathbf{u})}{dt} = c\gamma(\mathbf{u})^2\xi,\tag{2.5.2a}$$

$$\frac{d^2\mathbf{r}}{d\tau^2} = \frac{d}{d\tau}\frac{d\mathbf{r}}{d\tau} = \gamma(\mathbf{u})\frac{d}{dt}[\gamma(\mathbf{u})\mathbf{u}] = \gamma(\mathbf{u})^2(\mathbf{a} + \mathbf{u}\xi),\tag{2.5.2b}$$

where

$$\xi \equiv \gamma(\mathbf{u})^2\frac{\mathbf{a} \cdot \mathbf{u}}{c^2}.\tag{2.5.2c}$$

The 4-acceleration $a^\mu = d^2x^\mu/d\tau^2$ is a 4-vector and transforms like x^μ. Hence, following Eqs. (2.2.12), we have the transformation equations for the components of a^μ,

$$[\gamma(\mathbf{u}')]^2(a' + \mathbf{u}'\xi') = \gamma(\mathbf{v})[\gamma(\mathbf{u})]^2(a^\dagger + \mathbf{u}^\dagger\xi - \mathbf{v}\xi),\tag{2.5.3a}$$

$$[\gamma(\mathbf{u}')]^2\xi' = \gamma(\mathbf{v})[\gamma(\mathbf{u})]^2[\xi - \frac{\mathbf{v} \cdot (\mathbf{a} + \mathbf{u}\xi)}{c^2}],\tag{2.5.3b}$$

where $\mathbf{u}^\dagger$ is defined by (2.4.10c), and

$$\mathbf{a}^\dagger \equiv \frac{1}{\gamma(\mathbf{v})}\left[\mathbf{a} - (1 - \gamma(\mathbf{v}))\frac{(\mathbf{v} \cdot \mathbf{a})\mathbf{v}}{v^2} \right].\tag{2.5.4a}$$

By using Einstein's law of the addition of velocities (2.4.11), and then eliminating ξ from (2.5.3), we obtain

$$\mathbf{a}' = \frac{(1 - \mathbf{u} \cdot \mathbf{v}/c^2)\mathbf{a}^\dagger + (\mathbf{u}^\dagger - \mathbf{v})(\mathbf{a} \cdot \mathbf{v})/c^2}{\gamma(\mathbf{v})(1 - \mathbf{u} \cdot \mathbf{v}/c^2)^3}. \qquad (2.5.4b)$$

This is the transformation law of accelerations for a moving particle. In particular, if we choose the instantaneous frame in where the particle is at rest, i.e., $\mathbf{u} = 0$, then Eq. (2.5.4b) reduces to

$$\mathbf{a}' = \frac{\mathbf{a}^\dagger - \mathbf{v}(\mathbf{a} \cdot \mathbf{v})/c^2}{\gamma(\mathbf{v})}. \qquad (2.5.5)$$

In Newtonian limit, $c \to \infty$, the transformation (2.5.4) or (2.5.5) reduces to the Newtonian result $\mathbf{a}' = \mathbf{a}$.

2.6. Infinitesimal Lorentz Transformation

It is easy to see that when the constant c goes to infinity the Lorentz transformations (2.2.4) reduce to the Galilean transformations (2.1.1). Of course, this result is trivial, because the difference between Newtonian frames and Einstein frames is just the different definitions of simultaneity as mentioned in Sec. 1.3; In other words, Einstein frames would become Newtonian frames if $c \to \infty$. However, the value of the velocity of light c is finite. Therefore, Lorentz transformations differ from the Galilean transformations, even if the first-order approximation in v/c is considered. In fact, the infinitesimal Lorentz transformations follow from Eqs. (2.2.4)

$$x' = x - vt, \qquad (2.6.1a)$$

$$y' = y, \qquad (2.6.1b)$$

$$z' = z, \qquad (2.6.1c)$$

$$t' = t - \frac{v}{c^2}x. \qquad (2.6.1d)$$

By dividing the derivatives of Eqs. (2.6.1a,b,c) by the derivative of Eq. (2.6.1d), we have the law of the addition of velocities:

$$u'_x = \frac{u_x - v}{1 - vu_x/c^2}, \qquad (2.6.2a)$$

$$u'_y = \frac{u_y}{1 - vu_x/c^2}, \qquad (2.6.2b)$$

$$u'_z = \frac{u_z}{1 - vu_x/c^2}. \qquad (2.6.2c)$$

Of course, Eqs. (2.6.2) is just an approximation of Eqs. (2.4.2) to the first-order in v/c.

On the right-hand side of the infinitesimal Lorentz transformation (2.6.1d), the second term vx/c^2 is called *Einstein's simultaneity factor* that comes from the definition of Einstein simultaneity. If $(x/t)/c$ or $(dx/dt)/c$ has the order of unity, then the simultaneity factor cannot be neglected. For example, if the space–time coordinates describe the motion of a particle, then dx/dt is just the x-component u_x of its velocity. In case of $u_x \sim c$, Einstein's simultaneity factor vx/c^2 is of the same order with v/c and, hence, cannot be neglected. An extreme case is just the electromagnetic interaction. Therefore, Lorentz transformations to the first-order will generally differ from the Galilean transformations. Similarly, the first-order approximation (2.6.2) of Einstein's law of the addition of velocities is different from the Newtonian law of the addition of velocities (2.1.2). This is just the reason that the electrodynamic equations of a moving body with a low velocity v are covariant to the first-order in v/c under the infinitesimal Lorentz transformations (2.6.1), but not under the Galilean transformations (2.1.1). In other words, this is just the reason that the electromagnetism experiments for moving bodies with low velocities can be explained by the infinitesimal Lorentz transformations (2.6.1), but not by the Galilean transformations (2.1.1). However, we want to stress that this does not mean that the electromagnetism experiments for moving bodies with low velocities provide the evidence of the vanishing directional parameter $q_r = 0$ (i.e., the constancy of the one-way velocity of light). We shall see in chapter 6 that a non-zero value of q_r would not give any observable effects in physical experiments.

If $(x/t)/c$ or u_x/c has the same order as v/c, then the infinitesimal Lorentz transformations (2.6.1) reduce to Galilean transformations (2.1.1) because vx/c^2 should be neglected as the second-order quantity and, at the same time, equations (2.6.2) reduce to the Newtonian law of the addition of velocities (2.1.2). Thus, we conclude that Galilean transformations (2.1.1) are the first-order approximation of Lorentz transformations (2.6.1) under the following conditions: Both v/c and $(dx/dt)/c$ are much less than unity so that the second-order of them can be neglected. Therefore, only for the macroscopic mechanics, can we say that Newtonian space–time theory is an approximation to Einstein's theory of special relativity.

2.7. Simultaneity and Causality

In Newtonian frames, as well known, the simultaneity is independent of motion, and therefore, is absolute: if two events occur simultaneously at two separate points in a given Newtonian frame, then they should occur simultaneously in any other Newtonian frames. However, it is not the case in Einstein's theory.

Consider two events, E_I and E_{II}, the space–time coordinates of which in the Einstein frame F are, respectively, $(x_I, 0, 0, t_I)$ and $(x_{II}, 0, 0, t_{II})$. By use of the Lorentz transformations (2.2.6), the differences between their space–time coordinates in the Einstein frame F' are obtained as follows:

$$\Delta x' = \frac{1}{\sqrt{1 - v^2/c^2}}(\Delta x - v\Delta t), \qquad (2.7.1a)$$

$$\Delta t' = \frac{1}{\sqrt{1 - v^2/c^2}}\left(\Delta t - \frac{v}{c^2}\Delta x\right), \qquad (2.7.1b)$$

where $\Delta x = x_{II} - x_I$, $\Delta t = t_{II} - t_I$, and so on. Assume that these two events occur simultaneously in the frame F at two different points, i.e.,

$$\Delta t = 0, \quad \Delta x \neq 0. \qquad (2.7.2)$$

Substituting the initial conditions (2.7.2) into (2.7.1), we have

$$\Delta x' = \frac{\Delta x}{\sqrt{1 - v^2/c^2}} \neq 0, \qquad (2.7.3a)$$

$$\Delta t' = -\frac{1}{\sqrt{1 - v^2/c^2}}\frac{v}{c}\frac{\Delta x}{c} \neq 0. \qquad (2.7.3b)$$

Equations (2.7.3) show that the two events occur at different times in the frame F'. Therefore, we conclude that *simultaneity is a relative conception: Any two events occurring simultaneously at two different space points in a given Einstein frame will occur no longer simultaneously in any other Einstein frames.*

We now discuss the causality problem. Causality can be stated as follows: *The time sequence of two events must be preserved.* This implies that the time differences Δt and $\Delta t'$ must have the same sign. However, the transformation (2.7.1b) shows that the relative sign of Δt and $\Delta t'$ depends upon the rate $\Delta x/\Delta t$. Now let us rewrite Eq. (2.7.1b) in the form

$$\Delta t' = \frac{\Delta t}{\sqrt{1 - v^2/c^2}}\left(1 - \frac{v}{c}\frac{\Delta x/\Delta t}{c}\right). \qquad (2.7.4)$$

Consider the following three cases:

(i) *Two time-like events.* In this case, $ds^2 = c^2(\Delta t)^2 - (\Delta x)^2 > 0$, i.e., $\Delta x/\Delta t < c$. Furthermore, owing to $v < c$, we have $1 - v(\Delta x/\Delta t)/c^2 > 0$. Therefore, we see from (2.7.4) that the time differences $\Delta t'$ and Δt have the same sign. This is to say that the causality will be preserved for the time-like events;

(ii) *Two light-like events.* We have $ds^2 = c^2(\Delta t)^2 - (\Delta x)^2 = 0$, i.e., $\Delta x/\Delta t = c$. Then, $1 - v(\Delta x/\Delta t)/c^2 = 1 - v/c > 0$ owing to $v < c$. We get the same result: $\Delta t'$ and Δt have the same sign, i.e., the causality is also preserved.

(iii) *Two space-like events.* We have $ds^2 = c^2(\Delta t)^2 - (\Delta x)^2 < 0$, i.e., $\Delta x/\Delta t > c$. In the case of

$$v < c^2\frac{\Delta t}{\Delta x}, \qquad (2.7.5)$$

i.e., $1 - v(\Delta x/\Delta t)/c^2 > 0$, then $\Delta t'$ and Δt have the same sign, and hence the causality is preserved. However, when the value of the relative velocity v satisfies the condition,

$$c > v > c^2 \frac{\Delta t}{\Delta x}, \qquad (2.7.6)$$

i.e., $1 - v(\Delta x/\Delta t)/c^2 < 0$, then Eq. (2.7.4) shows that $\Delta t'$ and Δt have different signs. This implies that the time sequence of the two space-like events in the frame F' is inverse. However, a connection between two space-like events must be realized by a superluminal light signal, which does not exist in the nature. This is to say that the time sequence of two space-like events has no any physical meaning. Therefore, the inversion of the time sequence is not relevant to the causality.

2.8. Contraction of a Moving Body

The length measurement of a rest rod is simple. Let a rod be rigidly connected with the frame F', the end points of which have the space coordinates (x'_2, y'_2, z'_2) and (x'_1, y'_1, z'_1). Its length in its rest frame, *proper length*, is simply given by

$$l'_0 = \sqrt{(\Delta x')^2 + (\Delta y')^2 + (\Delta z')^2}, \qquad (2.8.1)$$

where $\Delta x' = (x'_2 - x'_1)$, and so on. We now consider how to measure the length of the rod in the other frame F moving with the velocity $-\mathbf{v} = (-v, 0, 0)$ relative to F'. According to Eqs. (2.2.6a,b,c), the transformations of the space coordinates are

$$\Delta x' = \frac{1}{\sqrt{1 - v^2/c^2}}(\Delta x - v\Delta t), \qquad (2.8.2a)$$

$$\Delta y' = \Delta y, \qquad (2.8.2b)$$

$$\Delta z' = \Delta z, \qquad (2.8.2c)$$

where Δt is the difference between the time coordinates t_1 and t_2 at which the two end points of the rod are measured in the frame F, respectively. Equation (2.8.2a) shows that the value of Δx depends on Δt. Thus, such a coordinate difference cannot generally represent the projection of length of the rod in the x-axis. The length of the rigid rod measured by the observer in F is defined to be such a space distance that its corresponding time coordinate difference vanishes, i.e., $\Delta t = 0$. In other words, we must simultaneously measure the two end points of the rod in F. Thus, from Eqs. (2.8.2) we have the relation between the lengths of the rod in the two frames,

$$l_x = l'_x \sqrt{1 - \frac{v^2}{c^2}}, \qquad (2.8.3a)$$

$$l_y = l'_y, \qquad (2.8.3b)$$

$$l_z = l'_z, \tag{2.8.3c}$$

where $l_x \equiv \Delta x, l'_x \equiv \Delta x'$, and so on.

If the rigid rod is perpendicular to the direction of the relative velocity $\mathbf{v}$ of F' and F, then it has the same length in either frame.

On the other hand, if the rigid rod is parallel with the direction of the relative velocity $\mathbf{v}$, then the length of the rod measured in F is from (2.8.3a)

$$l = l'_0\sqrt{1 - \frac{v^2}{c^2}}. \tag{2.8.4a}$$

Inversely, if the rigid rod is fixed with the frame F and its rest length (i.e. proper length) is l_0, then its length as seen in F' is given by

$$l' = l_0\sqrt{1 - \frac{v^2}{c^2}}. \tag{2.8.4b}$$

These results show that the dimension of a moving rigid body in its moving direction is smaller than its rest dimension. This effect is called *contraction of a moving body* or simply *length contraction*.

Concept of "contraction" was firstly introduced by Fitzgerald and Lorentz for explaining the null result of the Michelson–Morley experiment (see subsection 8.2.1 for detail). Although their formula has the same form as Einstein's formula (2.8.4), the physical meanings are essentially different: The "contraction" in special relativity is a relative conception, while the contraction of a moving body in the ether theory is absolute.

2.9. Time Dilation of a Moving Clock

In section 2.3, we have given the expression for the *proper time* of any moving particle, i.e., Eq. (2.3.20),

$$\frac{d\tau}{dt} = \sqrt{1 - \frac{\mathbf{u}^2}{c^2}}. \tag{2.9.1}$$

This relation is valid for accelerated as well as unaccelerated particles. If the particle is just a moving clock, then this equation represents the reduction of the rate of the moving clock relative to the rate of a rest clock. This is the *time dilation* or *retardation of time*.

We now derive this effect from the Lorentz transformations (2.2.6). Let a standard clock be at rest in the frame F' and, therefore, its location be fixed, i.e., $\Delta x^i = 0$ for $i = 1, 2, 3$. Applying these initial conditions to Eqs. (2.2.6), we have the transformation between the time coordinates,

$$\Delta\tau' = \Delta t\sqrt{1 - \frac{v^2}{c^2}}, \tag{2.9.2a}$$

where $\Delta\tau' \equiv \Delta t'$ is the difference between two readings of the clock fixed at the frame F', i.e., its *proper time interval*.

Similarly, if a clock is at rest in F, then its proper time interval $\Delta\tau$ is related to its coordinate time interval $\Delta t'$ through the equation:

$$\Delta\tau = \Delta t'\sqrt{1 - \frac{v^2}{c^2}}, \tag{2.9.2b}$$

Equation (2.9.1) or (2.9.2) is the formula for the *time dilation* or *time retardation* of a moving clock. If the motion of the clock is not uniform, one should use the integral (2.2.21) to get a finite proper time interval.

We see that the *proper time interval* is *smaller* than its *coordinate time interval*. In other words, a moving clock runs slowly comparing to a series of rest clocks synchronized according to Einstein's definition of simultaneity. This retardation of a moving clock (or time dilation) is also a relative conception, just as the contraction of a moving rod does.

2.10. Aberration and Doppler Effect

Both the aberration and Doppler effect are the results coming directly from the Lorentz transformations. In particular, the effect of aberration comes from the transformations of the direction of a light ray, and thus we can derive this effect from Einstein's law of the addition of velocities. We first give the derivation of both effects in a 4-vector form.

Consider a single frequency electromagnetic wave that is described by the plane wave,

$$\mathbf{E} = \mathbf{E}_0 e^{i\psi}, \tag{2.10.1}$$

where

$$\psi = \mathbf{k} \cdot \mathbf{r} - \omega t \tag{2.10.2}$$

is the phase function of the plane wave, $\mathbf{k}$ and ω are, respectively, its wave vector and angular frequency, and $\mathbf{r}$ is the radial vector of the space point (x, y, z). The phase ψ is an invariant with respect to Lorentz transformations (see, e.g., [13]). Therefore, by introducing the four-dimensional covariant wave vector,

$$(k_0, k_1, k_2, k_3) \equiv \left(-\frac{\omega}{c}, \mathbf{k}\right), \tag{2.10.3}$$

we can now rewrite the phase function in a four-dimensional form as follows:

$$\psi = k_\mu x^\mu, \tag{2.10.4}$$

where

$$x^\mu = (ct, \mathbf{r}) \tag{2.10.5}$$

is a contravariant 4-vector.

The wave 4-vector k_μ transforms like a covariant 4-vector, i.e.,

$$k'_{x'} = \frac{1}{\sqrt{1 - v^2/c^2}} \left(k_x - \frac{v\omega}{c^2} \right), \tag{2.10.6a}$$

$$k'_{y'} = k_y, \tag{2.10.6b}$$

$$\omega' = \frac{\omega - vk_x}{\sqrt{1 - v^2/c^2}}, \tag{2.10.6c}$$

where the path of the light ray is assumed, for simplicity, to lie in the xy-plane, i.e., $\mathbf{k'} = (k'_{x'}, k'_{y'}, 0)$, and ω' is the frequency of light as measured in the frame F'. Introducing the angles θ and θ' between directions of the light ray and relative velocity $\mathbf{v}$, i.e., $k_x = k\cos\theta = (\omega/c)\cos\theta$, $k^y = (\omega/c)\sin\theta$, $k'_{x'} = k'\cos\theta' = (\omega'/c)\cos\theta'$, $k'_{y'} = (\omega'/c)\sin\theta'$, we can write Eqs. (2.10.6) as

$$\omega'\cos\theta' = \omega \frac{\cos\theta - v/c}{\sqrt{1 - v^2/c^2}} \tag{2.10.7a}$$

$$\omega'\sin\theta' = \omega\sin\theta, \tag{2.10.7b}$$

$$\omega' = \omega \frac{1 - (v/c)\cos\theta}{\sqrt{1 - v^2/c^2}}. \tag{2.10.7c}$$

The last equation (2.10.7c) is just the *relativistic Doppler effect* formula. By substituting Eq. (2.10.7c) into Eqs. (2.10.7a,b) for ω', we get the *aberration* formula,

$$\cos\theta' = \frac{\cos\theta - v/c}{1 - (v/c)\cos\theta}, \tag{2.10.8a}$$

$$\sin\theta' = \frac{\sin\theta}{1 - (v/c)\cos\theta} \sqrt{1 - \frac{v^2}{c^2}}. \tag{2.10.8b}$$

The difference between the above relativistic Doppler effect (or aberration effect) and that in classical physics is due to the contraction factor $\sqrt{1 - v^2/c^2}$. In particular, in case of $\theta = 90°$ the Doppler effect (2.10.7c) becomes

$$\omega' = \frac{\omega}{\sqrt{1 - v^2/c^2}}. \tag{2.10.9}$$

This is the so-called *transverse Doppler effect*, which differs essentially from the status in classical physics where there is not this kind of Doppler effect. The transverse Doppler effect (2.10.9) shows

$$\omega < \omega', \tag{2.10.10}$$

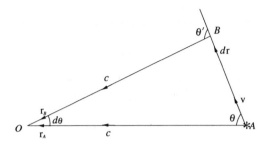

Fig. 4. A sketch for derivation of Doppler effect.

where ω' is the *proper frequency* (or the rest frequency). This indicates that the frequency of light as measured by a observer moving perpendicularly with respect to the light ray is smaller than its proper frequency. In other words, the wave length of light as measured by the moving observer is larger than its *proper wave length*. This is called *red-shift* of a spectral line.

The transverse Doppler effect is a relativistic effect, which comes just from the time dilation. In other wards, the contraction factor $\sqrt{1 - v^2/c^2}$ in Eq. (2.10.9) follows just from the time dilation. To make this explicit, let us use another method (see Fig. 4) to derive the Doppler effect formula (2.10.7c).

Figure 4 is similar to figure 6.1 which is a special case where the velocity of Jupiter is constant and the average direction of light rays is perpendicular to the path of motion of the Jupiter.

In figure 4, a light source S, which is at rest in the Einstein frame F', has the instantaneous velocity $\mathbf{v}$ as seen in another Einstein frame (i.e., the laboratory frame) F in where an observer is at rest at the point O. Let the source emit the n-th crest of light wave at the point A at the coordinate time t_A as seen in F, while the corresponding time as timed by a clock connected rigidly with the source is τ'_A. Then, at the coordinate time $t_A + dt$ the source reaches the point B as seen in F, and emits the $(n + dn)$-th crest, while the corresponding time with respect to the source is τ'_B. We see that τ'_A and τ'_B are the two readings of the same clock connected with the source, and thus the difference

$$d\tau' = \tau'_B - \tau'_A \qquad (2.10.11a)$$

is a proper time interval. However, t_A is given by a clock at A, and while $t_A + dt$ is the reading of another clock at B. So the difference

$$dt = t_B - t_A \qquad (2.10.11b)$$

is a coordinate time interval.

We now assume that the n-th and $(n + dn)$-th crests are received by the observer at the point O at the times τ_n and τ_{n+dn}, respectively. The difference,

$$d\tau \equiv \tau_{n+dn} - \tau_n, \tag{2.10.12}$$

is also a proper time interval because they are timed by the same clock at O.

According to the geometry shown in Fig. 4, we have,

$$d\mathbf{r} = \mathbf{v}dt, \qquad \theta' = \theta + d\theta, \qquad \mathbf{r}_B = \mathbf{r}_A - d\mathbf{r}, \tag{2.10.13}$$

where θ and θ' are the angles between $\mathbf{v}$ and $\mathbf{r}_A$ as well as $\mathbf{v}$ and $\mathbf{r}_B$, and $d\theta$ is the angle among $\mathbf{r}_A$ and $\mathbf{r}_B$.

Let us now calculate the relationship between $d\tau$ and dt. From the geometry we have

$$\tau_n = t_A + \frac{r_A}{c}, \qquad \tau_{n+dn} = t_B + \frac{r_B}{c} = t_A + dt + \frac{r_B}{c} \tag{2.10.14}$$

with $r_A = |\mathbf{r}_A|, r_B = |\mathbf{r}_B|$. By putting Eqs. (2.10.14) into Eq. (2.10.12), we get

$$d\tau = dt + \frac{r_B}{c} - \frac{r_A}{c}, \tag{2.10.15}$$

From the last one of equations (2.10.13) we arrive at

$$r_B{}^2 = (\mathbf{r}_A - d\mathbf{r})^2 = r_A^2 - 2\mathbf{r}_A \cdot d\mathbf{r} = r_A^2 - 2r_A \cos\theta |d\mathbf{r}|,$$

or

$$r_B = r_A - v \cos\theta dt, \tag{2.10.16}$$

where $dr = vdt$ is used.

Using Eq. (2.10.16), equation (2.10.15) becomes

$$d\tau = dt\left(1 - \frac{v}{c}\cos\theta\right). \tag{2.10.17}$$

By definition the proper frequency ω' of source is given by

$$\omega' = \frac{dn}{d\tau'}. \tag{2.10.18a}$$

Similarly, the frequency ω as measured by the observer is defined by

$$\omega = \frac{dn}{d\tau}. \tag{2.10.18b}$$

Then we obtain

$$\omega = \frac{dn}{d\tau} = \frac{dn}{d\tau'}\frac{d\tau'}{d\tau} = \omega'\frac{d\tau'}{d\tau}. \tag{2.10.19}$$

Substituting Eq. (2.10.17) into Eq. (2.10.19), we have

$$\omega = \omega' \frac{(d\tau'/dt)}{1 - (v/c)\cos\theta}. \tag{2.10.20}$$

In particular, for transverse motion ($\theta = 90°$), equation (2.10.20) reduces to

$$\omega = \omega' \frac{d\tau'}{dt}. \tag{2.10.21}$$

On the right–hand side of Eq. (2.10.20), the denominator is a directional effect which is the same as that in classical physics, while the numerator is the ratio of the proper time interval to the corresponding coordinate time interval. In classical physics, this ratio is equal to unity. However, in Einstein's theory of special relativity, this ratio is nothing but the time dilation, which is given by Eq. (2.9.1),

$$\frac{d\tau'}{dt} = \sqrt{1 - \frac{v^2}{c^2}}.$$

By putting the time dilation into Eq. (2.10.20), one arrives at the relativistic Doppler frequency shift formula,

$$\omega = \omega' \frac{\sqrt{1 - v^2/c^2}}{1 - (v/c)\cos\theta}. \tag{2.10.22}$$

The transverse Doppler effect, Eq. (2.10.21), is then given by

$$\omega = \omega' \sqrt{1 - \frac{v^2}{c^2}}. \tag{2.10.23}$$

Therefore, we see that the transverse Doppler effect is nothing but just the time dilation effect.

2.11. The Thomas Precession

In non-relativistic kinematics the Galilean transformations without spatial rotation of the Cartesian axes form a group. Consider three systems N, N' and N'' which are Newtonian inertial frames. The Cartesian axes in N' are chosen to be parallel to the corresponding axes in N; Similarly, The Cartesian axes in N'' are also parallel to the corresponding axes in N, i.e., without spatial rotation. Let the relative velocity of N' (or N'') to N be $\mathbf{v}$ (or $\mathbf{u}$), and in general the directions of $\mathbf{v}$ and $\mathbf{u}$ are not parallel to each other. The connection between the coordinates (x, y, z, t) in N and coordinates (x', y', z', t') in N' [or coordinates (x'', y'', z'', t'') in N''] is then given by a Galilean transformation without spatial rotation. It is well known that the transformation between N' and N'' is also a Galilean transformation without spatial rotation. This is, however, not the case in relativistic kinematics, for if we combine two Lorentz transformations without rotation the resultant Lorentz transformation will in general correspond to a change of orientation of the Cartesian axes. In 1927, Thomas [14] applied this effect to the electron in an atom and found for first time the precession phenomenon of the magnetic moment of electron. This is called the *Thomas precession.*

Let us now derive the Thomas precession formula (see, e.g., Ref. [13]). To do this, consider two successive (general) Lorentz transformations. Let the Einstein frame $F'(\mathbf{r}', t')$ move with the velocity $\mathbf{v}$ relative to the Einstein frame $F(\mathbf{r}, t)$, and their corresponding Cartesian axes be chosen to be parallel to each other. The connection of their coordinates is then given by the general Lorentz transformation, i.e., Eq. (2.2.12),

$$\mathbf{r}' = \left[\mathbf{r} + (\gamma - 1)\frac{\mathbf{v}(\mathbf{v} \cdot \mathbf{r})}{v^2}\right] - \gamma \mathbf{v} t, \tag{2.11.1a}$$

$$t' = \gamma\left(t - \frac{\mathbf{v} \cdot \mathbf{r}}{c^2}\right), \tag{2.11.1b}$$

where $\gamma = 1/\sqrt{1 - v^2/c^2}, \mathbf{r} = (x, y, z), \mathbf{r}' = (x', y', z')$, and $\mathbf{v} \cdot \mathbf{r} = v_x x + v_y y + v_z z$.

The inverse of Eqs. (2.11.1) can be obtained by making the changes $(\mathbf{r}', t') \leftrightarrow (\mathbf{r}, t)$ and $\mathbf{v} \to -\mathbf{v}$,

$$\mathbf{r} = \left[\mathbf{r}' + (\gamma - 1)\frac{\mathbf{v}'(\mathbf{v}' \cdot \mathbf{r}')}{v^2}\right] - \gamma \mathbf{v}' t', \tag{2.11.2a}$$

$$t = \gamma\left[t' - \frac{\mathbf{v}' \cdot \mathbf{r}'}{c^2}\right], \tag{2.11.2b}$$

where $\mathbf{v}' = -\mathbf{v}$ is the relative velocity of F to F'.

The component forms of Eqs. (2.11.1) are

$$x' = \left[1 + (\gamma - 1)\frac{v_x^2}{v^2}\right]x + (\gamma - 1)\frac{v_x v_y}{v^2}y + (\gamma - 1)\frac{v_x v_z}{v^2}z - \gamma v_x t, \tag{2.11.3a}$$

$$y' = (\gamma - 1)\frac{v_x v_y}{v^2}x + \left[1 + (\gamma - 1)\frac{v_y^2}{v^2}\right]y + (\gamma - 1)\frac{v_y v_z}{v^2}z - \gamma v_y t, \tag{2.11.3b}$$

$$z' = (\gamma - 1)\frac{v_z v_x}{v^2}x + (\gamma - 1)\frac{v_z v_y}{v^2}y + \left[1 + (\gamma - 1)\frac{v_z^2}{v^2}\right]z - \gamma v_z t, \tag{2.11.3c}$$

$$t' = -\gamma\frac{v_x x}{c^2} - \gamma\frac{v_y y}{c^2} - \gamma\frac{v_z z}{c^2} + rt. \tag{2.11.3d}$$

The inverse transformations can be got by making the changes of $(x, y, z, t) \leftrightarrow (x', y', z', t')$ and $(v_x, v_y, v_z) \to (v'_x, v'_y, v'_z)$.

We now consider the third Einstein frame F'', the origin of which coincides with that of F' and is at rest relative to F', and assume that the corresponding axes of F'' and F' are not parallel to each other. The relative velocity of F'' to F is also $\mathbf{v}$.

The Cartesian axes of F''' can be rotated to be coincided with the axes of F', and this spatial rotation is denoted by the *rotation operator* D^{-1}. A 3-vector, e.g., $\mathbf{r}'$ in F' can be transformed into the corresponding vector $\mathbf{r}''$ in F''', i.e., $D\mathbf{r}'' = \mathbf{r}'$. So the velocity of the origin of F relative to F''', $\mathbf{v}''$, is not equal to $-\mathbf{v}$, but should be

$$D\mathbf{v}'' = -\mathbf{v}. \tag{2.11.4}$$

By putting $D\mathbf{r}'' = \mathbf{r}'$ into Eqs. (2.11.1), we have

$$D\mathbf{r}'' = \left[\mathbf{r} + (\gamma - 1)\frac{\mathbf{v}(\mathbf{v} \cdot \mathbf{r})}{v^2}\right] - \gamma \mathbf{v} t, \tag{2.11.5a}$$

$$t'' = \gamma \left(t - \frac{\mathbf{v} \cdot \mathbf{r}}{c^2}\right). \tag{2.11.5b}$$

By multiplying both sides of Eq. (2.11.5a) by the inverse operator D^{-1} and using Eq. (2.11.4), then Eqs. (2.11.5) become the transformations between F''' and F,

$$\mathbf{r}'' = D^{-1}\mathbf{r} - \mathbf{v}''\left[(\gamma - 1)\frac{\mathbf{v} \cdot \mathbf{r}}{v^2} - \gamma t\right], \tag{2.11.6a}$$

$$t'' = \gamma \left[t - \frac{\mathbf{v} \cdot \mathbf{r}}{c^2}\right]. \tag{2.11.6b}$$

Equations (2.11.6) are the *Lorentz transformations with the spatial rotation*.

The Lorentz transformations (2.11.1) without rotation give the law of the addition of velocities, i.e., Eq. (2.4.11),

$$\mathbf{u}' = \frac{\gamma^{-1}\mathbf{u} + \{[1 - \gamma^{-1}](\mathbf{u} \cdot \mathbf{v})/v^2 - 1\}\,\mathbf{v}}{1 - (\mathbf{u} \cdot \mathbf{v})/c^2}, \tag{2.11.7}$$

where $\mathbf{u} = d\mathbf{r}/dt$, and $\mathbf{u}' = d\mathbf{r}'/dt'$. The inverse transformation of (2.11.7) is

$$\mathbf{u} = \frac{\gamma^{-1}\mathbf{u}' + \{[1 - \gamma^{-1}](\mathbf{u}' \cdot \mathbf{v})/v^2 + 1\}\,\mathbf{v}}{1 + (\mathbf{u}' \cdot \mathbf{v})/c^2}. \tag{2.11.8}$$

We now assume that the third Einstein frame $F'''(\mathbf{r}''t'')$ moves with the velocity $\mathbf{u}'$ relative to F', and its axes are, respectively, parallel to the axes in F'. Therefore, the transformations between F''' and F' is the Lorentz transformations without rotation, which can be obtained from Eqs. (2.11.1) by use of the replacements, $(\mathbf{r}, t, \mathbf{v}) \rightarrow (\mathbf{r}', t', \mathbf{u}')$ and $(\mathbf{r}', t') \rightarrow (\mathbf{r}'', t'')$. By eliminating $(\mathbf{r}', t')$ from the transformations between F and F' and the ones between F' and F''', one arrives at the transformation among F''' and F,

$$\mathbf{r}'' = D^{-1}\mathbf{r} - \mathbf{w}''\left\{\frac{\mathbf{r} \cdot \mathbf{w}}{w^2}\left[\frac{1}{\sqrt{1 - w^2/c^2}} - 1\right] - \frac{t}{\sqrt{1 - w^2/c^2}}\right\}, \tag{2.11.9}$$

which have the same form as Eq. (2.11.6a). We see that the axes in F'' are generally not parallel to the ones in F. Therefore, The resultant Lorentz transformation obtained from combination of two Lorentz transformations without rotation will in general correspond to a change of orientation of the Cartesian axes.

In Eq. (2.11.9), $\mathbf{w}$ is the velocity of F'' relative to F, $w = |\mathbf{w}|$, and $\mathbf{w}''$ is the relative velocity of F to F''. From Eqs. (2.11.7) and (2.11.8) they are given by

$$\mathbf{w} = \frac{\sqrt{1 - v^2/c^2}\,\mathbf{u}' + \mathbf{v}\left\{\left(1 - \sqrt{1 - v^2/c^2}\right)(\mathbf{u}' \cdot \mathbf{v})/v^2 + 1\right\}}{1 + (\mathbf{u}' \cdot \mathbf{v})/c^2}, \tag{2.11.10}$$

$$\mathbf{w}'' = \frac{\sqrt{1 - u'^2/c^2}\,\mathbf{v} + \mathbf{u}'\left\{\left(1 - \sqrt{1 - u'^2/c^2}\right)(\mathbf{u}' \cdot \mathbf{v})/u'^2 + 1\right\}}{1 + (\mathbf{u}' \cdot \mathbf{v})/c^2}. \tag{2.11.11a}$$

Equation (2.11.4) shows

$$D\mathbf{w}'' = -\mathbf{w}. \tag{2.11.11b}$$

We now consider the case where the transformations among F' and F'' is an infinitesimal transformation, i.e., where $\mathbf{u}'/c$ is infinitesimal. Neglecting all terms of higher than first-order in $\mathbf{u}'/c$, the transformations between F' and F'' reduce to

$$\mathbf{r}'' = \mathbf{r}' - \mathbf{u}'t', \tag{2.11.12a}$$

$$t'' = t' - \frac{\mathbf{u}' \cdot \mathbf{r}'}{c^2}. \tag{2.11.12b}$$

So that Eqs. (2.11.10) and (2.11.11a) become

$$\mathbf{w} = \mathbf{v} + \sqrt{1 - \frac{v^2}{c^2}}\left\{\mathbf{u}' + \mathbf{v}\frac{\mathbf{v} \cdot \mathbf{u}'}{v^2}\left[\sqrt{1 - \frac{v^2}{c^2}} - 1\right]\right\}, \tag{2.11.13a}$$

$$\mathbf{w}'' = -\left(\mathbf{v} + \mathbf{u}' - \mathbf{v}\frac{\mathbf{v} \cdot \mathbf{u}'}{c^2}\right), \tag{2.11.13b}$$

$$w^2 = w''^2 = v^2 + 2(\mathbf{v} \cdot \mathbf{u}')\left(1 - \frac{v^2}{c^2}\right). \tag{2.11.13c}$$

By substituting Eq. (2.11.1) for $\mathbf{r}'$ and t' we obtain the transformation which can be written in the form (2.11.9) with

$$D^{-1}\mathbf{r} = \mathbf{r} + \frac{1}{v^2}\left\{\frac{1}{\sqrt{1 - v^2/c^2}} - 1\right\}[(\mathbf{v} \times d\mathbf{v}) \times \mathbf{r}],$$

$$d\mathbf{v} = \mathbf{w} - \mathbf{v}. \tag{2.11.14}$$

To the first-order in $d\mathbf{v}$, equation (2.11.14) can be expressed as

$$D\mathbf{r} = \mathbf{r} + (\boldsymbol{\Omega} \times \mathbf{r}), \tag{2.11.15a}$$

where

$$\Omega = -\frac{1}{v^2} \left\{ \frac{1}{\sqrt{1 - v^2/c^2}} - 1 \right\} (\mathbf{v} \times d\mathbf{v}). \qquad (2.11.15b)$$

The rotation operator D represents an infinitesimal rotation around the direction of the vector Ω, the angle of rotation being equal to the magnitude $|\Omega|$ of the vector Ω.

Let us now consider motion of a classical electron with spin. If the velocity of the electron relative to F is $\mathbf{v} = \mathbf{v}(t)$, and if we put $d\mathbf{v} = \mathbf{a}dt$ in Eq. (2.11.15), where $\mathbf{a}(t) = \dot{\mathbf{v}}(t)$ is the acceleration of the electron, then the frames F' and F'' will be momentary rest frames for the electron at the times t and $t + dt$, respectively. Since the transformation from F' to F'' is an infinitesimal Lorentz transformation without rotation, it is natural to assume that the direction shown by the spin of the electron at the time $t + dt$ has the same orientation relative to the axes in F'' as it has at the time t relative to the axes in F', provided that the forces on the electron do not exert any torque on it.

If we put $d\mathbf{v} = \mathbf{a}dt$, the rotation vector Ω defined by Eq. (2.11.15) represents the rotation which has to be applied to the axes in F at the time $t + dt$ in order to give them the same orientation as the axes in F'''. Since, furthermore, the direction of the electron relative to the rest frame is constant, this means that the direction of the electron relative to F is turned through an angle corresponding to the rotation vector Ω. In other words, the electron performs a precession relative to F with the velocity of precession

$$\omega_T = \frac{d\Omega}{dt} = -\frac{1}{v^2}(\gamma - 1)(\mathbf{v} \times \mathbf{a}), \qquad (2.11.16)$$

where $\gamma = 1/\sqrt{1 - v^2/c^2}$, and $\mathbf{a} = d\mathbf{v}/dt$ is the acceleration of the electron. When $v \ll c$, the velocity of precession, Eq. (2.11.16), reduces to

$$\omega_T = -\frac{\mathbf{v} \times \mathbf{a}}{2c^2} \qquad (2.11.17)$$

to the lowest order approximation. This is the so-called *Thomas precession*. (see, e.g., Ref. [15] for another method of deriving the Thomas precession). The tests of Thomas precession will be given in Chap. 13.

CHAPTER 3

RELATIVISTIC MECHANICS

3.1. Mass, Momentum, Force, Work and Energy

Let us recall the relativistic invariance of a physical system. According to the principle of relativity the action of a physical system is invariant under the inhomogeneous Lorentz transformations

$$x^\mu \to x'^\mu = \Lambda^\mu_{\ \nu} x^\nu + b^\mu, \tag{3.1.1}$$

where b^μ is a constant 4-vector. All the inhomogeneous Lorentz transformations form a group which is called Poincaré group.

The first term on the right-hand side of Eq. (3.1.1) represents the general homogeneous Lorentz transformations with spatial rotation, i.e., Eqs. (2.2.15), the coefficients of which being denoted by $\Lambda^\mu_{\ \nu}$. The invariance of the action of a physical system under the homogeneous Lorentz transformations gives the *conservation law of angular momentum*; The second term represents a relative displacement, the *Poincaré displacement*, for the locations of the origins of the frames F and F'. The invariance of the action with respect to Poincaré displacement implies the homogeneity of the space and time and gives the *conservation law of energy–momentum*. These conservation laws are some of the foundations of relativistic mechanics.

The momentum of a material particle moving with the velocity $\mathbf{u}$ as measured in the frame F is regarded as to be proportional to $\mathbf{u}$,

$$\mathbf{p} = m\mathbf{u}. \tag{3.1.2}$$

The proportionality factor m is called the *inertial mass* of the particle, which is not constant but a universal function of the magnitude $u = |\mathbf{u}|$ of the velocity vector, i.e.,

$$m = m(u) = f(u). \tag{3.1.3}$$

In other frame F', according to the principle of relativity, the momentum of the particle, $\mathbf{p}'$, should have the same form

$$\mathbf{p}' = m'\mathbf{u}', \tag{3.1.4}$$

where $\mathbf{u}'$ represents the velocity of this particle relative to the frame F', and

$$m' = m'(u') = f(u') \tag{3.1.5}$$

is the same function of $u' = |\mathbf{u}'|$ as m is of u.

This function $f(u)$ is uniquely determined when we require that the conservation law of momentum should hold in any inertial frame. By applying this requirement to an elastic collision between two identical particles, one can obtain the function $f(u)$ as follows [16]:

$$f(u) = \frac{f(0)}{\sqrt{1 - u^2/c^2}},$$

or

$$m = \frac{m_0}{\sqrt{1 - u^2/c^2}}, \tag{3.1.6}$$

where $m_0 = f(0)$ is the value of $m(u)$ in case of $u = 0$. Thus m_0 is called the *rest mass* of the particle, and $m(u)$ with $u \neq 0$ is called the *relativistic mass* or the total mass of the particle. The momentum of the particle defined by Eq. (3.1.2) is then written as

$$\mathbf{p} = m\mathbf{u} = \frac{m_0 \mathbf{u}}{\sqrt{1 - u^2/c^2}}. \tag{3.1.7}$$

Equation (3.1.6) is the relation between mass and velocity or simply the *mass–velocity relation*.

Newtonian second law of mechanics states that the force $\mathbf{F}$ acting upon a material particle is directly proportional to the acceleration $\mathbf{a}$ of the particle, the proportional coefficient being called the inertial mass m not depending on the motion of the particle. This law, $\mathbf{F} = m\mathbf{a}$, is covariant under the Galilean transformation (2.1.1) but not under the Lorentz transformations (2.2.4). Therefore, Newtonian second law should be modified.

According to the conservation law of momentum, the momentum for a free particle is constant in time (i.e., $d\mathbf{p}/dt = 0$). On the other hand, when the particle is acted upon by a force, its momentum changes (i.e., $d\mathbf{p}/dt \neq 0$) . These show that a change in time of the momentum for a material particle is related to the force acting upon it. Therefore, a force $\mathbf{F}$ can be defined by the change of momentum per unit time:

$$\mathbf{F} = \frac{d\mathbf{p}}{dt}, \tag{3.1.8}$$

where the momentum $\mathbf{p}$ is given by Eq. (3.1.7). Thus, we see that for small particle velocities, equation (3.1.8) is identical with Newtonian second law of mechanics. By use of Eq. (3.1.7), equation (3.1.8) becomes

$$\mathbf{F} = m\frac{d\mathbf{u}}{dt} + \frac{dm}{dt}\mathbf{u} = m\frac{d\mathbf{u}}{dt} + \frac{\mathbf{F} \cdot \mathbf{u}}{c^2}\mathbf{u},$$

or

$$m\frac{d\mathbf{u}}{dt} = \mathbf{F} - \frac{\mathbf{F} \cdot \mathbf{u}}{c^2}\mathbf{u}, \tag{3.1.9}$$

where the relation $c^2 dm/dt = \mathbf{F} \cdot \mathbf{u}$, i.e., Eq. (3.1.16b), is used. This is the equation of motion for a particle acted by a force, which shows that the directions of the acceleration $d\mathbf{u}/dt$ and force $\mathbf{F}$ are in general not parallel to each other unless the direction of F is parallel with or perpendicular to the direction of the velocity $\mathbf{u}$.

In his first paper, Einstein considered the motion of a charged particle in an electromagnetic field. He assumed that the force acting upon the particle by the electromagnetic field is proportional to the acceleration of the particle, and then introduced the so-called "longitudinal mass" and "transverse mass". At the same time Einstein mentioned that a different definition of force would lead to a different value of mass. For example, the definition of force $\mathbf{F} = d(m\mathbf{u})/dt$ would give the mass formula (3.1.6). In 1909 Lewis and Tolman [16] started just from the view point of mechanics and considered the elastic collision between two material spheres. By means of the conservation law of momentum and Einstein's law of the addition of velocities, they obtained the mass–velocity relation (3.1.6). This derivation is independent of the electromagnetic field theory.

The relation between mass and energy can be derived from the mass–velocity relation (3.1.6) and the definition of force (3.1.8). The change of kinetic energy of the particle is equal to the rate of work W done by the force per unit time, which is defined by

$$\frac{dT}{dt} = W = \mathbf{F} \cdot \mathbf{u}, \tag{3.1.10}$$

where T is the kinetic energy of the particle.

By using Eqs. (3.1.7) and (3.1.8), equation (3.1.10) becomes

$$\frac{dT}{dt} = \mathbf{u} \cdot \frac{d}{dt}\left(\frac{m_0\mathbf{u}}{\sqrt{1 - u^2/c^2}}\right) = \frac{m_0}{\sqrt{1 - u^2/c^2}}\mathbf{u} \cdot \frac{d\mathbf{u}}{dt} + \frac{m_0 u^2}{c^2(1 - v^2/c^2)^{3/2}}u\frac{du}{dt}. \tag{3.1.11}$$

Since

$$\mathbf{u} \cdot \frac{d\mathbf{u}}{dt} = \frac{1}{2}\frac{d}{dt}(\mathbf{u} \cdot \mathbf{u}) = \frac{1}{2}\frac{d}{dt}(u^2) = \frac{1}{2}\frac{d}{dt}(u^2) = u\frac{du}{dt},$$

equation (3.1.11) reduces to

$$\frac{dT}{dt} = \frac{d}{dt}\left(\frac{m_0 c^2}{\sqrt{1 - u^2/c^2}}\right) = c^2\frac{dm}{dt}. \tag{3.1.12}$$

By integrating Eq. (3.1.12) we get the kinetic energy of a particle

$$T = \frac{m_0 c^2}{\sqrt{1 - u^2/c^2}} + C. \tag{3.1.13}$$

Here C is an integration constant to be determined by the boundary condition: the kinetic energy T vanishes when $u = 0$. This leads to $C = -m_0 c^2$. Thus, the kinetic energy is expressed as

$$T = \frac{m_0 c^2}{\sqrt{1 - u^2/c^2}} - m_0 c^2. \tag{3.1.14a}$$

By expanding the first term on the right-hand side of Eq. (3.1.14a), we have

$$T = \frac{1}{2}m_0 u^2 \left(1 + \frac{3}{4}\frac{u^2}{c^2} + \cdots\right). \tag{3.1.14b}$$

In case of $u/c \ll 1$, to the first-order, equation (3.2.14b) reduces to the Newtonian expression for the kinetic energy

$$T = \frac{1}{2}m_0 u^2. \tag{3.1.15}$$

Denoting the first term on the right-hand side of Eq. (3.1.14a) by E, then we have

$$E = T + m_0 c^2 = \frac{m_0 c^2}{\sqrt{1 - u^2/c^2}} = mc^2, \tag{3.1.16a}$$

or

$$\frac{dE}{dt} = \frac{dm}{dt}c^2 = \mathbf{F} \cdot \mathbf{u}, \tag{3.1.16b}$$

$$\Delta E = c^2 \Delta m, \tag{3.1.16c}$$

where m is the total mass of a particle defined by Eq. (3.1.6). Equation (3.1.16a) or Eq. (3.1.16c) is called the relation between mass and energy or simply the *mass–energy relation*.

Similarly, we can denote the second term by E_0, i.e.,

$$E_0 = m_0 c^2. \tag{3.1.17}$$

The quantities E and E_0 can be called the *total energy* and *rest energy* (or *proper energy*) of a particle, respectively.

We see that for $u \to c$ all the mass m, kinetic energy T and total energy E go to infinity. Thus, in order to guarantee the energy of a free photon being definite the rest mass of a photon should be zero.

By the definitions (3.1.7) and (3.1.16a) we can get the relation among E and $p = |\mathbf{p}|$

$$p^2 c^2 + m_0^2 c^4 = E^2,$$

or

$$p^2 c^2 - E^2 = -m_0^2 c^4, \tag{3.1.18}$$

i.e.,

$$E = c\sqrt{m_0^2 c^2 + p^2}. \tag{3.1.19}$$

Thus, the velocity of the particle can be written as

$$u = \frac{pc^2}{E} = \frac{dE}{dp}. \tag{3.1.20}$$

3.2. Transformations of Mass, Momentum, Energy and Force

The definitions and relations given in the previous section can be regarded as being valid within the Einstein frame F. In another frame F' the same definitions and relations for the corresponding quantities should be held:

$$m' = \frac{m_0}{\sqrt{1 - u'^2/c^2}}, \tag{3.2.1}$$

$$\mathbf{p}' = m'\mathbf{u}' = \frac{m_0 \mathbf{u}'}{\sqrt{1 - u'^2/c^2}}, \tag{3.2.2}$$

$$E' = m'c^2, \tag{3.2.3}$$

$$\frac{d\mathbf{p}'}{dt'} = \mathbf{F}', \tag{3.2.4}$$

$$\frac{dE'}{dt'} = \mathbf{F}' \cdot \mathbf{u}'. \tag{3.2.5}$$

The transformation equations between the primed quantities (m', p', E', F') and the unprimed quantities (m, p, E, F) can be obtained by using Einstein's law of the addition of velocities and Lorentz transformation.

Using Einstein's addition law Eq. (2.4.2), the dilation factor becomes

$$\frac{1}{\sqrt{1 - u'^2/c^2}} = \frac{1}{\sqrt{1 - \left(u_x'^2 + u_y'^2 + u_z'^2\right)/c^2}} = \frac{1}{\sqrt{1 - u^2/c^2}} \frac{1 - vu_x/c^2}{\sqrt{1 - v^2/c^2}}. \tag{3.2.6}$$

Then, we have the transformation equations for m and E,

$$m' = \frac{m_0}{\sqrt{1 - u^2/c^2}} \frac{1 - vu_x/c^2}{\sqrt{1 - v^2/c^2}} = m\frac{1 - vu_x/c^2}{\sqrt{1 - v^2/c^2}} = \frac{m - vp_x/c^2}{\sqrt{1 - v^2/c^2}}, \tag{3.2.7}$$

$$\left(\frac{E'}{c^2}\right) = \frac{(E/c^2) - (v/c^2)\, p_x}{\sqrt{1 - v^2/c^2}}. \tag{3.2.8}$$

Putting Eqs. (3.2.7) and (2.4.2) into (3.2.2) for m' and $\mathbf{u} = (u_x, u_y, u_z)$ respectively, we get

$$p_x' = \frac{p_x - vE/c^2}{\sqrt{1 - v^2/c^2}}, \tag{3.2.9a}$$

$$p_y' = p_y, \tag{3.2.9b}$$

$$p_z' = p_z. \tag{3.2.9c}$$

It is shown from the transformation equations (3.2.9) and (3.2.8) that the four quantities $(E/c, p_x, p_y, p_z)$ transform like the space–time coordinates (ct, x, y, z). Thus, $(p^\mu) = (E/c, p_x, p_y, p_z)$ form a 4-vector, and then its norm is an invariant,

$$\eta_{\mu\nu} p'^\mu p'^\nu = \eta_{\mu\nu} p^\mu p^\nu = m_0^2 c^4,$$

or

$$p'^2 c^2 - E'^2 = p^2 c^2 - E^2 = -m_0^2 c^4, \tag{3.2.10}$$

where the rest mass m_0 is a scalar under Lorentz transformations.

Under the more general Lorentz transformations (2.2.12) the transformation equations for $(E/c, \mathbf{p})$ can be obtained by the replacements $\mathbf{r} \to \mathbf{p}, \mathbf{r}' \to \mathbf{p}'$ and $t \to (E/c^2)$,

$$\mathbf{p}' = \gamma(\mathbf{p}^\dagger - \mathbf{v}\frac{E}{c^2}), \tag{3.2.11a}$$

$$E' = \gamma(E - \mathbf{v} \cdot \mathbf{p}), \tag{3.2.11b}$$

with

$$\mathbf{p}^\dagger \equiv \frac{1}{\gamma}\left[\mathbf{p} - (1 - \gamma)\frac{(\mathbf{v} \cdot \mathbf{p})\mathbf{v}}{v^2}\right], \tag{3.2.11c}$$

where γ is given by Eq. (2.2.10d).

By use of the transformations for $\mathbf{p}'$ and t' and the definition (3.1.16b) we can derive the transformation of $\mathbf{F}'$. Since $d\mathbf{p}'$ transforms like $\mathbf{r}'$, then $d\mathbf{p}'/dt$ do not form the components of a 4-vector. On the other hand, in relativistic mechanics the conception of force has no longer any absolute meaning. For these reasons, a force can be redefined by

$$\mathbf{f} = \dot{\mathbf{p}} = \frac{d\mathbf{p}}{d\tau}, \tag{3.2.12}$$

where $d\tau$ is a proper time interval, which is an invariant. Then, from (3.1.16b) we have

$$\dot{E} \equiv \frac{dE}{d\tau} = \mathbf{f} \cdot \mathbf{u}. \tag{3.2.13}$$

Therefore the quantities

$$\frac{\dot{E}}{c^2}, \quad \dot{\mathbf{p}} \tag{3.2.14}$$

transform like the space–time coordinates $t, \mathbf{r}$, so that the quantities

$$\left(\frac{\mathbf{f} \cdot \mathbf{u}}{c^2}, \quad \mathbf{f}\right) \tag{3.2.15}$$

form a 4-vector.

CHAPTER 4

ELECTRODYNAMICS IN MEDIA

The first complete theory describing the electromagnetic phenomena in moving media was developed by Hertz in about 1890. For the induction of electric current in conductors, this theory gives the same prediction as the modern theory, which was tested by experiments. When, however, applied to the motion of the dielectric and magnetic media in electromagnetic fields, Hertz's theory gives the predictions in contrast with the results given by the Röntgen, Eichenwald, and Wilson–Wilson [17] experiments. Later, Lorentz's electron theory could explain the magnetic induction of moving dielectric with the exception of the electric induction of magnetizable media. In his first paper, Einstein (1905) [1] gave the relativistic transformation of the electromagnetic fields in vacuum, but did not give the general construction of electromagnetic field equation in moving massive bodies. In 1908, Minkowski [18] proved that the electromagnetic field equation in moving massive bodies could be easily written in case that the covariance of the field equation with respect to Lorentz transformations is required. This kind of Maxwell–Minkowski electrodynamics of moving media could give the predictions in agreement with the electromagnetism experiments.

4.1. The Fundamental Equations

Consider a medium being at rest in the Einstein frame F'. In the frame, electromagnetic field equations in the rest medium are given by the usual *Maxwell's field equations* (in Gaussian units):

$$\nabla' \times \mathbf{E}' = -\frac{1}{c}\frac{\partial \mathbf{B}'}{\partial t'}, \tag{4.1.1a}$$

$$\nabla' \cdot \mathbf{B}' = 0, \tag{4.1.1b}$$

$$\nabla' \times \mathbf{H}' = \frac{1}{c}\frac{\partial \mathbf{D}'}{\partial t'} + \frac{4\pi}{c}\mathbf{J}', \tag{4.1.1c}$$

$$\nabla' \cdot \mathbf{D}' = 4\pi\rho', \tag{4.1.1d}$$

where the *current density* $\mathbf{J}'$ and the *charge density* ρ' satisfy the *charge conservation law*,

$$\nabla' \cdot \mathbf{J}' + \frac{\partial \rho'}{\partial t'} = 0. \tag{4.1.1e}$$

It is well known that equations (4.1.1) cannot completely determine the field variables $\mathbf{E}'$, $\mathbf{B}'$, $\mathbf{D}'$ and $\mathbf{H}'$. Thus, the *constitutive relations* are needed, which connect $\mathbf{D}'$ and

J' with **E'**, and **H'** with **B'**. For an isotropic, permeable, and conducting dielectric the *constitutive relations* are given by

$$\mathbf{D}' = \varepsilon \mathbf{E}', \quad \mathbf{B}' = \mu \mathbf{H}', \tag{4.1.2a}$$

$$\mathbf{J}' = \sigma \mathbf{E}', \tag{4.1.2b}$$

where ε, μ, and σ are the *dielectric constant* (the relative permittivity), the *permeability* and the *conductivity*, respectively. The vacuum values of ε and μ are unity. Equation (4.1.2b) is called *Ohm's law*.

Furthermore, the *polarization vector* **P'** and *magnetization vector* **M'** are defined by

$$\mathbf{E}' = \mathbf{D}' - 4\pi \mathbf{P}', \quad \mathbf{B}' = \mathbf{H}' + 4\pi \mathbf{M}'. \tag{4.1.3}$$

4.2. Relativistic Transformations of Electromagnetic Quantities

The equations for the electromagnetic field quantities given in the previous section are valid in the rest frame F'. Consider now any other frame F which has the velocity $-\mathbf{v}$ relative to F'. We assume that the frame F is the laboratory frame, and thus $\mathbf{v}$ represents the velocity of the medium as measured in the laboratory. Following the assumption by Minkowski, the electromagnetic field equations in the moving medium have the same form as Eqs. (4.1.1), then we have

$$\nabla \times \mathbf{E} = -\frac{1}{c}\frac{\partial \mathbf{B}}{\partial t}, \tag{4.2.1a}$$

$$\nabla \cdot \mathbf{B} = 0, \tag{4.2.1b}$$

$$\nabla \times \mathbf{H} = \frac{1}{c}\frac{\partial \mathbf{D}}{\partial t} + \frac{4\pi}{c}\mathbf{J}, \tag{4.2.1c}$$

$$\nabla \cdot \mathbf{D} = 4\pi\rho, \tag{4.2.1d}$$

and

$$\nabla \cdot \mathbf{J} + \frac{\partial \rho}{\partial t} = 0. \tag{4.2.1e}$$

Equations (4.1.1) should be, thus, changed to Eqs. (4.2.1) under a general Lorentz transformation from F' to F with the relative velocity $\mathbf{v}$. This covariance requires the following transformations of the field quantities:

$$\mathbf{E}'_{\parallel} = \mathbf{E}_{\parallel}, \quad \mathbf{E}'_{\perp} = \gamma\left(\mathbf{E}_{\perp} + \frac{1}{c}\mathbf{v} \times \mathbf{B}_{\perp}\right),$$

$$\mathbf{B}'_{\parallel} = \mathbf{B}_{\parallel}, \quad \mathbf{B}'_{\perp} = \gamma\left(\mathbf{B}_{\perp} - \frac{1}{c}\mathbf{v} \times \mathbf{E}_{\perp}\right),$$

$$\mathbf{D}'_{\parallel} = \mathbf{D}_{\parallel}, \quad \mathbf{D}'_{\perp} = \gamma\left(\mathbf{H}_{\perp} + \frac{1}{c}\mathbf{v} \times \mathbf{H}_{\perp}\right),$$

$$\mathbf{H}'_\parallel = \mathbf{H}_\parallel, \quad \mathbf{H}'_\perp = \gamma\left(\mathbf{H}_\perp - \frac{1}{c}\mathbf{v}\times\mathbf{D}_\perp\right), \tag{4.2.2}$$

and

$$\mathbf{J}'_\perp = \mathbf{J}_\perp, \quad \mathbf{J}'_\parallel = \gamma(\mathbf{J}_\parallel - \frac{1}{c}\mathbf{v}\rho), \tag{4.2.3a}$$

$$\rho' = \gamma(\rho - \frac{1}{c}\mathbf{v}\cdot\mathbf{J}). \tag{4.2.3b}$$

Here $\parallel$ and $\perp$ mean parallel and perpendicular to the velocity $\mathbf{v}$, and γ is defined by Eq. (2.2.10d). The inverse transformations can be obtained from Eqs. (4.2.2) and (4.2.3) by the inter-change of primed and unprimed quantities and $\mathbf{v}\to-\mathbf{v}$. The transformation equations (4.2.2) show the symmetry among the electric and magnetic fields: $\mathbf{E}$ and $\mathbf{B}$, as well as $\mathbf{D}$ and $\mathbf{H}$ have no independent existence; A purely electric or magnetic field in one frame will appear as a mixture of electric and magnetic fields in another frame. Of course certain restriction apply so that, for example, a purely electrostatic field in one frame cannot be transformed into a purely magnetostatic field in another.

It is similar to the case in F' that the *constitutive relations* in F are also needed for the uniqueness of solutions to Eqs. (4.2.1). Equations (4.2.2) indicate that the constitutive relations (4.1.2) in the medium are not covariant under Lorentz transformations. By putting Eq. (4.2.2) into Eq. (4.1.2a), one arrives at

$$\mathbf{D} + \frac{1}{c}\mathbf{v}\times\mathbf{H} = \varepsilon(\mathbf{E} + \frac{1}{c}\mathbf{v}\times\mathbf{B}), \tag{4.2.4a}$$

$$\mathbf{B} - \frac{1}{c}\mathbf{v}\times\mathbf{E} = \mu(\mathbf{H} - \frac{1}{c}\mathbf{v}\times\mathbf{D}). \tag{4.2.4b}$$

Solutions to Eqs. (4.2.4) for $\mathbf{D}$ and $\mathbf{B}$, the *constitutive relations* in the moving medium, are given by

$$\mathbf{D} = \frac{1}{1-\varepsilon\mu(v^2/c^2)}\left\{\varepsilon\mathbf{E}\left(1-\frac{v^2}{c^2}\right) + (\varepsilon\mu-1)\left[\frac{\mathbf{v}}{c}\times\mathbf{H} - \varepsilon\frac{\mathbf{v}}{c}\left(\frac{\mathbf{v}}{c}\cdot\mathbf{E}\right)\right]\right\}, \tag{4.2.5a}$$

$$\mathbf{B} = \frac{1}{1-\varepsilon\mu(v^2/c^2)}\left\{\mu\mathbf{H}\left(1-\frac{v^2}{c^2}\right) - (\varepsilon\mu-1)\left[\frac{\mathbf{v}}{c}\times\mathbf{E} + \mu\frac{\mathbf{v}}{c}\left(\frac{\mathbf{v}}{c}\cdot\mathbf{H}\right)\right]\right\}. \tag{4.2.5b}$$

Here ε and μ are determined by Eq. (4.1.2a), which are, in general, functions of the frequency ω', i.e., $\varepsilon = \varepsilon(\omega')$ and $\mu = \mu(\omega')$ where ω' is the frequency of light wave as measured in the rest frame F'. This is called the *dispersion*.

Similarly, Ohm's law, Eq. (4.1.2b), is also not covariant under Lorentz transformations. To find Ohm's law in the moving medium, write the inverse transformations of Eq. (4.2.3) as follows:

$$\mathbf{J}_\parallel = \gamma(\mathbf{J}'_\parallel + \frac{1}{c}\mathbf{v}\rho'), \quad \mathbf{J}_\perp = \mathbf{J}'_\perp, \tag{4.2.6a}$$

$$\rho = \gamma(\rho' + \frac{1}{c}\mathbf{v} \cdot \mathbf{J}').\tag{4.2.6b}$$

Then, we have

$$\mathbf{J} = \mathbf{J}_\parallel + \mathbf{J}_\perp = \mathbf{J}_c + \frac{1}{c}\gamma\mathbf{v}\rho',\tag{4.2.7a}$$

where

$$\mathbf{J}_c = \gamma\mathbf{J}'_\parallel + \mathbf{J}'_\perp.\tag{4.2.7b}$$

By using Ohm's law (4.1.2b) and the transformation Eqs. (4.2.2), equation (4.2.7b) can be written as

$$\mathbf{J}_c = \gamma\sigma\mathbf{E}'_\parallel + \sigma\mathbf{E}'_\perp = \sigma\gamma(\mathbf{E} + \frac{1}{c}\mathbf{v} \times \mathbf{B}).\tag{4.2.7c}$$

This shows that the current $\mathbf{J}_c$ is directly proportional to the conductivity σ, and hence is called the *conduction current density*. The second term, $\gamma\mathbf{v}\rho'/c$, is related to the motion of the density ρ', and therefore called the *convection current density*.

The definitions of polarization vector $\mathbf{P}$ and magnetization vector $\mathbf{M}$ in F are the same as those in F', i.e.,

$$\mathbf{E} = \mathbf{D} - 4\pi\mathbf{P}, \qquad \mathbf{B} = \mathbf{H} + 4\pi\mathbf{M}.\tag{4.2.8}$$

The equations should be obtained from Eq. (4.1.3) in which the primed quantities are changed to the unprimed quantities by using the transformations (4.2.2). This requires the following relations

$$\mathbf{P}'_\parallel = \mathbf{P}_\parallel, \qquad \mathbf{P}'_\perp = \gamma(\mathbf{P}_\perp - \frac{1}{c}\mathbf{v} \times \mathbf{M}_\perp),\tag{4.2.9a}$$

$$\mathbf{M}'_\parallel = \mathbf{M}_\parallel, \qquad \mathbf{M}'_\perp = \gamma(\mathbf{M}_\perp + \frac{1}{c}\mathbf{v} \times \mathbf{P}_\perp).\tag{4.2.9b}$$

4.3. Propagation of Electromagnetic Waves in a Moving Medium [19,20]

By making use of the field equations (4.2.1) and the constitutive relations (4.2.5) we can discuss propagation of electromagnetic waves in a moving medium. By putting Eq. (4.2.5) into the field equations (4.2.1) with $\mathbf{J} = 0$ and $\rho = 0$, and then eliminating $\mathbf{H}, \mathbf{B}$ and $\mathbf{D}$, we obtain the propagating equation for a free wave in the moving medium as follows:

$$\left[-\nabla^2 + \frac{n^2 - 1}{c^2 - v^2}(\mathbf{v} \cdot \nabla)^2\right]\mathbf{E} + 2\frac{n^2 - 1}{c^2 - v^2}(\mathbf{v} \cdot \nabla)\frac{\partial\mathbf{E}}{\partial t} + \left(\frac{n^2 - v^2/c^2}{c^2 - v^2}\right)\frac{\partial^2\mathbf{E}}{\partial t^2} = 0,\tag{4.3.1}$$

where $n = (\varepsilon\mu)^{1/2}$ is the *index of refraction* of the rest medium, which is a function of the frequency ω' in F'. By putting the plane-wave solution $\mathbf{E} = \mathbf{E}_0 exp[i(\mathbf{k} \cdot \mathbf{r} - \omega t)]$ into Eq. (4.3.1), we get the equation of the wave vector $\mathbf{k}$,

$$k^2 - \frac{n^2 - 1}{c^2 - v^2}(\mathbf{v} \cdot \mathbf{k})^2 + 2\frac{n^2 - 1}{c^2 - v^2}(\mathbf{v} \cdot \mathbf{k})\omega - \frac{n^2 - v^2/c^2}{c^2 - v^2}\omega^2 = 0.\tag{4.3.2}$$

We know that all the experiments test Eq. (4.3.2) only to first-order in v/c. For this reason we now simply consider the first-order approximation of Eq. (4.3.2), which is

$$F \equiv \mathbf{k}^2 + 2\{[n(\omega')]^2 - 1\}\mathbf{v} \cdot \mathbf{k}\frac{w}{c^2} - [n(\omega')]^2\frac{\omega^2}{c^2} = 0. \qquad (4.3.3)$$

Note that in Eq. (4.3.3) the wave vector $\mathbf{k}$ and the frequency ω are measured in the laboratory frame F, while the index of fraction $n(\omega')$ is related to ω' measured in F'. In the experiments the frequency, however, is always measured in the laboratory frame F as well as in vacuum. For instance, consider the case shown in Fig. 5. In the laboratory frame F the frequency of incident wave is assumed to be ω_0, while the frequency of refracted wave, ω, differs from ω_0 because of the motion of the surface of medium relative to the laboratory. On the other hand, the index of refraction $n(\omega')$ in Eq. (4.3.3) is a function of ω' which is the frequency of refraction wave as measured in the frame F'. The connection between ω_0' and ω' is given by the refraction law (4.4.1b), while the relation among ω and ω' is given by the Doppler formula (2.10.6c).

The first-order approximation of Eqs. (2.10.6) gives

$$\mathbf{k}' = \mathbf{k} - \frac{\omega}{c^2}\mathbf{v}, \qquad (4.3.4a)$$

$$\omega' = \omega - \mathbf{v} \cdot \mathbf{k}. \qquad (4.3.4b)$$

By expanding $n(\omega')$ in power series of $\Delta\omega = \omega' - \omega$, and then substituting (4.3.4b) for $\Delta\omega$, we have

$$n(\omega') = n(\omega) - \mathbf{v} \cdot \mathbf{k}\left(\frac{dn}{d\omega}\right) \qquad (4.3.5a)$$

to the first-order. Here

$$\frac{dn}{d\omega} \equiv \left(\frac{dn(\omega')}{d\omega'}\right)_{\omega'=\omega}. \qquad (4.3.5b)$$

Substituting Eq. (4.3.5a) for $n(\omega')$ in Eq. (4.3.3), we get

$$F(\mathbf{k}, \omega) = \mathbf{k}^2 + 2[n(\omega)]^2 f_1\frac{\omega}{c^2}(\mathbf{v} \cdot \mathbf{k}) - [n(\omega)]^2\frac{\omega^2}{c^2} = 0, \qquad (4.3.6a)$$

where

$$f_1 \equiv 1 - \frac{1}{[n(\omega)]^2} + \frac{\omega}{n(\omega)}\frac{dn}{d\omega}. \qquad (4.3.6b)$$

Solution to Eq. (6.3.6a) for $k \equiv |\mathbf{k}|$ is

$$k = n(\omega)\frac{\omega}{c}\left[1 - n(\omega)f_1\frac{1}{c}\mathbf{v} \cdot \hat{\mathbf{k}}\right], \qquad (4.3.7)$$

with $\hat{\mathbf{k}} = \mathbf{k}/k$. Then, the *phase velocity* u of the electromagnetic wave is given by

$$u = \frac{\omega}{k} = \frac{c}{n(\omega)}\left[1 + n(\omega)f_1\frac{1}{c}\mathbf{v} \cdot \hat{\mathbf{k}}\right] \qquad (4.3.8)$$

to the first-order. The frequency ω in Eqs. (4.3.6)–(4.3.8) is measured in F.

We now consider the *group velocity* $\mathbf{W} = (W_i, i = x, y, z)$. The group velocity in the moving medium is defined by

$$W_i = -\frac{\partial F}{\partial k_i} \left(\frac{\partial F}{\partial \omega} \right)^{-1}.$$

From Eq. (4.3.6) we obtain

$$W_i = \left(\frac{k_i}{n^2} + f_1 \frac{\omega}{c^2} v_i \right) \left[\left(1 + \frac{\omega}{n} \frac{dn}{d\omega} \right) \omega - \left(f_1 + 3 \frac{\omega}{n} \frac{dn}{d\omega} \right) \mathbf{v} \cdot \mathbf{k} \right]^{-1}, \qquad (4.3.9)$$

where $n = n(\omega)$, and the terms $d^2n/d\omega^2$ and $(dn/d\omega)^2$ are neglected (i.e., we assume that $n(\omega)$ is a slow varying function of ω). From Eq. (4.3.9) we have

$$\tan \phi = \frac{W_x}{W_z} = \frac{\sin \theta + n f_1 v_x/c}{\cos \theta + n f_1 v_x/c}. \qquad (4.3.10)$$

Here we assume that the vectors $\mathbf{k}$ and $\mathbf{v}$ lie in the xz-plane, i.e., $\sin \theta = k_x/k$ and $\cos \theta = k_z/k$. From Eq. (4.3.9) we know that the vector $\mathbf{W}$ lies in the xz-plane too, and thus ϕ is the angle between $\mathbf{W}$ and the z-axis, and θ is the angle among $\mathbf{k}$ and the z-axis (see Fig. 5).

4.4. Reflection and Refraction of Electromagnetic Waves [19,20]

Let us consider the special case as shown in Fig. 5. In the frame F', in where the medium is at rest, we have the usual boundary conditions for an electromagnetic wave: On the surface of the medium the tangent component of the refraction and reflection wave vectors $\mathbf{k}'$ and $\mathbf{k}'_r$ are equal to that of the incident wave vector $\mathbf{k}'_0$; The frequencies ω', ω'_r and ω'_0 of the refraction, reflection and incident waves are the same. Thus, we write

$$k'_x = (k'_0)_x, \qquad (4.4.1a)$$

$$\omega' = \omega'_0, \qquad (4.4.1b)$$

$$(k'_r)_x = (k'_0)_x, \qquad (4.4.2a)$$

$$\omega'_r = \omega'_0, \qquad (4.4.2b)$$

with

$$(k'_0)_x = (k'_0) \sin \theta'_i, \qquad (k'_0) = \frac{\omega'_0}{c}, \qquad (4.4.3a)$$

$$k'_x = k' \sin \theta', \qquad k' = n(\omega') \frac{\omega'_0}{c}, \qquad (4.4.3b)$$

$$(k'_r)_x = (k'_r) \sin \theta'_r, \qquad (k'_r) = \frac{\omega'_r}{c}. \qquad (4.4.3c)$$

Here θ'_i, θ' and θ'_r are, respectively, the incident, refraction and reflection angles.

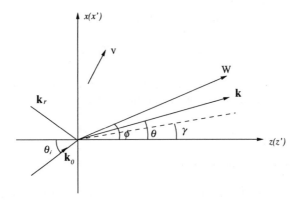

Fig. 5. Refraction and reflection of light ray. The region $z' \geq 0$ in the frame $F'(x'y'z't')$ is filled with an isotropic transparent substance with the index of refraction $n(\omega')$, while the region $z' < 0$ is an empty space. The plane $z' = 0$ is the surface of the substance. The substance moves with the velocity $\mathbf{v} = (v_x, 0, v_z)$. $\mathbf{k_0} = (k_{0x}, 0, k_{0z})$, $\mathbf{k} = (k_x, 0, k_z)$ and $\mathbf{k_r} = (k_{rx}, 0, k_{rz})$ are, respectively, the wave vectors of the incident, refraction and reflection light rays. $\mathbf{W} = (W_x, 0, W_z)$ is the group velocity. γ is the refraction angle in case of $\mathbf{v} = 0$. All the quantities are defined in the laboratory frame $F(xyzt)$.

Equations (4.4.1)–(4.4.3) are the usual refraction and reflection laws for the rest medium (i.e., in the comoving frame F'). By using Lorentz transformation one can obtain the corresponding laws in the laboratory frame F (i.e., in the moving medium).

Using the transformations (4.3.4) in Eq. (4.4.1), we get

$$k_x - \frac{\omega}{c}\frac{v_x}{c} = k_{0x} - \frac{\omega_0}{c}\frac{v_x}{c}, \tag{4.4.4a}$$

$$\omega - \mathbf{v} \cdot \mathbf{k} = \omega_0 - \mathbf{v} \cdot \mathbf{k_0}, \tag{4.4.4b}$$

with

$$(k_0)_x = \frac{\omega_0}{c} \sin \theta_i, \tag{4.4.5a}$$

$$k_x = k \sin \theta, \tag{4.4.5b}$$

where θ_i and θ are, respectively, the incident and refraction angles as measured in the laboratory frame F (see Fig. 5).

By using Eqs. (4.4.3), equation (4.4.4a) becomes

$$k \sin \theta = \frac{\omega_0}{c} \sin \theta_i + \frac{v_x}{c} \frac{\omega - \omega_0}{c},$$

where the second term on the right-hand side is of second-order in v/c owing to Eq. (4.4.4b), and thus should be neglected. So that we have

$$\sin \theta = \frac{\omega_0}{ck} \sin \theta_i. \tag{4.4.6a}$$

Equation (4.4.4b) can be written as

$$\omega = \omega_0 + \mathbf{v} \cdot (\mathbf{k} - \mathbf{k_0}) = \omega_0 + v_x(k_x - k_{0x}) + v_z(k_z - k_{0z}).$$

Equation (4.4.6a) shows $k_x = k_{0x}$ to the first-order, so that the second term on the right-hand side of the above equation vanishes. The third term is of the first-order, and therefore k_z can be replaced by k_z' (the wave vector of the refraction wave in the rest frame F'). Thus, the above equation reduces to

$$\omega = \omega_0 \left[1 + \frac{v_z}{c}(n \cos \gamma - \cos \theta_i)\right], \tag{4.4.6b}$$

where $k_z \to k_z' \to (n\omega_0/c) \cos \gamma$ and $k_{0z} = (\omega_0/c) \cos \theta_i$ are used. Equations (4.4.6) are the *refraction law* in the laboratory frame F (i.e., in the moving medium). In equation (4.4.6b), the angle γ is the refraction angle as measured in the rest frame F' of the medium. The refraction law in the rest frame F' (4.4.1a) gives $\sin \gamma = (1/n) \sin \theta_i'$, where θ_i' is the incident angle as measured in the rest frame F', and thus differs from the incident angle θ_i as seen in F by a first-order quantity, i.e., $\sin \theta_i' = \sin \theta_i + O(v/c)$. So that we have

$$\cos \gamma = \sqrt{1 - n^{-2} \sin^2 \theta_i} + O\left(\frac{v}{c}\right).$$

Since the second term on the right-hand side of Eq. (4.4.6b) is of the first-order, the second term on the right-hand side of the above equation, $O(v/c)$, can be neglected. Then, we write

$$\cos \gamma = \sqrt{1 - n^{-2} \sin^2 \theta_i}. \tag{4.4.7}$$

Similarly, from Eqs. (4.4.2) and by using the transformations (4.3.4) we can get the *reflection law* on the surface of the moving medium as seen in F:

$$\sin \theta_r = \left(1 + 2\frac{v_z}{c} \cos \theta_i\right) \sin \theta_i, \tag{4.4.8a}$$

$$\omega_r = \omega_0 \left[1 - 2\frac{v_z}{c} \cos \theta_i\right]. \tag{4.4.8b}$$

By use of Eq. (4.4.6b) the quantity $n(\omega)$ in Eq. (4.3.8) can be expanded in the power series of $\omega - \omega_0$ as follows:

$$n(\omega) = n(\omega_0) \left\{ 1 + [n(\omega_0) \cos \gamma - \cos \theta_i] \frac{\omega_0}{n(\omega_0)} \frac{dn}{d\omega_0} \frac{v_z}{c} \right\}. \tag{4.4.9}$$

Substituting Eq. (4.4.9) for $n(\omega)$ in Eq. (4.3.8), the phase velocity in the moving medium as seen in F can be rewritten as

$$u = \frac{\omega}{k} = \frac{c}{n(\nu_0)} + f_1 v_x \sin \gamma + f_2 v_z \cos \gamma, \tag{4.4.10a}$$

$$f_1 = 1 - \frac{1}{[n(\nu_0)]^2} + \frac{\nu_0}{n(\nu_0)} \frac{dn}{d\nu_0}, \tag{4.4.10b}$$

$$f_2 = 1 - \frac{1}{[n(\nu_0)]^2} + \frac{\nu_0}{[n(\nu_0)]^2} \frac{dn}{d\nu_0} \left(\frac{\cos \theta_i}{\cos \gamma} \right). \tag{4.4.10c}$$

Here $\nu_0 = \omega_0/2\pi$ is used, and $n(\nu_0) = n(\omega_0)$ is the index of refraction of the medium with ν_0 being the frequency of incident light in vacuum as measured in the laboratory frame F. Equations (4.4.10) are the same as those obtained by Parks and Dowell (1974) [21] from Fresnel's formula.

The quantity f_1 defined by Eq. (4.4.10b) is called *Einstein's drag coefficient*. *Laub coefficient* is the special case of f_2 defined by Eq. (4.4.10c) where $\theta_i = 0$. For a medium without dispersion, i.e., $dn/d\nu_0 = 0$, equations (4.4.10b,c) reduce to

$$f_1 = f_2 = 1 - \frac{1}{n^2} \tag{4.4.11a}$$

that is called *Fresnel's drag coefficient*. Another special case is $\theta_i = $ Brewster's angle, i.e.,

$$\frac{\cos \theta_i}{\cos \gamma} = \frac{1}{n}. \tag{4.4.11b}$$

In this case the coefficient f_2 becomes

$$f_{(Brewster)} = 1 - \frac{1}{n^2} + \frac{\nu_0}{n^3} \frac{dn}{d\nu_0}. \tag{4.4.11c}$$

By using Eqs. (4.4.6b) and (4.4.10), the refraction law for the moving medium, Eq. (4.4.6a), can be expressed as

$$\sin \theta = \frac{\sin \theta_i}{n} \left\{ 1 + f_1 \frac{v_x}{c} \sin \theta_i + \frac{v_z}{c} \left[\left(1 + \frac{\nu_0}{n} \frac{dn}{d\nu_0} \right) \cos \theta_i - \frac{1}{n} \cos \gamma \right] \right\}. \tag{4.4.12}$$

CHAPTER 5

THE PROCA VECTOR FIELD

It is well known that the electrodynamic constant c in Maxwell's electromagnetic field equations in vacuum is the ratio of the electromagnetic units and the electrostatic units. The constant c represents the velocity of electromagnetic waves propagating in vacuum. This means that the one-way velocity of light is isotropic at least in a certain inertial frame. According to the second postulate in Einstein's theory of special relativity, the light rays always propagate in vacuum with the constant velocity c which is independent of inertial frames. Thus, there is no such an inertial frame in where a photon would be at rest, so that the rest mass of a photon must be zero. This result also follows from the mass–velocity–energy relations and the four-dimensional formulation of electromagnetic field action. This theoretical prediction has been tested by the experiments, e.g., in particular the quantum electrodynamic experiments. However many experiments and analyses directly testing the rest mass of a photon have been performed. All the tests are based on the massive electromagnetic field (Proca field) theory that will be introduced in this chapter.

5.1. Covariant Form of Maxwell's Field Equations

Let us firstly recall Maxwell's electromagnetic field theory. Maxwell's electromagnetic field equations in vacuum (in Gaussian units) are

$$\nabla \times \mathbf{E} = -\frac{1}{c}\frac{\partial \mathbf{B}}{\partial t}, \tag{5.1.1a}$$

$$\nabla \cdot \mathbf{B} = 0, \tag{5.1.1b}$$

$$\nabla \times \mathbf{B} = \frac{1}{c}\frac{\partial \mathbf{E}}{\partial t} + \frac{4\pi}{c}\mathbf{J}, \tag{5.1.1c}$$

$$\nabla \cdot \mathbf{E} = 4\pi\rho, \tag{5.1.1d}$$

and the charge conservation law is

$$\nabla \cdot \mathbf{J} + \frac{\partial \rho}{\partial t} = 0. \tag{5.1.1e}$$

We next rewrite the equations in the four-dimensional form, the covariant form. By introducing the charge-current 4-vector J_μ defined by

$$J_\mu = (\mathbf{J}, ic\rho) \tag{5.1.2}$$

with $\mu = 1, 2, .3, 4,$ then the continuity equation (5.1.1e) can be expressed in the covariant form:

$$\frac{\partial J_\mu}{\partial x_\mu} = 0. \tag{5.1.3}$$

Note that in this chapter for simplicity we employ such a formulation that the fourth component of the four-dimensional coordinate vector is a pure imaginary quantity, i.e.,

$$x_\mu = (x, y, z, ict). \tag{5.1.4}$$

The 4-interval ds^2 is defined by

$$ds^2 = \eta_{\mu\nu} dx_\mu dx_\nu = dx^2 + dy^2 + dz^2 - c^2 dt^2, \tag{5.1.5a}$$

where the metric takes

$$\eta_{\mu\nu} = \text{diag}\ (+1, +1, +1, +1). \tag{5.1.5b}$$

The inverse metric is the same:

$$\eta^{\mu\nu} = \text{diag}\ (+1, +1, +1, +1) = \eta_{\mu\nu}.$$

So that any covariant tensors do not differ from their contravariant form, for example $x^\mu = x_\mu, A^\mu = A_\mu, J^\mu = J_\mu, F^{\mu\nu} = F_{\mu\nu}$, etc. So that we do not distinguish the superscripts from the subscripts.

The wave equations for the 3-vector potential $\mathbf{A}$ and the 3-scalar potential ϕ are given by (in Gaussian units)

$$\nabla^2 \mathbf{A} - \frac{1}{c^2} \frac{\partial^2 \mathbf{A}}{\partial t^2} = -\frac{4\pi}{c} \mathbf{J}, \tag{5.1.6a}$$

$$\nabla^2 \phi - \frac{1}{c^2} \frac{\partial^2 \phi}{\partial t^2} = -4\pi \rho, \tag{5.1.6b}$$

with the Lorentz gauge condition

$$\nabla \cdot \mathbf{A} + \frac{1}{c} \frac{\partial \phi}{\partial t} = 0. \tag{5.1.7}$$

By introducing the 4-vector potential A_μ defined by

$$A_\mu = (\mathbf{A}, i\phi), \tag{5.1.8}$$

then the wave equations can be written

$$\Box^2 A_\mu = -\frac{4\pi}{c} J_\mu, \tag{5.1.9}$$

where $\Box^2 = \partial_\mu \partial_\mu = \nabla^2 - c^{-2} \partial_t^2$, and the Lorentz gauge condition reduces to

$$\frac{\partial A_\mu}{\partial x_\mu} = 0. \tag{5.1.10}$$

To write Maxwell's field equations in a covariant form, let us introduce the second-rank, antisymmetric, field-strength tensor (the electromagnetic field tensor) $F_{\mu\nu}$ defined by

$$F_{\mu\nu} = \frac{\partial A_\nu}{\partial x_\mu} - \frac{\partial A_\mu}{\partial x_\nu}. \tag{5.1.11}$$

Explicitly, the field-strength tensor is

$$(F_{\mu\nu}) = \begin{pmatrix} 0 & B_z & -B_y & -iE_x \\ -B_z & 0 & B_x & -iE_y \\ B_y & -B_x & 0 & -iE_z \\ iE_x & iE_y & iE_z & 0 \end{pmatrix}, \tag{5.1.12}$$

which satisfies the identity

$$\frac{\partial F_{\mu\nu}}{\partial x_\lambda} + \frac{\partial F_{\lambda\mu}}{\partial x_\nu} + \frac{\partial F_{\nu\lambda}}{\partial x_\mu} = 0, \tag{5.1.13}$$

where $\mu, \nu, \lambda = 1, 2, 3, 4$. The explicit form of this identity are just Maxwell's field equations (5.1.1a,b).

By using the tensor $F_{\mu\nu}$, the other two Maxwell's equations (5.1.1c,d) can be written in the covariant form:

$$\frac{\partial F_{\mu\nu}}{\partial x_\nu} = \frac{4\pi}{c} J_\mu. \tag{5.1.14}$$

Under Lorentz transformations the tensor $F_{\mu\nu}$ defined in the Einstein frame F would be changed to the tensor $F'_{\sigma\rho}$ defined in another Einstein frame F', i.e.,

$$F'_{\sigma\rho} = a_{\sigma\mu} a_{\rho\nu} F_{\mu\nu}, \tag{5.1.15}$$

where the transformation coefficient matrix $(a_{\sigma\mu})$ for a Lorentz transformation without rotation from F to F' with a relative velocity $\mathbf{v}$, e.g., parallel to the x-axis is given by

$$(a_{\sigma\mu}) = \begin{pmatrix} \gamma & 0 & 0 & i\gamma\beta \\ 0 & 1 & 0 & 0 \\ 0 & 0 & 1 & 0 \\ -i\gamma\beta & 0 & 0 & \gamma \end{pmatrix}, \tag{5.1.16}$$

with

$$\beta = \frac{v}{c}, \qquad \gamma = \frac{1}{\sqrt{1 - \beta^2}}, \tag{5.1.17}$$

and the definition of $F'_{\sigma\mu}$ is given by Eq. (5.1.12) with the replacements from the unprimed quantities to primed quantities. The explicit form of Eq. (5.1.15) is just equation (4.2.2).

Similarly, under Lorentz transformations the charge-current 4-vector J_μ should be changed to

$$J'_\lambda = a_{\lambda\mu} J_\mu. \tag{5.1.18}$$

The explicit form is just equation (4.2.3).

Maxwell's field equations without a current 4-vector (i.e., $J_\mu = 0$) can be obtained from the Lagrangian

$$L_0 = -\frac{1}{16\pi} \int d^4x F_{\mu\nu} F_{\mu\nu}. \tag{5.1.19}$$

In a general case, the Lagrangian representing interaction between the electromagnetic potential A_μ and matter field,

$$L_i = \frac{1}{c} \int d^4x A_\mu J_\mu, \tag{5.1.20}$$

as well as Lagrangian of the matter field should be added to Eq. (5.1.19).

The Lagrangian (5.1.19) is invariant under $U(1)$ gauge transformation

$$A_\mu \rightarrow A'_\mu = A_\mu + \partial_\mu \alpha, \tag{5.1.21}$$

where $\alpha = \alpha(x)$ is an arbitrary scalar function of space–time coordinate $x = (x_\mu)$. A mass term for a photon, $\mu^2 A_\mu A_\mu$, would violate the $U(1)$ gauge invariant (the phase invariant).

5.2. Proca's Vector Field Equations

If the gauge invariant is abandoned, the mass term can be added to Eq. (5.1.19). Then, Lagrangian for a massive electromagnetic field is

$$L_{tot} = \int d^4x \left[-\frac{1}{16\pi} F_{\mu\nu} F_{\mu\nu} - \frac{1}{8\pi} \mu^2 A_\mu A_\mu + \frac{1}{c} A_\mu J_\mu \right], \tag{5.2.1}$$

where μ is the rest mass of photon[1].

The variation of L_{tot} with respective to A_μ gives the Proca equations in Gaussian units (A. Proca first developed the missive field theory during 1930–1936; see, e.g., Ref. [197] for the original references)

$$\frac{\partial F_{\mu\nu}}{\partial x_\nu} + \mu^2 A_\mu = \frac{4\pi}{c} J_\mu, \tag{5.2.2}$$

where the tensor $F_{\mu\nu}$ still satisfies the identity (5.1.13).

The current density J_μ is required to satisfy the charge conservation law (5.1.3). By making derivative of the Proca equation (5.2.2) with respect to x_μ and using the definition (5.1.11) and the conversation law (5.1.3), we obtain the Lorentz gauge condition (5.1.10). This implies that in Proca theory the Lorentz gauge condition is equivalent to the conservation law.

---

[1]The rest mass of a photon μ is in the wave number unit (dimension is cm^{-1}), which is related to the mass m_0 in gram as follows: $\mu = m_0 c/\hbar$, i.e., 1 cm$^{-1} = 3.50 \times 10^{-38}$ gm.

Substituting the definition of $F_{\mu\nu}$ in Eq. (5.2.2), we have the wave equation of Proca vector field A_μ

$$\left(\Box^2 - \mu^2\right) A_\mu = -\frac{4\pi}{c} J_\mu. \tag{5.2.3}$$

The three-dimensional forms of the above equations are collected as follows:

$$\nabla \times \mathbf{H} - \frac{1}{c}\frac{\partial \mathbf{E}}{\partial t} = \frac{4\pi}{c}\mathbf{J} - \mu^2\mathbf{A}, \tag{5.2.4a}$$

$$\nabla \cdot \mathbf{E} = 4\pi\rho - \mu^2\phi, \tag{5.2.4b}$$

$$\nabla \times \mathbf{E} = -\frac{1}{c}\frac{\partial \mathbf{H}}{\partial t}, \tag{6.2.5a}$$

$$\nabla \cdot \mathbf{H} = 0, \tag{5.2.5b}$$

$$\mathbf{H} = \nabla \times \mathbf{A}, \tag{5.2.6a}$$

$$\mathbf{E} = -\nabla\phi - \frac{1}{c}\frac{\partial \mathbf{A}}{\partial t}, \tag{5.2.6b}$$

$$\nabla \cdot \mathbf{J} + \frac{\partial \rho}{\partial t} = 0, \tag{5.2.7a}$$

$$\nabla \cdot \mathbf{A} + \frac{1}{c}\frac{\partial \phi}{\partial t} = 0, \tag{5.2.7b}$$

$$(\Box - \mu^2)\mathbf{A} = -\frac{4\pi}{c}\mathbf{J}, \tag{5.2.8a}$$

$$(\Box - \mu^2)\phi = -4\pi\rho. \tag{5.2.8b}$$

Obviously, when $\mu = 0$ Proca's equations reduce to Maxwell's field equations. Equations (5.2.2) were developed by Proca in 1930's, which is the unique generalization of Maxwell's field equations with preserving the Lorentz invariant. Equations (5.2.4)–(5.2.8) are the foundations for testing the rest mass μ of photon by the experiments (see Chap. 12).

5.3. Dispersion in Vacuum

One of the predictions given by the massive electromagnetic theory is the dispersion for the velocity of a massive photon ($\mu \neq 0$) in vacuum. The plane wave solution to Eq. (5.2.3) without current ($J_\mu = 0$) is

$$A_\nu \sim e^{i(\mathbf{k}\cdot\mathbf{r}-\omega t)}, \tag{5.3.1}$$

where the wave vector $\mathbf{k}$, angular frequency ω and rest mass μ should satisfy the relationship

$$\mathbf{k}^2 - \frac{\omega^2}{c^2} = -\mu^2. \tag{5.3.2}$$

This is the dispersion relation for a massive electromagnetic wave propagating in vacuum. The phase velocity of a free massive wave is

$$u = \frac{\omega}{k} = c\left(1 - \frac{\mu^2 c^2}{\omega^2}\right)^{-\frac{1}{2}}, \tag{5.3.3}$$

where $k = |\mathbf{k}| = 2\pi/\lambda$ with λ being the wave length. This shows that the phase velocity depends on the frequency, and thus the group velocity (the velocity of energy flow) would differ from the phase velocity. By definition the group velocity is

$$v_g = \frac{d\omega}{dk} = c\left(1 - \frac{\mu^2 c^2}{\omega^2}\right)^{\frac{1}{2}}. \tag{5.3.4}$$

Since the mass μ has a finite value, so that in the limit $\omega \to \infty$ the phase and group velocities will go to the constant c, i.e.,

$$\lim_{\omega \to \infty} u = \lim_{\omega \to \infty} v_g = c. \tag{5.3.5}$$

It is shown from Eqs. (5.3.2) and (5.3.1) that $k = 0$ for $\omega = \mu c$, namely the massive wave does not propagate; When $\omega < \mu c$, we have $k^2 < 0$, i.e., k is a imaginary quantity. So that there is an exponential reduction factor in the vector potential given by Eq. (5.3.1), i.e., the amplitude of a free massive wave would be evanescent during its propagating in vacuum; Only the waves with $\omega > \mu c$ could propagate in vacuum without reduction, and therefore, the phase and group velocities of which are given by Eqs. (5.3.3) and (5.3.4).

Part II

Test Theories of Special Relativity

CHAPTER 6

EDWARDS' THEORY

6.1. Introduction

As mentioned in Sec. 1.3, in Einstein's theory of special relativity, the constancy of the speed of light is the second postulate. With this postulate, a clock located at any position in an inertial frame can be synchronized with a clock at the origin of the frame by means of a light pulse. Since that time, the clock synchronization problem has been discussed by many authors. Such as Reichenbach (1958) [22] and Grunbaum (1960) [23] discussed this problem in detail, and pointed out that no observable difference would result if the speed of light really were anisotropic. Ruderfer (1960) [24] held that special relativity contains an important assumption which has not and possibly cannot be tested. Edwards (1963) [2] and Winnie (1970) [5] obtained a kind of generalized Lorentz transformations starting from the constancy of the two-way speed of light. It was concluded that the generalized Lorentz transformations predict the same observable effects as the standard Lorentz transformation [25].

Another kind of generalized Lorentz transformations were proposed by H. P. Robertson in 1949 [3]. Twenty-eight years late, Mansouri and Sexl (MS) (1977) [4] suggested a kind of more general transformations. Since that time, many papers on this topic, such as Bertotti (1979) [26], MacArthur (1986) [27], Haugan and Will (1987) [28], Abolghasem, Khajehpour and Mansouri (1988) [29], Riis et al. (1988, 1989) [30,31], Bay and White (1989) [32], Gabriel and Haugan (1990) [33], Krisher et al. (1990) [34], and Will (1992) [35], have been published. However, some ambiguities still exist in explaining the physical meaning of the so-called one-way experiments by use of the test theory of special relativity [4,35]. Thus it is necessary to reanalyze these kinds of test theories in detail.

The difference between Einstein's theory and test theories is simply the different definitions of simultaneity, i.e., the different postulates concerning the velocity of light. In Einstein's theory, the velocity of light c is a universal constant independent of the relative motion of inertial frames, as well as of the motion of the light source. The value of this parameter c takes a value of the two-way speed of light, which has already given by measurements, such as [36] $c=299\ 792\ 458$ m s^{-1}. In test theories there are some other parameters, for instance, the directional parameter $\mathbf{q}$ and the two-way speeds ($c_{\parallel}$ and $c_{\perp}$). Among these parameters, we will see that $\mathbf{q}$ has no any effect on physical experiments, while the difference among the two-way velocities $c_{\parallel}$ and $c_{\perp}$ has not yet observed by experiments. Thus, the test theories involve these two parameters awaiting tests by experiments.

In this chapter, we shall discuss relationships between Edwards' theory and Einstein's theory. In particular, we will clarify how to compare the test theory with data of experiments, which will show that the directional parameter **q** cannot be observed if we use a light signal to synchronize clocks [25].

In chapter 7, we will analyze relationship between the MS transformations and Robertson transformations. We can see that the MS transformations are trivial in the sense that they are a generalization of the Robertson transformations, just as the Edwards transformations are a generalization of the Lorentz transformations. In other words, the MS transformations differ from the Robertson transformations by a directional parameter **q**, just as the Edwards transformations do from the Lorentz transformations. So that the MS transformations predict the same observable effects as the Robertson transformations, just as the Edwards transformations do as the Lorentz transformations.

It will be shown that our conclusion is as follows: *Robertson's theory is a test theory of Einstein's special relativity because the two theories would give different predictions to physical experiments; while Edwards' theory cannot be called a test theory due to the physical equivalence with Einstein's special relativity; The MS theory is not a new test theory due to its equivalence with Robertson's theory.* Thus, we only need to use Robertson's theory for yielding theoretical predictions and then to compare with physical experiments. In other words, it is not needed to use the MS theory for getting theoretical predictions because the directional parameter **q** is not observable physically.

6.2. One-Way and Two-Way Velocities of Light

In Sec. 1.3, we have given the relation between the two-way velocity and one-way velocity, and introduced a directional parameter. They are

$$\bar{c}_r = \frac{2c_r c_{-r}}{c_r + c_{-r}} \qquad (6.2.1)$$

and

$$c_r = \frac{\bar{c}_r}{1 - q_r}, \qquad c_{-r} = \frac{\bar{c}_r}{1 + q_r}, \qquad (6.2.2)$$

with

$$-1 \le q_r \le +1, \qquad (6.2.3)$$

where $\bar{c}_r$ is the two-way speed of light along the path $l_{OP} + l_{PO}$, $c_{\pm r}$ are, respectively, the one-way velocities in the directions $\pm \mathbf{r}/r$ with r being the length of the path l_{OP} (or l_{PO}), and q_r is the projection of the directional parameter **q** in the direction $\mathbf{r}/r$, i.e., $q_r = \mathbf{q} \cdot \mathbf{r}/r$.

In particular, along the x-, y-, and z-axis, we have

$$c_i = \frac{\bar{c}_i}{1 - q_i}, \qquad c_{-i} = \frac{\bar{c}_i}{1 + q_i}, \qquad (6.2.4a)$$

$$-1 \le q_i \le +1, \qquad i = x, y, z. \qquad (6.2.4b)$$

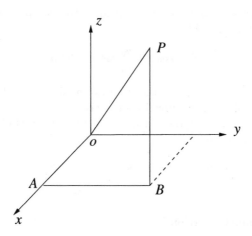

Fig. 6. The paths of light rays: $OABPO$ and $OPBAO$.

Let us discuss the relation between q_r and q_i. Consider the following "loops" of light (see Fig. 6):

$$l_+ = l_{OA} + l_{AB} + l_{BP} + l_{PO},$$

$$l_- = l_{OP} + l_{PB} + l_{BA} + l_{AO}, \tag{6.2.5}$$

where l_{OA} is the distance between O and A, and so on. Coordinates of the points O, A, B and P are $(0, 0, 0), (x, 0, 0), (x, y, 0)$ and (x, y, z), respectively. Let t_+ and t_- denote the time intervals spent by the light pulse traversing the distances l_+ and l_-, respectively, i.e.,

$$t_+ = t_{OA} + t_{AB} + t_{BP} + t_{PO},$$

$$t_- = t_{OP} + t_{PB} + t_{BA} + t_{AO}. \tag{6.2.6}$$

Substituting $t_{OA} = x/c_x, t_{AO} = x/c_{-x}, t_{AB} = y/c_y, t_{BA} = y/c_{-y}, t_{BP} = z/c_z, t_{PB} = z/c_{-z}, t_{OP} = r/c_r$, and $t_{PO} = r/c_{-r}$ into Eqs. (6.2.6), we have

$$t_+ = \frac{x}{c_x} + \frac{y}{c_y} + \frac{z}{c_z} + \frac{r}{c_{-r}},$$

$$t_- = \frac{x}{c_{-x}} + \frac{y}{c_{-y}} + \frac{z}{c_{-z}} + \frac{r}{c_r}. \tag{6.2.7}$$

By using the definitions (6.2.2) and (6.2.4a) in Eqs. (6.2.7), we obtain

$$t_+ - t_- = 2r \left[\frac{q_r}{\bar{c}_r} - \left(\frac{q_x}{\bar{c}_x} \cos \alpha + \frac{q_y}{\bar{c}_y} \cos \beta + \frac{q_z}{\bar{c}_z} \cos \gamma \right) \right], \tag{6.2.8}$$

where $\cos\alpha = x/r, \cos\beta = y/r, \cos\gamma = z/r$, and $\cos^2\alpha + \cos^2\beta + \cos^2\gamma = 1$. Assuming $t_+ = t_-$, equation (6.2.8) yields

$$\frac{q_r}{\bar{c}_r} = \frac{q_x}{\bar{c}_x}\cos\alpha + \frac{q_y}{\bar{c}_y}\cos\beta + \frac{q_z}{\bar{c}_z}\cos\gamma. \tag{6.2.9}$$

Note that $\bar{c}_i$ with $i = x$ (or y, z) is the two-way speed of light along the i-axis, but not the component of $\bar{c}_r$ on the direction of i-axis.

In particular, if we assume the constancy of two-way speed of light, i.e., $\bar{c}_r = \bar{c}_i$, equation (6.2.9) reduces to [2]

$$q_r = q_x\cos\alpha + q_y\cos\beta + q_z\cos\gamma. \tag{6.2.10}$$

6.3. Edwards Transformation

Edwards (1963) [2] assumed the *constancy of two-way speed of light*, and changed Einstein's second postulate to the following: *The two-way speed of light in a vacuum as measured in two coordinate systems moving with constant relative velocity is the same constant regardless of any assumptions concerning the one-way speed.*

The constancy of two-way speed of light has been used for defining the Edwards frames in Sec. 1.3.3. For simplicity, let $\mathbf{q} \equiv (q, 0, 0)$ and $\mathbf{q}' = (q', 0, 0)$, so the one-way velocities of light as seen in the Edwards frames $\tilde{F}$ and $\tilde{F}'$ are expressed as (see the previous section)

$$c_x = \frac{c}{1-q}, \quad c_{-x} = \frac{c}{1+q}, \quad c_y = c_{-y} = c_z = c_{-z} = c, \tag{6.3.1a}$$

$$c_x' = \frac{c}{1-q'}, \quad c_{-x}' = \frac{c}{1+q'}, \quad c_y' = c_{-y}' = c_z' = c_{-z}' = c, \tag{6.3.1b}$$

where the constant c denotes the two-way speed of light within both $\tilde{F}$ and $\tilde{F}'$, i.e., $c = \bar{c}_r = \bar{c}_{r'}'$. The one-way speeds of light in the directions of the vectors $\pm\mathbf{r}$ (or $\pm\mathbf{r}'$) as measured in $\tilde{F}$ (or $\tilde{F}'$) are from (6.2.2)

$$c_r = \frac{c}{1-q_r}, \quad c_{-r} = \frac{c}{1+q_r}, \tag{6.3.2a}$$

or

$$c_{r'}' = \frac{c}{1-q_{r'}'}, \quad c_{-r'}' = \frac{c}{1+q_{r'}'}. \tag{6.3.2b}$$

where the projection q_r (or $q_{r'}'$) of $\mathbf{q}$ (or $\mathbf{q}'$) in the direction of $\mathbf{r}$ (or $\mathbf{r}'$) is from (6.2.10)

$$q_r = q_x\cos\alpha + q_y\cos\beta + q_z\cos\gamma = q\cos\alpha, \tag{6.3.3a}$$

or

$$q_{r'}' = q_x'\cos\alpha' + q_y'\cos\beta' + q_z'\cos\gamma' = q'\cos\alpha', \tag{6.3.3b}$$

with α (or α') being the angle among the directions of **r** and x-axis (or **r'** and x'-axis).

From the postulate about constancy of the two-way speed of light, (6.3.1), and the principle of relativity, Edwards (1963) [2] obtained his generalized Lorentz transformations among the two Frames $\tilde{F}$ and $\tilde{F}'$ as follows:

$$\tilde{x}' = \xi(\tilde{x} - \tilde{v}\tilde{t}), \tag{6.3.4a}$$

$$\tilde{y}' = \tilde{y}, \tag{6.3.4b}$$

$$\tilde{z}' = \tilde{z}, \tag{6.3.4c}$$

$$\tilde{t}' = \xi\left\{\left[1 + (q+q')\frac{\tilde{v}}{c}\right]\tilde{t} - \left[(1-q^2)\frac{\tilde{v}}{c} + (q'-q)\right]\frac{\tilde{x}}{c}\right\}, \tag{6.3.4d}$$

with

$$\xi \equiv \frac{1}{\sqrt{(1+q\tilde{v}/c)^2 - \tilde{v}^2/c^2}}, \tag{6.3.4e}$$

where $\tilde{v}$ is the velocity of the Edwards frames $\tilde{F}'$ relative to $\tilde{F}$. The relative velocity of $\tilde{F}'$ to $\tilde{F}$ can be obtained by putting $d\tilde{x}' = d\tilde{y}' = d\tilde{z}' = 0$ into Eqs. (6.3.4a,b,c),

$$\tilde{\mathbf{v}} = (\tilde{v}, 0, 0)$$

with

$$\tilde{v} = \frac{d\tilde{x}}{d\tilde{t}}. \tag{6.3.4f}$$

Note that the *coordinate time interval* $d\tilde{t}$ here is measured by use of *Edwards clocks*, and thus we call $\tilde{v}$ the *Edwards velocity* as stated in Sec. 1.5.

The inverse of the transformations (6.3.4) are

$$\tilde{x} = \xi\left\{\left[1 + (q+q')\frac{\tilde{v}}{c}\right]\tilde{x}' + \tilde{v}'\tilde{t}'\right\}, \tag{6.3.5a}$$

$$\tilde{y} = \tilde{y}', \tag{6.3.5b}$$

$$\tilde{z} = \tilde{z}', \tag{6.3.5c}$$

$$\tilde{t} = \xi\left\{\tilde{t}' + \left[(1-q^2)\frac{\tilde{v}}{c} + (q'-q)\right]\frac{\tilde{x}'}{c}\right\}, \tag{6.3.5d}$$

Introducing the quantity

$$\tilde{v}' = \frac{\tilde{v}}{1 + (q+q')\tilde{v}/c}, \tag{6.3.6a}$$

then Eqs. (6.3.5) become

$$\tilde{x} = \xi'(\tilde{x}' + \tilde{v}'\tilde{t}'), \tag{6.3.6b}$$

$$\tilde{y} = \tilde{y}', \tag{6.3.6c}$$

$$\tilde{z} = \tilde{z}', \tag{6.3.6d}$$

$$\tilde{t} = \xi' \left\{ \left[1 - (q + q') \frac{\tilde{v}'}{c} \right] \tilde{t}' + \left[(1 - q'^2) \frac{\tilde{v}'}{c} + (q' - q) \right] \frac{\tilde{x}'}{c} \right\}, \qquad (6.3.6e)$$

with

$$\xi' \equiv \frac{1}{\sqrt{(1 - q'\tilde{v}'/c)^2 - \tilde{v}'^2/c^2}}. \qquad (6.3.6f)$$

Wee see from Eqs. (6.3.6) that the quantity $-\tilde{v}'$ represents the velocity of $\tilde{F}$ relative to $\tilde{F}'$.

In the case of $q = 0$ the frame $\tilde{F}$ reduces to the Einstein frame $F(xyzt)$ (see Sec. 1.3.4), which is sometimes called a "preferred" reference system. In this case the Edwards transformations (6.3.4) reduce to

$$\tilde{x}' = \frac{1}{\sqrt{1 - v^2/c^2}} (x - vt), \qquad (6.3.7a)$$

$$\tilde{y}' = y, \qquad (6.3.7b)$$

$$\tilde{z}' = z. \qquad (6.3.7c)$$

$$\tilde{t}' = \frac{1}{\sqrt{1 - v^2/c^2}} \left[\left(1 + \frac{v}{c} q' \right) t - \left(\frac{v}{c} + q' \right) \frac{x}{c} \right], \qquad (6.3.7d)$$

It is worth to note that the connection (1.3.23) of the Edwards frame with the Einstein frame as well as the relation (1.5.6) between the Edwards and Einstein velocities represent the relationship between Edwards' and Einstein's definitions of simultaneity. In other words, putting the relations (1.3.23) and (1.5.6) into Eqs. (6.3.4) is equivalent to putting $q = 0$ into Eqs. (6.3.4). In fact, substituting equations (1.3.23) and (1.5.6) into (6.3.4) for $\tilde{x}, \tilde{y}, \tilde{z}, \tilde{t}$ and $\tilde{v}$, one arrives at the same results with Eqs. (6.3.7).

We now give the derivations of Edwards transformations by use of two different methods.

The first method is similar to that by which the Lorentz transformations are obtained in section 2.2. We start from the Edarwds frame $\tilde{F}'(\tilde{x}'\tilde{y}'\tilde{z}'\tilde{t}')$ and the Einstein frame $F(xyzt)$, which are defined in Secs. 1.3.3 and 1.3.2, respectively. As done in Sec. 2.2, we choose the initial conditions: The corresponding axes in the two frames coincide with each other at the initial time $t = \tilde{t}' = 0$, and the frame $\tilde{F}'$ moves relative to the frame F with the constant velocity v in the direction of x-axis. Using these initial conditions and the principle of special relativity, the coordinate transformations between $\tilde{F}'$ and F take the linear forms:

$$\tilde{x}' = \tilde{a}_{11}(x - vt), \qquad (6.3.8a)$$

$$\tilde{y}' = y, \qquad (6.3.8b)$$

$$\tilde{z}' = z, \qquad (6.3.8c)$$

$$\tilde{t}' = \tilde{a}_{01}x + \tilde{a}_{00}t. \tag{6.3.8d}$$

In order to establish the constants $\tilde{a}_{11}, \tilde{a}_{00}$ and $\tilde{a}_{01}$, we recall the fact that the Einstein frame F and the Edwards frame $\tilde{F}$ are defined in Secs. 1.3.3 and 1.3.2 in such a way that the one-way velocities of light as measured in these two frames are, respectively, the constant c and $c'_{r'}$. This implies

$$(\tilde{x}')^2 + (\tilde{y}')^2 + (\tilde{z}')^2 - (c'_{r'}\tilde{t}')^2 = x^2 + y^2 + z^2 - (ct)^2 = 0. \tag{6.3.9}$$

Substituting (6.3.8) into (6.3.9) and considering the special cases of equations (6.3.1b), we can determine $\tilde{a}_{11}, \tilde{a}_{00}$ and $\tilde{a}_{01}$. In this way, equations (6.3.8) become the Edwards transformations (6.3.7).

The second method for deriving the Edwards transformations is based on the consideration concerning the difference among the definitions of simultaneity in the frames $\tilde{F}$ and F, which we have already discussed in Sec. 1.3.4. Therefore, we start from the Lorentz transformations (2.2.4),

$$x' = \frac{1}{\sqrt{1 - v^2/c^2}}(x - vt), \tag{6.3.10a}$$

$$y' = y, \tag{6.3.10b}$$

$$z' = z, \tag{6.3.10c}$$

$$t' = \frac{1}{\sqrt{1 - v^2/c^2}}\left(t - \frac{v}{c^2}x\right). \tag{6.3.10d}$$

Then, one can get the general Edwards transformations (6.3.4) from the Lorentz transformations (6.3.10) by the method: Change the Einstein frames F and F' into the Edwards frames $\tilde{F}$ and $\tilde{F}'$, respectively.

To do this, let us recall the transformations (1.3.23) among F and $\tilde{F}$:

$$x = \tilde{x}, \quad y = \tilde{y}, \quad z = \tilde{z}, \tag{6.3.11a}$$

$$t = \tilde{t} + q\frac{\tilde{x}}{c}. \tag{6.3.11b}$$

Similarly, the transformations between F' and $\tilde{F}'$ are

$$x' = \tilde{x}', \quad y' = \tilde{y}', \quad z' = \tilde{z}', \tag{6.3.12a}$$

$$t' = \tilde{t}' + q'\frac{\tilde{x}'}{c} \tag{6.3.12b}$$

On the other hand, as stated in Sec. 1.5, the *Einstein velocity v* in (6.3.10) should also be changed into the Edwards velocity $\tilde{v}$ by use of equation (1.5.6b), i.e.,

$$v = \frac{\tilde{v}}{1 + q\tilde{v}/c}. \tag{6.3.13}$$

By substituting (6.3.11)–(6.3.13) for the coordinates $(x', y', z', t'), (x, y, z, t)$ and the Einstein velocity v in the Lorentz transformations (6.3.10), we obtain the general Edwards transformations (6.3.4).

We have seen in the above derivations that after the Einstein frames have been changed into the Edwards frames the Lorentz transformations would become the Edwards transformations. This shows that the difference among Edwards' theory and Einstein's theory is just from the difference between definitions of simultaneity. It will shows, however, that this difference has no any physical effect in experiments if Einstein's simultaneity is used in measuring the quantities relevant to coordinate time intervals just as it does usually in laboratories.

6.4. Anisotropic Four-Dimensional Space–Time

Introduce the 4-vectors,

$$dX = \begin{pmatrix} cdt \\ dx \\ dy \\ dz \end{pmatrix}, \qquad dX' = \begin{pmatrix} cdt' \\ dx' \\ dy' \\ dz' \end{pmatrix}, \tag{6.4.1a}$$

and the transpose of dX and dX',

$$dX^\dagger = (cdt, dx, dy, dz), \qquad (dX')^\dagger = (cdt', dx', dy', dz'). \tag{6.4.1b}$$

Then the 4-interval (2.3.4) becomes

$$ds^2 = dX^\dagger \eta dX, \tag{6.4.1c}$$

where the Minkowski metric is

$$\eta \equiv (\eta_{\mu\nu}) = \begin{pmatrix} 1 & 0 & 0 & 0 \\ 0 & -1 & 0 & 0 \\ 0 & 0 & -1 & 0 \\ 0 & 0 & 0 & -1 \end{pmatrix}. \tag{6.4.1d}$$

The Lorentz transformations (2.2.4) can be expressed as

$$dX' = \Lambda dX, \tag{6.4.2a}$$

where Λ is the coefficient matrix of the right-hand side of Eq. (2.2.4):

$$\Lambda = \begin{pmatrix} \gamma & -\gamma v/c & 0 & 0 \\ -\gamma v/c & \gamma & 0 & 0 \\ 0 & 0 & 1 & 0 \\ 0 & 0 & 0 & 1 \end{pmatrix}, \tag{6.4.2b}$$

with

$$\gamma = \frac{1}{\sqrt{1 - v^2/c^2}}. \qquad (6.4.2c)$$

The Lorentz transformations are those linear coordinate transformations which carry the metric tensor $\eta_{\mu\nu}$ over into itself:

$$\eta = \Lambda^\dagger \eta \Lambda. \qquad (6.4.3)$$

We now return to the space–time for the Edwards frames $\tilde{F}$ and $\tilde{F}'$. It is easy to know that the coordinate transformations between the Edwards frame $\tilde{F}$ and the Einstein frame F, Eqs. (1.3.23), are also valid for the space and time derivatives:

$$dt = d\tilde{t} + q\frac{dx}{c}, \qquad (6.4.4a)$$

$$dx = d\tilde{x}, \quad dy = d\tilde{y}, \quad dz = d\tilde{z}. \qquad (6.4.4b)$$

where the $\mathbf{q} = (q, 0, 0)$ and Eq. (6.2.10) are used. These transformations can be expressed in the compact form:

$$dX = Q d\tilde{X}, \qquad (6.4.5a)$$

where dX is defined by Eq. (6.4.1a), $d\tilde{X}$ has the similar definition

$$d\tilde{X} = \begin{pmatrix} cd\tilde{t} \\ d\tilde{x} \\ d\tilde{y} \\ d\tilde{z} \end{pmatrix}, \qquad (6.4.5b)$$

and

$$Q = \begin{pmatrix} 1 & q & 0 & 0 \\ 0 & 1 & 0 & 0 \\ 0 & 0 & 1 & 0 \\ 0 & 0 & 0 & 1 \end{pmatrix}. \qquad (6.4.6)$$

The transpose of Eq. (6.4.5) is

$$dX^\dagger = d\tilde{X}^\dagger Q^\dagger, \qquad (6.4.7)$$

where $dX^\dagger$ is given by (6.4.1b), similarly

$$d\tilde{X}^\dagger = (cd\tilde{t}, d\tilde{x}, d\tilde{y}, d\tilde{z}), \qquad (6.4.8)$$

and

$$Q^\dagger = \begin{pmatrix} 1 & 0 & 0 & 0 \\ q & 1 & 0 & 0 \\ 0 & 0 & 1 & 0 \\ 0 & 0 & 0 & 1 \end{pmatrix}. \qquad (6.4.9)$$

The inverse transformation of (6.4.5) is

$$d\tilde{X} = Q^{-1}dX. \tag{6.4.10a}$$

Here Q^{-1} is the inverse matrix of Q, which is given by

$$Q^{-1}(q) = \begin{pmatrix} 1 & -q & 0 & 0 \\ 0 & 1 & 0 & 0 \\ 0 & 0 & 1 & 0 \\ 0 & 0 & 0 & 1 \end{pmatrix} = Q(-q). \tag{6.4.10b}$$

The transpose of (6.4.10a) is

$$d\tilde{X}^{\dagger} = dX^{\dagger}\left(Q^{-1}\right)^{\dagger}. \tag{6.4.11}$$

The matrix Q is the transformation matrix from the isotropic Minkowski space–time (M_4) to a new space–time. Because Q is not orthogonal, so that the new space–time is an anisotropic space–time to be denoted by A_4. We now determine the metric tensor $g_{\alpha\beta}$ from the transformation (6.4.5) or (6.4.10). The 4-interval in the space A_4 takes the form

$$d\tilde{s}^2 = g_{\alpha\beta}d\tilde{X}^{\alpha}d\tilde{X}^{\beta} = d\tilde{X}^{\dagger}g d\tilde{X}, \tag{6.4.12}$$

On the other hand, under the transformation (6.4.5) the 4-interval in M_4, Eq. (6.4.1c), will be transformed into

$$\begin{aligned} ds^2 &= dX^{\dagger}\eta dX \\ &= d\tilde{X}^{\dagger}\left[Q^{\dagger}\eta Q\right]d\tilde{X}. \end{aligned}$$

$$\tag{6.4.13}$$

We will give the argument that the 4-intervals should be equal to each other:

$$ds^2 = d\tilde{s}^2. \tag{6.4.14}$$

This leads from Eqs. (6.4.13) and (6.4.12) to

$$g \equiv (g_{\alpha\beta}) = Q^{\dagger}\eta Q. \tag{6.4.15a}$$

By using Eqs. (6.4.6), (6.4.9), and (6.4.1d), we can write explicitly the metric:

$$g = \begin{pmatrix} 1 & q & 0 & 0 \\ q & -1+q^2 & 0 & 0 \\ 0 & 0 & -1 & 0 \\ 0 & 0 & 0 & -1 \end{pmatrix}. \tag{6.4.15b}$$

Then the 4-interval, Eq. (6.4.12), can be explicitly written as

$$ds^2 = d\tilde{s}^2 = (cd\tilde{t} + qd\tilde{x})^2 - d\tilde{x}^2 - d\tilde{y}^2 - d\tilde{z}^2. \tag{6.4.16}$$

The above deals with the frames F and $\tilde{F}$. Similarly, for the primed Einstein frame F' and the primed Edwards frame $\tilde{F}'$ we have the similar results:

$$dX' = Q'd\tilde{X}', \tag{6.4.17a}$$

$$d\tilde{X}' = (Q')^{-1}dX', \tag{6.4.17b}$$

$$Q' = \begin{pmatrix} 1 & q' & 0 & 0 \\ 0 & 1 & 0 & 0 \\ 0 & 0 & 1 & 0 \\ 0 & 0 & 0 & 1 \end{pmatrix}, \tag{6.4.17c}$$

$$(Q')^{-1} = \begin{pmatrix} 1 & -q' & 0 & 0 \\ 0 & 1 & 0 & 0 \\ 0 & 0 & 1 & 0 \\ 0 & 0 & 0 & 1 \end{pmatrix} = Q'(-q'), \tag{6.4.17d}$$

$$(ds')^2 = (d\tilde{s}')^2 = (d\tilde{X}')^\dagger g' d\tilde{X}', \tag{6.4.18a}$$

where definitions of $d\tilde{X}'$ and $(d\tilde{X}')^\dagger$ are similar to those of $d\tilde{X}$ and $d\tilde{X}^\dagger$, and

$$(ds')^2 = (dX')^\dagger \eta dX', \tag{6.4.18b}$$

$$g' \equiv (g'_{\alpha\beta}) = (Q')^\dagger \eta Q', \tag{6.4.19a}$$

or explicitly

$$g' = \begin{pmatrix} 1 & q' & 0 & 0 \\ q' & -1+(q')^2 & 0 & 0 \\ 0 & 0 & -1 & 0 \\ 0 & 0 & 0 & -1 \end{pmatrix}. \tag{6.4.19b}$$

Then the 4-interval, Eq. (6.4.18a), is explicitly written as

$$(d\tilde{s}')^2 = (cd\tilde{t}' + q'd\tilde{x}')^2 - (d\tilde{x}')^2 - (d\tilde{y}')^2 - (d\tilde{z}')^2. \tag{6.4.20}$$

It is obvious from the discussion below that the 4-interval $d\tilde{s}^2$ in the anisotropic space A_4 is an invariant under the Edwards transformations due to the invariance of the 4-interval ds^2 under the Lorentz transformations.

The Edwards transformations (6.3.4) between the two frames $\tilde{F}'$ and $\tilde{F}$ can be expressed in the compact form:

$$d\tilde{X}' = \tilde{\Lambda}d\tilde{X}, \tag{6.4.21}$$

where the transformation matrix $\tilde{\Lambda}$ is the coefficient matrix on the right-hand side of (6.3.4),

$$\tilde{\Lambda} = (\tilde{\Lambda}^{\mu}{}_{\nu}) = \left(\frac{\partial \tilde{x}'^{\mu}}{\partial \tilde{x}^{\nu}}\right)$$

$$= \begin{pmatrix} \xi\left[1 + (q + q')\tilde{v}/c\right] & -\xi\left[(1 - q^2)\tilde{v}/c + (q' - q)\right] & 0 & 0 \\ -\xi\tilde{v}/c & \xi & 0 & 0 \\ 0 & 0 & 1 & 0 \\ 0 & 0 & 0 & 1 \end{pmatrix}$$

$$(6.4.22a)$$

with

$$\xi \equiv \frac{1}{\sqrt{(1 + q\tilde{v}/c)^2 - \tilde{v}^2/c^2}}. \qquad (6.4.22b)$$

The inverse of $\tilde{\Lambda}$ is

$$\tilde{\Lambda}^{-1} = (\tilde{\Lambda}_{\mu}{}^{\nu}) = \left(\frac{\partial \tilde{x}^{\nu}}{\partial \tilde{x}'^{\mu}}\right)$$

$$= \begin{pmatrix} \xi & \xi\left[(1 - q^2)\tilde{v}/c + (q' - q)\right] & 0 & 0 \\ \xi\tilde{v}/c & \xi\left[1 + (q + q')\tilde{v}/c\right] & 0 & 0 \\ 0 & 0 & 1 & 0 \\ 0 & 0 & 0 & 1 \end{pmatrix}.$$

$$(6.4.22c)$$

From the derivation of (6.3.4) in the previous section we can know that the matrix $\tilde{\Lambda}$ should be some product of the Lorentz matrix Λ and the Q matrix. In fact, by putting (6.4.17b) and (6.4.10a) into (6.4.21), we have

$$(Q')^{-1}dX' = \tilde{\Lambda}Q^{-1}dX. \qquad (6.4.23)$$

Using (6.4.2a), this equation becomes

$$(Q')^{-1}\Lambda dX = \tilde{\Lambda}Q^{-1}dX. \qquad (6.4.24)$$

Owing to the arbitrariness of dX, this equation gives

$$(Q')^{-1}\Lambda = \tilde{\Lambda}Q^{-1}, \qquad (6.4.25)$$

or

$$\tilde{\Lambda} = (Q')^{-1}\Lambda Q. \qquad (6.4.26a)$$

The inverse transformation equation of (6.4.26a) then is

$$\Lambda = Q'\tilde{\Lambda}Q^{-1}. \qquad (6.4.26b)$$

It is easy to check that the expressions (6.4.2b), (6.4.6), (6.4.17c), (6.4.22), and (6.3.13) satisfy the relation (6.4.26).

Let us discuss relations among the matrixes g, g', Q, and $\tilde{\Lambda}$. Substituting Eq. (6.4.26b) for Λ in (6.4.3), one arrives at

$$\eta = (Q^{-1})^\dagger \tilde{\Lambda}^\dagger (Q')^\dagger \eta Q' \tilde{\Lambda} Q^{-1}. \tag{6.4.27}$$

By use of the definition of the matrix g', Eq. (6.4.19a), equation (6.4.27) becomes

$$\eta = (Q^{-1})^\dagger \tilde{\Lambda}^\dagger g' \tilde{\Lambda} Q^{-1}. \tag{6.4.28}$$

This equation can be expressed as

$$Q^\dagger \eta Q = \tilde{\Lambda}^\dagger g' \tilde{\Lambda}, \tag{6.4.29}$$

or

$$g = \tilde{\Lambda}^\dagger g' \tilde{\Lambda}, \tag{6.4.30}$$

where the definition of matrix g, equation (6.4.15a), is used. The relation (6.4.30) indicates that the Edwards transformations are those linear coordinate transformations which remain the form of the metric tensor $g_{\alpha\beta}$ invariant. Thus, this shows the invariance of the 4-interval under Edwards transformations:

$$(d\tilde{s}')^2 = d\tilde{s}^2,$$

or explicitly

$$(cd\tilde{t} + qd\tilde{x})^2 \quad - \quad d\tilde{x}^2 - d\tilde{y}^2 - d\tilde{z}^2$$
$$= \quad (cd\tilde{t}' + q'd\tilde{x}')^2 - (d\tilde{x}')^2 - (d\tilde{y}')^2 - (d\tilde{z}')^2.$$
$$\tag{6.4.31}$$

In the special case of $q = 0$ (i.e., $g = \eta$), Eq. (6.4.30) reduces to

$$\eta = \left[\tilde{\Lambda}(q=0)\right]^\dagger g' \left[\tilde{\Lambda}(q=0)\right], \tag{6.4.32}$$

or

$$g' = \left[\tilde{\Lambda}^{-1}(q=0)\right]^\dagger \eta \left[\tilde{\Lambda}^{-1}(q=0)\right]. \tag{6.4.33}$$

Similarly, for the matrix g we have

$$g = \left[\tilde{\Lambda}(q'=0)\right]^\dagger \eta \left[\tilde{\Lambda}(q'=0)\right]. \tag{6.4.34}$$

The quantity $\tilde{V} = (\tilde{V}^0, \tilde{V}^1, \tilde{V}^2, \tilde{V}^3)$ in the anisotropic space–time A_4 is called a contravariant 4-vector if transformed under the Edwards transformations like the coordinate derivative $d\tilde{X} = (cd\tilde{t}, d\tilde{x}, d\tilde{y}, d\tilde{z})$. The transformation law of a covariant

4-vector under the Edwards transformations is the same as the derivative operator $(\partial/\partial \tilde{c}t, \partial/\partial \tilde{x}, , \partial/\partial \tilde{y}, \partial/\partial \tilde{z})$. The definition of tensors is given by Eq. (2.3.15) where Λ should be replaced by $\tilde{\Lambda}$.

Equation (6.4.30) indicates that the metric tensor $g_{\alpha\beta}$ is a covariant tensor of rank 2 under the Edwards transformations. The inverse of $g_{\alpha\beta}$ can be easily found out

$$g^{-1} = (g^{\alpha\beta}) = \begin{pmatrix} 1-q^2 & q & 0 & 0 \\ q & -1 & 0 & 0 \\ 0 & 0 & -1 & 0 \\ 0 & 0 & 0 & -1 \end{pmatrix}, \tag{6.4.35}$$

which is a contravariant tensor of rank 2. The metrics $g_{\alpha\beta}$ and $g^{\alpha\beta}$ can be used to raise or to lower the indices of tensors.

The above discussion shows that the choice for the 4-interval $d\tilde{s}$ is consistent. We want to point out that the equality of ds^2 and $d\tilde{s}^2$ is indeed an identity (see Sec. 6.6).

6.5. Comparison Among Edwards' Theory and Experiments

It is shown from the above sections that the difference among Lorentz and Edwards transformations follows just from the different definitions of the *coordinate times*. So that Edwards' theory and Einstein's theory would give the same physical predictions, if a light signal is used in clock synchronizing. In this section we shall repeat the procedure of changing simultaneity, as done in the above sections, for some examples to reveal the physical equivalence between the two theories.

6.5.1. Edwards' law of the Addition of Velocities

Edwards' addition law of velocities can be obtained from dividing (6.3.4a,b,c) by (6.3.4d):

$$\tilde{u}'_x = \frac{\tilde{u}_x - \tilde{v}}{[1 + (\tilde{v}/c)(q'+q)] - [(\tilde{v}/c)(1-q^2) + q' - q](\tilde{u}_x/c)}, \tag{6.5.1a}$$

$$\tilde{u}'_y = \frac{\tilde{u}_y\sqrt{[1+(\tilde{v}/c)q]^2 - \tilde{v}^2/c^2}}{[1 + (\tilde{v}/c)(q'+q)] - [(\tilde{v}/c)(1-q^2) + q' - q](\tilde{u}_x/c)}, \tag{6.5.1b}$$

$$\tilde{u}'_z = \frac{\tilde{u}_z\sqrt{[1+(\tilde{v}/c)q]^2 - \tilde{v}^2/c^2}}{[1 + (\tilde{v}/c)(q'+q)] - [(\tilde{v}/c)(1-q^2) + q' - q](\tilde{u}_x/c)}, \tag{6.5.1c}$$

where $\tilde{v}$ is the Edwards velocity of $\tilde{F}'$ relative to $\tilde{F}$, $\tilde{u}'_x = d\tilde{x}'/d\tilde{t}'$ and $\tilde{u}_x = d\tilde{x}/d\tilde{t}$ are the *Edwards velocities* of a moving body as seen in $\tilde{F}'$ and $\tilde{F}$, respectively, and so on. If $q' \neq 0$ and/or $q \neq 0$, then Edwards' addition law is mathematically different from Einstein's addition law. As stressed in the previous sections, however, this difference comes just from the different definitions of clock synchronization. Thus, by using Eq. (1.5.6a) for changing Edwards velocities to the Einstein velocities, i.e.,

$$\tilde{u}_x = \frac{u_x}{1 - qu_x/c}, \tag{6.5.2a}$$

$$\tilde{u}'_x = \frac{u'_x}{1 - q'u'_x/c}, \tag{6.5.2b}$$

$$\tilde{v} = \frac{v}{1 - qu_x/c}, \tag{6.5.2c}$$

then Edwards' addition law (6.5.1) becomes Einstein' addition law (2.4.2). Therefore, the difference is not observable if we use a light signal in synchronizing.

6.5.2. On Reciprocity of Relative Velocities

Firstly consider the relative velocities among the two Einstein frames F and F'. By substituting $u'_x = u'_y = u'_z = 0$ into Einstein's addition law, Eqs. (2.4.3), we obtain the relative velocity of F' to F,

$$u_x = v, \quad u_y = u_z = 0. \tag{6.5.3}$$

Similarly, by putting $u_x = u_y = u_z = 0$ into Eqs. (2.4.2), we have the relative velocity of F to F',

$$u'_x = -v, \quad u'_y = u'_z = 0. \tag{6.5.4}$$

Hence, the reciprocity of relative velocities is valid for the two Einstein frames:

$$\mathbf{u}' = -\mathbf{u}. \tag{6.5.5}$$

We now discuss relative velocities of two Edwards frames $\tilde{F}'$ and $\tilde{F}$ by use of Edwards' addition law. Substituting $\tilde{u}_x = \tilde{u}_y = \tilde{u}_z = 0$ into Eqs. (6.5.1), we get the velocity of $\tilde{F}'$ relative to $\tilde{F}$,

$$\tilde{\mathbf{u}} = (\tilde{v}, 0, 0). \tag{6.5.6}$$

Similarly, putting $\tilde{u}'_x = \tilde{u}'_y = \tilde{u}'_z = 0$ into Eqs. (6.5.1), we obtain the velocity of $\tilde{F}$ relative to $\tilde{F}'$,

$$\tilde{u}'_x = \frac{-\tilde{v}}{1 + (\tilde{v}/c)(q' + q)}, \tag{6.5.7a}$$

$$\tilde{u}'_y = \tilde{u}'_z = 0. \tag{6.5.7b}$$

The results (6.5.6) and (6.5.7) show that we may have the reciprocity of the relative velocities,

$$\mathbf{u}' = -\mathbf{u}, \tag{6.5.8a}$$

if and only if

$$q' = -q. \tag{6.5.8b}$$

This is just the *reciprocity condition* obtained in Sec. 1.5. The reciprocity of relative velocities will be valid just for those frames which have the same simultaneity, Einstein frames being the special cases of $q = -q' = 0$.

However, we want to emphasize that reciprocity of relative velocities is not a universal rule. In particular, reciprocity of relative velocities is not valid for the Edwards frames with different directional parameters. As mentioned in Sec. 1.5, a value of a velocity has no physically absolute meaning in the sense that it depends on the definition of simultaneity. Therefore, we may say that velocities are not directly observable in experiments if we cannot fined any instantaneous signal in laboratories.

6.5.3. Time Dilation of a Moving Clock

Let the clock $\tilde{C}'$ be rigidly connected with the Edwards frame $\tilde{F}'$, i.e., its spatial coordinates in $\tilde{F}'$ do not change: $\Delta \tilde{x}' = \Delta \tilde{y}' = \Delta \tilde{z}' = 0$. Thus, the corresponding time coordinate is just the difference among two readings of the same clock $\tilde{C}'$, which is a proper time interval denoted by $\Delta \tilde{\tau}'$. To get the corresponding time interval $\Delta \tilde{t}$ as measured in the Edwards frame $\tilde{F}$, putting $\Delta \tilde{x}' = \Delta \tilde{y}' = \Delta \tilde{z}' = 0$ into the Edwards transformations (6.3.4), we have

$$\Delta \tilde{x} = \tilde{v} \Delta \tilde{t}, \tag{6.5.9a}$$

$$\Delta \tilde{\tau}' = \frac{1}{\sqrt{[1 + (\tilde{v}/c)q]^2 - \tilde{v}^2/c^2}} \left\{ \left[1 + \frac{\tilde{v}}{c}(q + q') \right] \Delta \tilde{t} - \left[\frac{\tilde{v}}{c}(1 - q^2) + (q' - q) \right] \frac{\Delta \tilde{x}}{c} \right\}. \tag{6.5.9b}$$

Substituting (6.5.9a) into (6.5.9b), we obtain Edwards' time dilation formula,

$$\Delta \tilde{\tau}' = \Delta \tilde{t} \sqrt{\left[1 + q\frac{\tilde{v}}{c} \right]^2 - \frac{\tilde{v}^2}{c^2}}, \tag{6.5.10}$$

where $\Delta \tilde{t}$ is Edwards' coordinate time interval as measured by two separate Edwards clocks in $\tilde{F}$. It is easily proved that the difference among Eq. (6.5.10) and (2.9.2a) comes just from the different definitions of the velocities ($\tilde{v}$ and v) and *coordinate time intervals* ($\Delta \tilde{t}$ and Δt). In fact, by using the transformations (6.3.13) and (6.3.11b), i.e.,

$$\tilde{v} = \frac{v}{1 - qv/c}, \quad or \quad \left(1 + q\frac{\tilde{v}}{c} \right) \left(1 - q\frac{v}{c} \right) = 1, \tag{6.5.11a}$$

$$\Delta \tilde{t} = \Delta t - q\frac{\Delta x}{c} = \Delta t \left(1 - q\frac{v}{c} \right), \tag{6.5.11b}$$

then Edwards' time dilation formula (6.5.10) reduces to Einstein's formula (2.9.2a). Let us now explain the physical meaning of this "reduction" as follows.

Let a clock C move with a constant velocity from a point A (at the local time t_A) to a point B (at the local time t_B). Assume that the difference $(t_B - t_A)$ is, e.g., 1×10^{-7} sec, the distance $AB = \Delta x = 3$m, and hence the velocity of the clock is $\Delta x/(t_B - t_A) = 3 \times 10^4$ km. It is needed to stress that both v and $(t_B - t_A)$ are measured by the Einstein clocks, as it always does in the physical experiments.

Now the question is in the following: How to compare the time dilation formulas with these measurements? Answer is obvious:

If the "reduction" formula, i.e., Einstein's formula (2.9.2a), is used, we can simply identify 1×10^{-7} sec and 3×10^4 km with Δt and v of Eq. (2.9.2a), respectively, and then obtain

$$\Delta \tilde{\tau}' = 0.3 \times 10^{-7} \text{sec},$$

because the clock synchronization in the frame F is the same as that in this experiment.

However, if one would like to use Edwards' formula (6.5.10), then 1×10^{-7}sec and 3×10^4 km *cannot* be identified with $\Delta \tilde{t}$ and $\tilde{v}$, respectively, because the clock synchronization in the formula (6.5.10) is Edwards' synchronization which differs from Einstein's synchronization as used in this experiment. One *should* substitute 1×10^{-7}sec and 3×10^4 km for Δt and v in Eq. (6.5.11a,b), respectively, and then get the values,

$$\tilde{v} = \frac{3 \times 10^4}{1 - 0.1q} \text{ km}, \qquad (6.5.11c)$$

$$\Delta \tilde{t} = \Delta t \left(1 - q\frac{v}{c} \right) = (1 - 0.1q) \times 3 \times 10^{-7}\text{sec}. \qquad (6.5.11d)$$

Then, putting these values of $\tilde{v}$ and $\Delta \tilde{t}$ into the Edwards formula (6.5.10), we still predict the same value of $\Delta \tilde{\tau}' = 0.3 \times 10^{-7}$sec as the above result from Einstein's time dilation formula.

The above example strongly shows that Edwards' and Einstein's formulas are equivalent in physical experiments. In other words, the directional parameter q is not observable, if a light signal is used in clock synchronizing in experiments.

6.5.4. The Römer Experiment

O. Römer [37,38] determined the speed of light from the occultation of the moons of Jupiter. The interval in which one of the moons of Jupiter enters into the shadow of this planet is constant, as seen from Jupiter. Seen from the earth irregularities appear due to the change in the Earth–Jupiter distance. As this distance increases, light signals would take longer to reach the earth. This permits a determination of the velocity of light. In this section We shall re-analyze the Römer experiment by making use of the Edwards transformations.

Jupiter involving its moons is regarded as a source (J) which emits light signals in the constant time interval towards the earth (E) where they are received. In principle

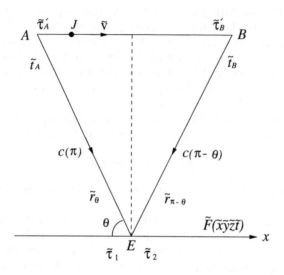

Fig. 7. Schematic of a Römer-type experiment. The Jupiter system J moves with the velocity $\tilde{v}$ as seen from the Earth E. J and E are at rest in $\tilde{F}'$ and $\tilde{F}$, respectively. The symbols "τ" and t represent "proper" and "coordinate" time, respectively. The primed and unprimed quantities are measured in the frames $\tilde{F}'$ and $\tilde{F}$, respectively.

this experiment can be shown in Fig. 7. Let the Edwards frame $\tilde{F}$ be rigidly connected with the earth. In $\tilde{F}$, Jupiter (J) moves with the velocity $\tilde{v}$ from the point A to the point B. The first light signal emitted by Jupiter at the time $\tilde{t}_A$ leaves the point A for the Earth E and then reaches E at the time $\tilde{\tau}_1$, as seen in $\tilde{F}$. Similarly, the second light signal emitted by Jupiter at the time $\tilde{t}_B$ leaves the point B for the Earth E and then reaches E at the time $\tilde{\tau}_2$, as seen in $\tilde{F}$. It is known that the interval

$$\Delta\tilde{\tau}_E = \tilde{\tau}_2 - \tilde{\tau}_1 \tag{6.5.12a}$$

is a proper time interval, while the interval

$$\Delta\tilde{t} = \tilde{t}_B - \tilde{t}_A \tag{6.5.12b}$$

is a coordinate time interval. Seen from Jupiter the two signals are emitted in the time interval

$$\Delta\tilde{\tau}'_J = \tilde{\tau}'_B - \tilde{\tau}'_A \tag{6.5.12c}$$

which is also a proper time interval.

It is known that the relation between $\Delta \tilde{t}$ and $\Delta \tilde{\tau}'_J$ is given by Edwards' time dilation formula (6.5.10), i.e.,

$$\Delta \tilde{\tau}'_J = \Delta \tilde{t} \sqrt{\left[1 + q\frac{\tilde{v}}{c}\right]^2 - \frac{\tilde{v}^2}{c^2}}. \tag{6.5.13}$$

Let us now derive the relationship between $\Delta \tilde{\tau}_E$ and $\Delta \tilde{t}$. The geometry in Fig. 7 shows

$$\tilde{r} \equiv \tilde{r}(\theta) = \tilde{r}(\pi - \theta), \qquad \Delta \tilde{x} \equiv AB = 2\tilde{r} \cos \theta \tag{6.5.12d}$$

and

$$\tilde{v} = \frac{\Delta \tilde{x}}{\Delta \tilde{t}}. \tag{6.5.12e}$$

By definition we have

$$\tilde{\tau}_2 = \tilde{t}_B + \frac{\tilde{r}(\pi - \theta)}{c(\pi - \theta)},$$

$$\tilde{\tau}_1 = \tilde{t}_A + \frac{\tilde{r}(\theta)}{c(\theta)}, \tag{6.5.14}$$

and then

$$\Delta \tilde{\tau}_E = \tilde{\tau}_2 - \tilde{\tau}_1 = \Delta \tilde{t} + \frac{\tilde{r}}{c(\pi - \theta)} - \frac{\tilde{r}}{c(\theta)}, \tag{6.5.15}$$

where the definitions (6.5.12) are used. The symbols $c(\theta)$ and $c(\pi - \theta)$ represent the velocities of the first and second signals along the paths AE and BE, respectively, which are given by (6.3.2) and (6.3.3), i.e.,

$$c(\theta) = \frac{c}{1 - q \cos \theta}, \tag{6.5.16a}$$

$$c(\pi - \theta) = \frac{c}{1 + q \cos \theta}, \tag{6.5.16b}$$

By putting (6.5.16) into (6.5.15), one arrives at

$$\Delta \tilde{\tau}_E = \Delta \tilde{t} + (2\tilde{r} \cos \theta)\frac{q}{c} = \Delta \tilde{t}\left(1 + \frac{\Delta \tilde{x}}{\Delta \tilde{t}}\frac{q}{c}\right) = \Delta \tilde{t}\left(1 + q\frac{\tilde{v}}{c}\right)$$

or

$$\Delta \tilde{t} = \Delta \tilde{\tau}_E \left(1 + q\frac{\tilde{v}}{c}\right)^{-1}, \tag{6.5.17}$$

where Eqs. (6.5.12d,e) are used.

Finally, by eliminating the coordinate time interval $\Delta \tilde{t}$ from Eqs. (6.5.17) and (6.5.13) we then obtain the expected relation between the two proper time intervals $\Delta \tilde{\tau}_E$ and $\Delta \tilde{\tau}'_J$ as follows:

$$\Delta \tilde{\tau}'_J = \frac{\sqrt{[1 + q(\tilde{v}/c)]^2 - \tilde{v}^2/c^2}}{1 + q(\tilde{v}/c)}\Delta \tilde{\tau}_E. \tag{6.5.18a}$$

By expanding the right-hand side of Eq. (6.5.18a) we can find that the first-order term in q does not appear. Furthermore while Eq. (6.5.18a) involves the second-order term, this does not implies the observability of q. The reason is the same as that mentioned in Sec. 6.5.3. In fact, changing the Edwards velocity $\tilde{v}$ to the Einstein velocity v by use of Eq. (6.5.11a), then equation (6.5.18a) reduces to the ordinary Einstein's time dilation formula (2.9.2a),

$$\Delta \tilde{\tau}_J' = \sqrt{1 - \frac{v^2}{c^2}} \Delta \tilde{\tau}_E. \tag{6.5.18b}$$

As stressed in the previous subsections, if we would like to use directly the data given by the experiment we must employ Eq. (6.5.18b) because the velocity v in this equation is defined by the same definition of simultaneity as that in the experiment. So the Römer-type experiment can not give a test of the one-way velocity of light. On the other hand, if you want to employ Eq. (6.5.18a) you must first identity the data of the velocity of Jupiter with v in Eq. (6.5.11a) to find out values of $\tilde{v}$ and then put the results in Eq. (6.5.18a). In this way you would have the same result as that given by Eq. (6.5.18b).

From the above analyses we can see that the Römer-type experiment is nothing but the measurement of the time dilation. So we have the conclusion: There is also not any effect of a non-vanishing q on the Römer-type experiment.

6.5.5. Aberration and Doppler Effect

To make physical meanings explicit, as done in Sec. 2.10, let us now employ the following methods to derive the aberration and Doppler effect.

Aberration comes from the change of direction of light ray when observed in other frame. Thus, we use Edwards' addition law to derive aberration formula. For simplicity, we assume that a light ray lies in the xy-plane: $c_z = 0, c_x = c_r \cos \theta, c_y = c_r \sin \theta$. The addition law (6.5.1) shows that the light ray also lies in the $x'y'$-plane: $c_z' = 0, c_x' = c_{r'}' \cos \theta', c_y' = c_{r'}' \sin \theta'$. From Eqs. (6.3.2) and (6.3.3) we have

$$c_r = \frac{c}{1 - q \cos \theta}, \qquad c_{r'}' = \frac{c}{1 - q' \cos \theta'}, \tag{6.5.19}$$

or

$$c_r = c + q c_r \cos \theta = c + q c_x, \tag{6.5.20a}$$

$$c_{r'}' = c + q' c_{r'}' \cos \theta' = c + q' c_x'. \tag{6.5.20b}$$

The addition law, Eq. (6.5.1a), gives the relation among c_x' and c_x,

$$c_x' = \frac{c_x - \tilde{v}}{[1 + (\tilde{v}/c)(q' + q)] - [(\tilde{v}/c)(1 - q^2) + q' - q](c_x/c)}, \tag{6.5.21a}$$

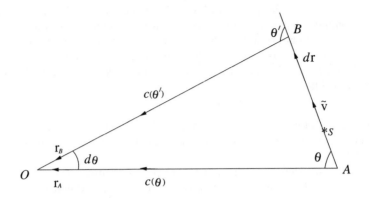

Fig. 8. Schematic for deriving Doppler effect in Edwards' theory.

where

$$c_x = c_r \cos\theta = \frac{c}{1 - q\cos\theta}\cos\theta. \tag{6.5.21b}$$

By definition, the cosine of the angle θ' between the directions of the light ray and x'-axis as seen in $\tilde{F}'$ is given by

$$\cos\theta' = \frac{c'_x}{c'_{r'}} = \frac{c'_x}{c + q'c'_x}, \tag{6.5.22}$$

where Eq. (6.5.20b) is used. Substituting Eq. (6.5.21a) for c'_x in Eq. (6.5.22) and then using (6.5.21b), we get Edwards' aberration formulas,

$$\cos\theta' = \frac{(c + \tilde{v}q)\cos\theta - \tilde{v}}{(c + \tilde{v}q) - \tilde{v}\cos\theta}, \tag{6.5.23a}$$

and

$$\sin\theta' = \frac{c'_y}{c'_{r'}} = \sqrt{1 - \cos^2\theta'} = \frac{\sqrt{[1 + (\tilde{v}/c)q]^2 - \tilde{v}^2/c^2}\sin\theta}{1 + (\tilde{v}/c)q - (\tilde{v}/c)\cos\theta}. \tag{6.5.23b}$$

We now derive Doppler effect by using the method shown in Fig. 8 (Fig. 7 is a special case of Fig. 8, where the relative velocity of Jupiter is constant and the average direction of light rays is perpendicular to the path of motion of Jupiter).

In Fig. 8, a light source S, which is at rest in the Edwards frame $\tilde{F}'$, has the instantaneous velocity $\tilde{\mathbf{v}}$, as seen in another Edwards frame (i.e., the laboratory frame) $\tilde{F}$ in where an observer is at rest at the point O. Let the source emit the Nth crest of light wave at the point A at the coordinate time $\tilde{t}_A$ as seen in $\tilde{F}$, while the corresponding time as timed by a clock connected rigidly with the source is $\tilde{\tau}'_A$. Then, at the coordinate time $\tilde{t}_A + d\tilde{t}$ the source reaches the point B, and emits the $(N + dN)$th crest, while the corresponding time with respect to the source is $\tilde{\tau}'_B$. We see that $\tilde{\tau}'_A$ and $\tilde{\tau}'_B$ are the two readings of the same clock connected with the source, and thus their difference

$$d\tilde{\tau}' = \tilde{\tau}'_A - \tilde{\tau}'_B \qquad (6.5.24)$$

is a proper time interval. However, $\tilde{t}_A$ is given by a clock at A, and while $\tilde{t}_A + d\tilde{t}$ is the reading of another clock at B. Then their difference $d\tilde{t}$ is an Edwards' coordinate time interval.

We now assume that the Nth and $(N + dN)$th crests are received by the observer at the point O at the times $\tilde{\tau}_N$ and $\tilde{\tau}_{N+dN}$, respectively. The difference

$$d\tilde{\tau} \equiv \tilde{\tau}_{N+dN} - \tilde{\tau}_N \qquad (6.5.25)$$

is also a proper time interval because they are timed by the same clock at O.

According to the geometry in Fig. 8, we have

$$\mathbf{r}_B = \mathbf{r}_A - d\mathbf{r}, \qquad (6.5.26a)$$

$$dr = \tilde{v}d\tilde{t}, \qquad (6.5.26b)$$

$$\theta' = \theta + d\theta, \qquad (6.5.26c)$$

$$d\theta = \frac{dr}{r_A} \sin\theta, \qquad (6.5.26d)$$

where θ and θ' are, respectively, the angles between $\tilde{v}$ and $\mathbf{r}_A$ as well as $\tilde{v}$ and $\mathbf{r}_B$, and $d\theta$ is the angle among $\mathbf{r}_A$ and $\mathbf{r}_B$. Note that the relative velocity $\tilde{v}$ is an Edwards velocity.

The one-way speeds of the Nth and $(N + dN)$th crests of light waves along the directions of $\mathbf{r}_A$ and $\mathbf{r}_B$ are from Eq. (6.3.2)

$$c(\theta) = \frac{c}{1 - q\cos\theta} \qquad (6.5.27a)$$

and

$$c(\theta') = c(\theta + d\theta) = \frac{c}{1 - q\cos(\theta + d\theta)} = \frac{c}{1 - q\cos\theta + qd\theta\sin\theta}, \qquad (6.5.27b)$$

respectively.

Let us now calculate the relationship between $d\tilde{\tau}$ and $d\tilde{t}$. From the geometry we have

$$\tilde{\tau}_1 = \tilde{t}_A + \frac{r_A}{c(\theta)},$$

$$\tilde{\tau}_2 = \tilde{t}_A + d\tilde{t} + \frac{r_B}{c(\theta + d\theta)} \qquad (6.5.28a)$$

with $r_A = |\mathbf{r}_A|, r_B = |\mathbf{r}_B|$, and then

$$d\tilde{\tau} = d\tilde{t} + \frac{r_B}{c(\theta + d\theta)} - \frac{r_A}{c(\theta)}. \qquad (6.5.28b)$$

From Eq. (6.5.26a) we get

$$r_B{}^2 = (\mathbf{r_A} - d\mathbf{r})^2 = r_A^2 - 2\mathbf{r}_A \cdot d\mathbf{r} = r_A^2 - 2r_A dr\cos\theta$$

or

$$r_B = r_A - dr\cos\theta. \qquad (6.5.29)$$

Substituting Eqs. (6.5.27) and (6.5.29) into (6.5.28b), and using Eqs. (6.5.26b,d), one arrives at

$$d\tilde{\tau} = d\tilde{t}\left(1 + q\frac{\tilde{v}}{c} - \frac{\tilde{v}}{c}\cos\theta\right). \qquad (6.5.30)$$

On the other hand, by definition, the proper frequency ω' of source is given by

$$\omega' = \frac{dN}{d\tilde{\tau}'}. \qquad (6.5.31a)$$

Similarly, the frequency ω as measured by the observer is defined by

$$\omega = \frac{dN}{d\tilde{\tau}}. \qquad (6.5.31b)$$

Then, we obtain

$$\omega = \frac{dN}{d\tilde{\tau}} = \frac{dN}{d\tilde{\tau}'}\frac{d\tilde{\tau}'}{d\tilde{\tau}} = \omega'\frac{d\tilde{\tau}'}{d\tilde{\tau}}. \qquad (6.5.32)$$

Substituting Eq. (6.5.30) into Eq. (6.5.32), we get

$$\omega = \omega'\frac{(d\tilde{\tau}'/d\tilde{t})}{1 + q\tilde{v}/c - (\tilde{v}/c)\cos\theta}. \qquad (6.5.33)$$

On the right-hand side of Eq. (6.5.33), the numerator is the ratio of the proper time interval to the corresponding Edwards' coordinate time interval, which is given by Edwards' time dilation formula Eq. (6.5.10). Then, by use of Eq. (6.5.10), we obtain the Doppler frequency shift formula in Edwards' theory,

$$\omega = \omega'\frac{\sqrt{[1 + (\tilde{v}/c)q]^2 - \tilde{v}^2/c^2}}{1 + q\tilde{v}/c - (\tilde{v}/c)\cos\theta}. \qquad (6.5.34)$$

Let us now consider how to compare Edwards' aberration and shift formulas, Eqs. (6.5.23) and (6,5,34), with physical experiments.

Note that the angle θ and frequencies ω and ω' are not related to the definition of simultaneity. However, the Edwards velocity $\tilde{v}$ in these formulas is defined by Edwards' simultaneity. In fact, by use of Eq. (6.5.11a), i.e.

$$\tilde{v} = \frac{v}{1 - qv/c},\qquad(6.5.35)$$

for changing the Edwards velocity $\tilde{v}$ to the Einstein velocity v, then Eqs. (6.5.23) and (6.5.34) reduce to the ordinary formulas.

Therefore, the status is similar to that in subsections 6.5.3 and 6.5.4: When the ordinary formulas (i.e., the formulas given by Einstein's theory) are used to compare with data of experiments, then we can simply identify the data given by experiments with the Einstein velocity v; However, if we want to employ Edwards' formulas, Eqs. (6.5.23) and (6.5.34), we should firstly substitute the data for the Einstein velocity v of Eq. (6.5.35) to get values of the Edwards velocity $\tilde{v}$, and then put these values in Edwards' formulas, Eqs. (6.5.23) and (6.5.34). In this way the formulas (6.5.23) and (6.5.34) would give the same results as that given by the ordinary formulas.

6.5.6. Contraction of a Moving Body

Let a rod be rigidly connected with the frame $\tilde{F}'$, the end points of which have the space coordinates $(\tilde{x}'_2, 0, 0)$ and $(\tilde{x}'_1, 0, 0)$. The length in its rest frame, i.e. the rest (proper) length, is simply given by

$$l'_0 = \Delta\tilde{x}' \equiv \tilde{x}'_2 - \tilde{x}'_1.\qquad(6.5.36)$$

According to Eq. (6.3.4a), the transformation of the space coordinate interval is

$$\Delta\tilde{x}' = \xi(\Delta\tilde{x} - \tilde{v}\Delta\tilde{t}),\qquad(6.5.37)$$

where $\Delta\tilde{x} = \tilde{x}_2 - \tilde{x}_1$ is the space coordinate interval corresponding to the coordinate time interval $\Delta\tilde{t} = \tilde{t}_2 - \tilde{t}_1$ as measured in the frame $\tilde{F}$.

Edwards' length $\tilde{l}$ of the moving rod measured by the observer in $\tilde{F}$ is defined by such a space distance $\Delta\tilde{x}$ that its corresponding coordinate time interval, $\Delta\tilde{t}$, vanishes. In other words, we must simultaneously (i.e., according to Edwards' definition of simultaneity) measure the two end points of the moving rod in $\tilde{F}$. Thus, from Eqs. (6.5.37) with $\Delta\tilde{t} = 0$ and (6.3.4e) we have the relation between the lengths of the rod in the two frames,

$$\tilde{l} = l'_0\sqrt{\left(1 + q\frac{\tilde{v}}{c}\right)^2 - \frac{\tilde{v}^2}{c^2}}.\qquad(6.5.38)$$

It is emphasized that the rest (proper) length $l'_0 = \Delta\tilde{x}'$ is independent of simultaneity, and while Edwards' "moving" length $\tilde{l} = \Delta\tilde{x}$ depends on Edwards' simultaneity, $\Delta\tilde{t} = 0$.

Let us now consider the relationship among Edwards' contraction formula (6.5.38) and Einstein's formula (2.8.4a).

In Einstein's theory the "moving" length l in Eq. (2.8.4a) is defined by Einstein's simultaneity, $\Delta t = 0$. From Eq. (6.3.11b) we know

$$\Delta t = \Delta \tilde{t} + q \frac{\Delta \tilde{x}}{c}. \tag{6.5.39}$$

Putting $\Delta \tilde{t} = 0$ and $\Delta \tilde{x} = \tilde{l}$ into Eq. (6.5.39), we have

$$\Delta t = q \frac{\tilde{l}}{c}. \tag{6.5.40}$$

This means that $\tilde{l}$ is not Einstein's "moving" length l because of $\Delta t \neq 0$. In fact, the path running by the rod at the time interval Δt is given by

$$\delta l = v \Delta t = q \frac{v}{c} \tilde{l}, \tag{6.5.41}$$

where v is the Einstein velocity. Then, the relationship among l and $\tilde{l}$ is

$$l = \tilde{l} - \delta l = \tilde{l} \left(1 - q \frac{v}{c} \right),$$

or

$$\tilde{l} = \frac{l}{1 - qv/c}. \tag{6.5.42}$$

Substituting (6.5.42) into (6.5.38), we get

$$\frac{l}{1 - qv/c} = l'_0 \sqrt{ \left(1 + q \frac{\tilde{v}}{c} \right)^2 - \frac{\tilde{v}^2}{c^2}}. \tag{6.5.43}$$

Using Eq. (6.5.11a) for the change on the right-hand side from the Edwards velocity $\tilde{v}$ to the Einstein velocity v, Edwards' formula (6.5.43) then becomes the usual Einstein's formula (2.8.4a),

$$l = l'_0 \sqrt{1 - \frac{v^2}{c^2}}.$$

Therefore, Edwards' formula (6.5.38) gives the same prediction as Einstein's formula (2.8.4a) in physical experiments, just as it does for the time dilation formulas in subsection 6.5.3.

6.5.7. Conclusion

The Edwards transformations Eqs. (6.3.4) are based on Edwards' simultaneity, and thus cannot be directly compared with data of physical experiments in which Einstein's simultaneity is always used. In fact the data for the time coordinates and velocities given in the physical experiments must be firstly identified with Einstein's time coordinates (t and t') and the Einstein velocity v, respectively. Then, the relations (6.3.11b), (6.3.12b) and (6.3.13) should be employed to get values of Edwards' times ($\tilde{t}$ and $\tilde{t}'$) and the Edwards velocity $\tilde{v}$. In this way the Edwards transformations (6.3.4) would give the same predictions in kinematics as Lorentz transformations for the physical experiments. This means that the directional parameter **q** has no any effect on experimental results. In other words, the one-way velocity of light cannot be tested by the experiments, just as it does in the above examples. We want to stress here that the relationship among Edwards transformations and Lorentz transformations is valid for all physical events involving the examples given in the above subsections. Therefore the analyses in the above subsections are trivial and not necessary because we have already had the general relation between Edwards and Lorentz transformations in Sec. 6.3.

Dynamic formulation in Edwards' theory may be constructed by using the four-dimensional formalism described in subsection 6.4. The key point is still that one must distinguish unobservable quantities from observable quantities.

6.6. On Dynamics of Edwards' Theory

In the previous sections we discuss the kinematics of Edwards' theory. This section will give some arguments on the dynamics of Edwards' theory.

In Sec. 6.4, we discuss the 4-interval $d\tilde{s}^2$ in the anisotropic space–time, and make it equal to that in Minkowski space–time, i.e.,

$$ds^2 = d\tilde{s}^2, \tag{6.6.1}$$

or explicitly

$$(cdt)^2 - dx^2 - dy^2 - dz^2 = (cd\tilde{t} + qd\tilde{x})^2 - d\tilde{x}^2 - d\tilde{y}^2 - d\tilde{z}^2. \tag{6.6.2}$$

Note that the relations between the coordinates dx^μ and $d\tilde{x}^\mu$ in this equation are given by Eq. (6.4.4), and that we are dealing with the same Cartesian frame $S(x, y, z)$. For simplicity, we choose $dy = dz = 0$. Then Eq. (6.6.2) reduces to

$$(cdt)^2 - dx^2 = (cd\tilde{t} + qd\tilde{x})^2 - d\tilde{x}^2. \tag{6.6.3}$$

Note also that the time coordinates t and $\tilde{t}$ are, respectively, the readings of the Einstein clock and Edwrads clock located at the same spatial point. We now give the implication of equation (6.6.3).

Let us recall the meanings of the coordinate time intervals $d\tilde{t}$ and dt given in Sec. 1.3. In the same Cartesian coordinate system $S(x, y, z)$, each point has an Einstein clock and, at the same time, has an Edwrads clock which have been already defined in Sec. 1.3. Consider two events E_1 and E_2 occur at two points $P_1 = (x_1, 0, 0)$ and

$P_2 = (x_2, 0, 0)$, the distance of which being $dl = dx = x_2 - x_1$. Now we ask the question: What is the time interval during which the two events occur? Answer is the following: If the Einstein clocks are used we then say that the time interval of the two events as seen in the frame $S(x, y, z)$ will be

$$dt = t_2 - t_1, \tag{6.6.4}$$

while if the Edwards clocks are used we then say that the time interval for the same two events as seen in the same frame $S(x, y, z)$ should be

$$d\tilde{t} = \tilde{t}_2 - \tilde{t}_1. \tag{6.6.5}$$

Relation between the values of the two intervals is just Eq. (6.4.4a). In other words, Eq. (6.4.4a) is just equation (6.6.3) due to Eq. (6.4.4b). This is to say that equation (6.6.3) is an identity for measurements of any events by using Edwards and Einstein clocks. Similarly, For measurements in another frame $S'(x', y'z')$ moving with a constant velocity relative to $S(x, y, z)$, the similar equation

$$(cdt')^2 - dx'^2 = (cd\tilde{t}' + q'd\tilde{x}')^2 - d\tilde{x}'^2 \tag{6.6.6}$$

is also an identity for measured values.

Lorentz transformations make the left-hand sides of Eqs. (6.6.3) and (6.6.6) equal to each other, while Edwards transformations make the right-hand sides equal to each other. It is well-known that Einstein's time dilation formula can be derived from the invariance of the 4-interval $ds^2 = ds'^2$. Similarly, Edwards' time dilation formula (6.5.10) can be derived from the invariance of $d\tilde{s}^2 = d\tilde{s}'^2$. Therefore, the 4-interval $d\tilde{s}$ must be defined by Eqs. (6.4.16) and (6.4.20).

Dynamics could be constructed by means of the four-dimensional formulation of the anisotropic space–time given in Sec. 6.4. The above arguments show that the invariance of the 4-interval is valid and gives the same predictions in kinematics in Einstein's and Edwards' theories. Because the events may be arbitrary, the invariance of the 4-interval then should be valid in dynamics. Thus, Edwards' dynamics based on this invariance should predict the same dynamic effects as that in Einstein's theory.

CHAPTER 7

THE GENERAL TEST THEORIES

7.1. Introduction

In the previous chapter we introduce and analyze Edwards' transformations, and then provide the correct method for comparing the transformations with experiments. It has been shown that Edwards' theory is physically equivalent to Einstein's theory of special relativity. Other kinds of test theories were proposed by Robertson (1949) [3] as well as by Mansouri and Sexl (MS) (1977) [4]. Since that time, many papers on this topic have been published [26–35]. However, some ambiguities still exist in comparing the test theories with physical experiments. Thus, we should re-analyze the test theories in detail. It is shown [25] that the MS transformations are trivial in the sense that they are a generalization of the Robertson transformations, just as the Edwards transformations are a generalization of the Lorentz transformations. In other words, the MS transformations differ from the Robertson transformations just by a directional parameter $\mathbf{q}$, just as the difference between the Edwards and Lorentz transformations. So that the MS transformations predict the same observable effects as the Robertson transformations, just as the Edwards transformations and Lorentz transformations do. Thus, MS' theory is not a new test theory, because (i) we have had Robertson's theory as a test theory of Einstein's special relativity, (ii) MS' and Robertson's theories are physically equivalent (thus, it is not necessary to use MS' theory for getting predictions to physical experiments).

7.2. Robertson's Test Theory

Robertson's test theory is based on the definition of Robertson's inertial frames of reference, i.e., the definition of Robertson's simultaneity, as stated in subsection 1.3.5, which is equivalent to the postulate that the velocity of light is given by Eqs. (1.3.24) and is independent of the motion of the light source.

7.2.1. Robertson Transformation

Robertson (1949) proposed the following transformations among the Robertson frame $\bar{F}(\bar{x}\bar{y}\bar{z}\bar{t})$ (see subsection 1.3.5 for definition) and Einstein frame $F(xyzt)$:

$$\bar{x} = a_1^{-1} \left(1 - \frac{v^2}{c^2} \right)^{-1} (x - vt), \qquad (7.2.1a)$$

$$\bar{y} = a_2^{-1} y, \qquad (7.2.1b)$$

$$\bar{z} = a_2^{-1} z, \tag{7.2.1c}$$

$$\bar{t} = a_0^{-1} \left(1 - \frac{v^2}{c^2}\right)^{-1} \left(t - \frac{v}{c^2}x\right), \tag{7.2.1d}$$

where v is the velocity of $\bar{F}$ as seen in F, and thus is an Einstein velocity (see subsection 1.3.2), a_0, a_1 and a_2 are arbitrary functions of v^2. The Einstein frame F can be called a "preferred" reference system in the sense that the one-way velocity of light is the constant c. It is easy from Eqs. (7.2.1) to get general Robertson transformations among any two Robertson frames $\bar{F}$ and $\bar{F}'$ by eliminating x, y, z and t.

For the convenience of comparing the Robertson transformations with the MS transformations (see Sec. 7.3), the parameters a_0, a_1, and a_2 are replaced by the following parameters:

$$a_0 = a^{-1}, \quad a_1 = b^{-1}\left(1 - \frac{v^2}{c^2}\right)^{-1}, \quad a_2 = d^{-1}. \tag{7.2.2}$$

In terms of these parameters, the *Robertson transformations* (7.2.1) become

$$\bar{x} = b(x - vt), \tag{7.2.3a}$$

$$\bar{y} = (d)y, \tag{7.2.3b}$$

$$\bar{z} = (d)z, \tag{7.2.3c}$$

$$\bar{t} = \frac{a}{1 - (v^2/c^2)}\left(t - \frac{v}{c^2}x\right). \tag{7.2.3d}$$

In order to understand physical meaning of the parameters, let us consider the velocity of light in the frame $\bar{F}$. In the Einstein frame F, the one-way speed of light is isotropic, i.e.,

$$c^2 t^2 - x^2 - y^2 - z^2 = 0. \tag{7.2.4}$$

Substituting the inverse transformations of (7.2.3) into (7.2.4), we have

$$c_{\bar{F}}^2\left(\frac{1}{\bar{c}_\|^2}\cos^2\alpha + \frac{1}{\bar{c}_\perp^2}\cos^2\beta + \frac{1}{\bar{c}_\perp^2}\cos^2\gamma\right) - 1 = 0, \tag{7.2.5}$$

where $\cos^2\alpha + \cos^2\beta + \cos^2\gamma = 1$, $\bar{x}/\bar{t} = c_{\bar{F}}\cos\alpha$, $\bar{y}/\bar{t} = c_{\bar{F}}\cos\beta$ and $\bar{z}/\bar{t} = c_{\bar{F}}\cos\gamma$ are used, and the parameters $\bar{c}_\|$ and $\bar{c}_\perp$ are defined by

$$\bar{c}_\| = \frac{cb}{a}\left(1 - \frac{v^2}{c^2}\right), \tag{7.2.6a}$$

$$\bar{c}_\perp = \frac{cd}{a}\sqrt{1 - \frac{v^2}{c^2}}. \tag{7.2.6b}$$

Thus, $c_{\bar{r}}$ is the one-way velocity of light along the direction of $\bar{\mathbf{r}}$ in the Robertson frame $\bar{F}$. If $\bar{\mathbf{r}} = (0, \bar{y}, \bar{z})$, i.e., the direction of light ray is perpendicular to the relative velocity $\mathbf{v}$, then from (7.2.5) we have $c_{\bar{r}} = \bar{c}_\perp$; If $\bar{\mathbf{r}} = (\bar{x}, 0, 0)$, i.e., the direction of light ray is parallel to the relative velocity $\mathbf{v}$, then from (7.2.5) we get $c_{\bar{r}} = \bar{c}_\parallel$. These show that the new parameters $\bar{c}_\parallel$ and $\bar{c}_\perp$ are nothing but the velocities of light in the parallel and perpendicular directions of $\mathbf{v}$, respectively.

A solution to Eq. (7.2.5) for $c_{\bar{r}}$ is given by

$$c_{\bar{r}} = \frac{\bar{c}_\parallel \bar{c}_\perp}{\sqrt{\bar{c}_\parallel^2 + (\bar{c}_\perp^2 - \bar{c}_\parallel^2) \cos^2 \alpha}}. \tag{7.2.7}$$

In particular, Eq. (7.2.7) leads to

$$c_{-\bar{r}} = c_{\bar{r}},$$

$$\bar{c}_{\bar{r}} \equiv \frac{2 c_{\bar{r}} c_{-\bar{r}}}{c_{\bar{r}} + c_{-\bar{r}}} = c_{\bar{r}},$$

$$c_{\bar{x}} = c_{-\bar{x}} = \bar{c}_\parallel, \qquad c_{\bar{y}} = c_{-\bar{y}} = c_{\bar{z}} = c_{-\bar{z}} = \bar{c}_\perp. \tag{7.2.8}$$

One can see from Eqs. (7.2.8) that in the Robertson frame $\bar{F}$ the one-way velocity of light in a given direction is equal to the one in its opposite direction, and that the two-way velocity of light, in general, is anisotropic. Therefore the new parameters $\bar{c}_\parallel$, $\bar{c}_\perp$, d (or a_0, a_1, a_2) can be determined by physical experiments. For instance, a limit on $\bar{c}_\parallel$ and $\bar{c}_\perp$ may come from the performed measurements of the two-way velocity of light, and then together with the time dilation experiments a limit on the parameter d can be obtained. Thus Robertson transformations are different physically from Lorentz transformations.

In terms of the new parameters defined by Eq. (7.2.6) the Robertson transformations (7.2.3) are written as

$$\bar{x} = (d) \frac{\bar{c}_\parallel}{\bar{c}_\perp} \frac{1}{\sqrt{1 - (v^2/c^2)}} (x - vt), \tag{7.2.9a}$$

$$\bar{y} = (d) y, \tag{7.2.9b}$$

$$\bar{z} = (d) z, \tag{7.2.9c}$$

$$\bar{t} = (d) \frac{c}{\bar{c}_\perp} \frac{1}{\sqrt{1 - (v^2/c^2)}} \left(t - \frac{v}{c^2} x \right). \tag{7.2.9d}$$

7.2.2. Comparison among Robertson and Lorentz Transformations

Robertson transformations (7.2.9) can be expressed in a standard Lorentz-like form. To do this, let us introduce the *Lorentz-like coordinates*:

$$\bar{x}_0 = d^{-1} \frac{\bar{c}_\perp}{\bar{c}_\parallel} \bar{x}, \tag{7.2.10a}$$

$$\bar{y}_0 = d^{-1}\bar{y}, \tag{7.2.10b}$$

$$\bar{z}_0 = d^{-1}\bar{z}, \tag{7.2.10c}$$

$$\bar{t}_0 = d^{-1}\frac{\bar{c}_\perp}{c}\bar{t}. \tag{7.2.10d}$$

By using the Lorentz-like coordinates, the Robertson transformations, Eqs. (7.2.9), become the Lorentz-like transformations:

$$\bar{x}_0 = \frac{1}{\sqrt{1 - v^2/c^2}}(x - vt), \tag{7.2.11a}$$

$$\bar{y}_0 = y, \tag{7.2.11b}$$

$$\bar{z}_0 = z, \tag{7.2.11c}$$

$$\bar{t}_0 = \frac{1}{\sqrt{1 - v^2/c^2}}\left(t - \frac{v}{c^2}x\right). \tag{7.2.11d}$$

It is easy to check that the *general Robertson transformations* between the two Robertson frames $\bar{F}(\bar{x}\bar{y}\bar{z}\bar{t})$ and $\bar{F}'(\bar{x}'\bar{y}'\bar{z}'\bar{t}')$ in terms of Lorentz-like coordinates take the Lorentz-like form:

$$\bar{x}'_0 = \frac{1}{\sqrt{1 - \bar{v}_0^2/c^2}}(\bar{x}_0 - \bar{v}_0\bar{t}_0), \tag{7.2.12a}$$

$$\bar{y}'_0 = \bar{y}_0, \qquad \bar{z}'_0 = \bar{z}_0, \tag{7.2.12b}$$

$$\bar{t}'_0 = \frac{1}{\sqrt{1 - \bar{v}_0^2/c^2}}\left(\bar{t}_0 - \frac{\bar{v}_0}{c^2}\bar{x}_0\right). \tag{7.2.12c}$$

This is the reason that the new variables defined by Eqs. (7.2.10) and (7.2.13) are referred to the Lorentz-like coordinates. Here the primed Lorentz-like coordinates are defined by

$$\bar{x}'_0 = (d')^{-1}\frac{\bar{c}'_\perp}{\bar{c}'_\parallel}\bar{x}', \tag{7.2.13a}$$

$$\bar{y}'_0 = (d')^{-1}\bar{y}', \tag{7.2.13b}$$

$$\bar{z}'_0 = (d')^{-1}\bar{z}', \tag{7.2.13c}$$

$$\bar{t}'_0 = (d')^{-1}\frac{\bar{c}'_\perp}{c}\bar{t}'. \tag{7.2.13d}$$

7.2.3. Four-Dimensional Space–Time

An invariant four-dimensional interval $d\bar{s}^2$ for the Robertson transformations can be easily introduced by means of the Lorentz-like coordinates:

$$d\bar{s}^2 = (cd\bar{t}_0)^2 - \left[(d\bar{x}_0)^2 + (d\bar{y}_0)^2 + (d\bar{z}_0)^2\right]. \tag{7.2.14}$$

The Robertson transformations (7.2.12) transform the 4-interval into the same form:

$$d\bar{s}^2 = (cd\bar{t}_0')^2 - \left[(d\bar{x}_0')^2 + (d\bar{y}_0')^2 + (d\bar{z}_0')^2\right].$$
(7.2.15)

By use of the definitions (7.2.10) the 4-interval is explicitly written in terms of the Robertson coordinates $(\bar{t}, \bar{x}, \bar{y}, \bar{z})$:

$$d\bar{s}^2 = (d^{-2}) \left[(\bar{c}_\perp d\bar{t})^2 - \left(\frac{\bar{c}_\perp^2}{\bar{c}_\parallel^2} d\bar{x}^2 + d\bar{y}^2 + d\bar{z}^2\right)\right].$$
(7.2.16)

This shows that the parameter d is nothing but a conformal factor.

For convenience in the remainder of this book, compare the metric (7.2.16) with the metric given by Robertson:

$$d\bar{s}^2 = (g_0 c d\bar{t})^2 - \left[(g_1 d\bar{x})^2 + g_2^2 \left(d\bar{y}^2 + d\bar{z}^2\right)\right].$$
(7.2.17)

Relations between Robertson's parameters and ours then are given by

$$g_0 = d^{-1}\frac{\bar{c}_\perp}{c}, \qquad g_1 = d^{-1}\frac{\bar{c}_\perp}{\bar{c}_\parallel}, \qquad g_2 = d^{-1}.$$
(7.2.18)

In particular, we have

$$\frac{g_2}{g_1} - 1 = \frac{\bar{c}_\parallel - \bar{c}_\perp}{\bar{c}_\perp}.$$
(7.2.19)

This means that the closed-path experiments, such as a laser test of the isotropy of space given by Brillet and Hall (1979) [40], tested the isotropy of the two-way speed of light (see Sec. 8.2.1).

Now introduce four-dimensional coordinate derivatives:

$$d\bar{X} \equiv (d\bar{X}^\mu) = \begin{pmatrix} \bar{c}_\parallel d\bar{t} \\ d\bar{x} \\ d\bar{y} \\ d\bar{z} \end{pmatrix}, \qquad d\bar{X}_0 \equiv (d\bar{X}_0^\mu) = \begin{pmatrix} cd\bar{t}_0 \\ d\bar{x}_0 \\ d\bar{y}_0 \\ d\bar{z}_0 \end{pmatrix}.$$
(7.2.20)

Equations (7.2.10) can be expressed as

$$d\bar{X}_0 = \bar{Q} d\bar{X},$$
(7.2.21)

where

$$\bar{Q} = \frac{1}{d} \begin{pmatrix} (\bar{c}_\perp/\bar{c}_\parallel) & 0 & 0 & 0 \\ 0 & (\bar{c}_\perp/\bar{c}_\parallel) & 0 & 0 \\ 0 & 0 & 1 & 0 \\ 0 & 0 & 0 & 1 \end{pmatrix}.$$
(7.2.22)

In this way, the 4-interval equations (7.2.14) and (7.2.16) can be written as

$$d\bar{s}^2 = d\bar{X}_0^\dagger \eta d\bar{X}_0$$
$$= d\bar{X}^\dagger \left[\bar{Q}^\dagger \eta \bar{Q}\right] d\bar{X} = d\bar{X}^\dagger \bar{g} d\bar{X},$$

(7.2.23)

where the metric $\bar{g}$ is defined by

$$\bar{g} \equiv (\bar{g}_{\mu\nu}) = \bar{Q}^\dagger \eta \bar{Q},$$

(7.2.24)

or explicitly

$$\bar{g} = \frac{1}{d^2}\begin{pmatrix} \left(\bar{c}_\perp/\bar{c}_\parallel\right)^2 & 0 & 0 & 0 \\ 0 & -\left(\bar{c}_\perp/\bar{c}_\parallel\right)^2 & 0 & 0 \\ 0 & 0 & -1 & 0 \\ 0 & 0 & 0 & -1 \end{pmatrix}.$$

(7.2.25)

Similarly, in the primed Robertson frame $\bar{F}'$ we have the similar formulas:

$$d\bar{X}_0' = \bar{Q}' d\bar{X}',$$

(7.2.26)

with

$$\bar{Q}' = \frac{1}{d'}\begin{pmatrix} \left(\bar{c}'_\perp/\bar{c}'_\parallel\right) & 0 & 0 & 0 \\ 0 & \left(\bar{c}'_\perp/\bar{c}'_\parallel\right) & 0 & 0 \\ 0 & 0 & 1 & 0 \\ 0 & 0 & 0 & 1 \end{pmatrix},$$

(7.2.27)

and

$$d\bar{s}^2 = d\bar{X}_0^{'\dagger} \eta d\bar{X}_0' = d\bar{X}^{'\dagger} \bar{g} d\bar{X}',$$

(7.2.28)

with

$$\bar{g}' \equiv (\bar{g}'_{\mu\nu}) = \bar{Q}'^\dagger \eta \bar{Q}',$$

(7.2.29)

or explicitly

$$\bar{g}' = \frac{1}{d'^2}\begin{pmatrix} \left(\bar{c}'_\perp/\bar{c}'_\parallel\right)^2 & 0 & 0 & 0 \\ 0 & -\left(\bar{c}'_\perp/\bar{c}'_\parallel\right)^2 & 0 & 0 \\ 0 & 0 & -1 & 0 \\ 0 & 0 & 0 & -1 \end{pmatrix}.$$

(7.2.30)

The general Robertson transformations (7.2.12) can be expressed as

$$d\bar{X}' = \bar{\Lambda} d\bar{X}.$$

(7.2.31)

Here the transformation matrix $\bar{\Lambda}$ can be obtained through the relation:

$$\bar{\Lambda} = (\bar{Q}')^{-1} \bar{\Lambda}_0 \bar{Q},$$

(7.2.32)

where $\bar{\Lambda}_0$ is the coefficient matrix on the right-hand side of the Lorentz-like transformation:

$$d\bar{X}'_0 = \bar{\Lambda}_0 d\bar{X}_0, \tag{7.2.33}$$

or explicitly

$$\bar{\Lambda}_0 = \begin{pmatrix} \bar{\gamma} & -\bar{\gamma}\bar{v}_0/c & 0 & 0 \\ -\bar{\gamma}\bar{v}_0/c & \bar{\gamma} & 0 & 0 \\ 0 & 0 & 1 & 0 \\ 0 & 0 & 0 & 1 \end{pmatrix}, \tag{7.2.34a}$$

with

$$\bar{\gamma} = \frac{1}{\sqrt{1 - \bar{v}_0^2/c^2}}. \tag{7.2.34b}$$

By use of Eqs. (7.2.22), (7.2.27), and (7.2.34a), the Robertson transformation matrix $\bar{\Lambda}$ in (7.2.32) is explicitly written as

$$\bar{\Lambda} = \left(\frac{d'}{d}\right) \begin{pmatrix} \bar{\gamma}(\bar{c}'_\parallel/\bar{c}_\parallel)(\bar{c}_\perp/\bar{c}'_\perp) & -\bar{\gamma}(\bar{v}_0/c)(\bar{c}'_\parallel/\bar{c}_\parallel)(\bar{c}_\perp/\bar{c}'_\perp) & 0 & 0 \\ -\bar{\gamma}(\bar{v}_0/c)(\bar{c}'_\parallel/\bar{c}_\parallel)(\bar{c}_\perp/\bar{c}'_\perp) & \bar{\gamma}(\bar{c}'_\parallel/\bar{c}_\parallel)(\bar{c}_\perp/\bar{c}'_\perp) & 0 & 0 \\ 0 & 0 & 1 & 0 \\ 0 & 0 & 0 & 1 \end{pmatrix}. \tag{7.2.35}$$

The quantity $\bar{v}_0$ in Eqs. (7.2.12), (7.2.34), and (7.2.35) is the *Lorentz-like velocity* of the origin of $\bar{F}'$ as seen from $\bar{F}$:

$$\frac{\bar{v}_0}{c} = \frac{\bar{x}_0}{c\bar{t}_0} = \frac{1}{\bar{c}_\parallel}\frac{\bar{x}}{\bar{t}} = \frac{\bar{v}}{\bar{c}_\parallel}, \tag{7.2.36}$$

where $\bar{x}$ and $\bar{t}$ are respectively the space and time coordinates of the origin of $\bar{F}'$ in $\bar{F}$, and

$$\bar{v} = \frac{\bar{x}}{\bar{t}} \tag{7.2.37}$$

is called the Robertson velocity, the velocity of the origin of $\bar{F}'$ measured by the Robertson clocks in $\bar{F}$.

7.2.4. Kinematical Effects

In the previous subsection we have seen that the general Robertson transformation and the 4-interval can be obtained from those of the Lorentz-like type. Therefore, all kinematical predictions given by the general Robertson transformations can be got from those given in Einstein's theory. Next we give some examples.

Robertson's law of the addition of velocities can be obtained from Eqs. (7.2.12), which has the same form as Einstein's law,

$$\bar{u}'_{0x} = \frac{\bar{u}_{0x} - \bar{v}_0}{1 - (\bar{v}_0\bar{u}_{0x}/c^2)}, \tag{7.2.38a}$$

$$\bar{u}'_{0y} = \bar{u}_{0y} \frac{\sqrt{1 - (\bar{v}_0^2/c^2)}}{1 - (\bar{v}_0 \bar{u}_{0x}/c^2)}, \tag{7.2.38b}$$

$$\bar{u}'_{0z} = \bar{u}_{0z} \frac{\sqrt{1 - (\bar{v}_0^2/c^2)}}{1 - (\bar{v}_0 \bar{u}_{0x}/c^2)}, \tag{7.2.38c}$$

where

$$\bar{u}'_{0x} = \frac{d\bar{x}'_0}{d\bar{t}'_0}, \quad \bar{u}_{0x} = \frac{d\bar{x}_0}{d\bar{t}_0}, \quad \bar{u}'_{0y} = \frac{d\bar{y}'_0}{d\bar{t}'_0},$$

$$\bar{u}_{0y} = \frac{d\bar{y}_0}{d\bar{t}_0}, \quad \bar{u}'_{0z} = \frac{d\bar{z}'_0}{d\bar{t}'_0}, \quad \bar{u}_{0z} = \frac{d\bar{z}_0}{d\bar{t}_0}, \tag{7.2.39}$$

and $\bar{v}_0$ is defined by Eq. (7.2.36). Using Eqs. (7.2.10) and (7.2.13), equations (7.2.39) become

$$\bar{u}'_{0x} = (c/\bar{c}'_\parallel)\frac{d\bar{x}'}{d\bar{t}'} = (c/\bar{c}'_\parallel)\bar{u}'_x, \quad \bar{u}_{0x} = (c/\bar{c}_\parallel)\frac{d\bar{x}}{d\bar{t}} = (c/\bar{c}_\parallel)\bar{u}_x,$$

$$\bar{u}'_{0y} = (c/\bar{c}'_\perp)\frac{d\bar{y}'}{d\bar{t}'} = (c/\bar{c}'_\perp)\bar{u}'_y, \quad \bar{u}_{0y} = (c/\bar{c}_\perp)\frac{d\bar{y}}{d\bar{t}} = (c/\bar{c}_\perp)\bar{u}_y,$$

$$\bar{u}'_{0z} = (c/\bar{c}'_\perp)\frac{dz'}{d\bar{t}'} = (c/\bar{c}'_\perp)\bar{u}'_z, \quad \bar{u}_{0z} = (c/\bar{c}_\perp)\frac{dz}{d\bar{t}} = (c/\bar{c}_\perp)\bar{u}_z. \tag{7.2.40}$$

By use of Eqs. (7.2.40) and (7.2.36), equations (7.2.38) are expressed as

$$(\bar{u}'_x/\bar{c}'_\parallel) = \frac{(\bar{u}_x/\bar{c}_\parallel) - (\bar{v}/\bar{c}_\parallel)}{1 - (\bar{v}\bar{u}_x/\bar{c}_\parallel^2)}, \tag{7.2.41a}$$

$$(\bar{u}'_y/\bar{c}'_\perp) = (\bar{u}_y/\bar{c}_\perp)\frac{\sqrt{1 - (\bar{v}^2/\bar{c}_\parallel^2)}}{1 - (\bar{v}\bar{u}_x/\bar{c}_\parallel^2)}, \tag{7.2.41b}$$

$$(\bar{u}'_z/\bar{c}'_\perp) = (\bar{u}_z/\bar{c}_\perp)\frac{\sqrt{1 - (\bar{v}^2/\bar{c}_\parallel^2)}}{1 - (\bar{v}\bar{u}_x/\bar{c}_\parallel^2)}. \tag{7.2.41c}$$

This is *Robertson's law of the addition of velocities* in terms of the Robertson velocities defined by the Robertson simultaneity.

The *condition of reciprocity of relative velocities*. Putting $\bar{u}'_x = \bar{u}'_y = \bar{u}'_z = 0$ into Eqs. (7.2.41), we get the relative velocity of $\bar{F}'$ to $\bar{F}$

$$\bar{u}_x = \bar{v}, \quad \bar{u}_y = \bar{u}_z = 0. \tag{7.2.42a}$$

Similarly, taking $\bar{u}_x = \bar{u}_y = \bar{u}_z = 0$ in (7.2.41), we then obtain the relative velocity of $\bar{F}$ to $\bar{F}'$

$$\bar{u}'_x = -\bar{v}\frac{\bar{c}'_\parallel}{\bar{c}_\parallel}, \quad \bar{u}'_y = \bar{u}'_z = 0. \tag{7.2.42b}$$

This shows that the *condition of reciprocity of relative velocities* is

$$\bar{c}_\parallel = \bar{c}'_\parallel. \tag{7.2.43}$$

Retardation of time (or *time dilation*). Let a standard clock be at rest in the frame $\bar{F}'$. Therefore, its location is fixed, i.e., $\Delta \bar{x}' = \Delta \bar{y}' = \Delta \bar{z}' = 0$. Applying these initial conditions in the Lorentz-type transformations, Eqs. (7.2.12), we have *Robertson's time dilation formula in terms of the new variables* as follows:

$$\Delta \bar{\tau}'_0 = \Delta \bar{t}_0 \sqrt{1 - \frac{\bar{v}_0^2}{c^2}}, \tag{7.2.44a}$$

where $\Delta \bar{\tau}'_0 \equiv \Delta \bar{t}'_0$. This has the same form as Einstein's formula. By using Eqs. (7.2.10d), (7.2.13d) and (7.2.36), equation (7.2.44a) becomes

$$\Delta \bar{\tau}' = \left(\frac{d'}{d} \right) \left(\frac{\bar{c}_\perp}{\bar{c}'_\perp} \right) \sqrt{1 - \frac{\bar{v}^2}{\bar{c}_\parallel^2}} \, \Delta \bar{t}, \tag{7.2.44b}$$

where $\Delta \bar{\tau}' = (cd)(\Delta \bar{\tau}'_0)/\bar{c}'_\perp$ is the difference between the two readings of the clock at rest in the frame $\bar{F}'$, i.e., *a proper time interval*, $\Delta \bar{t}$ is the corresponding coordinate time interval, and $\bar{v}$ is the velocity of the clock as measured in $\bar{F}$ by means of Robertson clocks. Equation (7.2.44b) is *Robertson's time dilation formula in terms of the physical quantities*.

7.2.5. Comparison between Robertson's Theory and Experiments

Test of the two-way velocities of light. The two-way velocities, $\bar{c}_\parallel$ and $\bar{c}_\perp$, can be directly measured by experiments. Up to now the value of the velocity of light given by experiments is a value of the two-way velocity of light, which show the constancy of the two-way velocity of light. In other wards, any possible deviation of $\bar{c}_\parallel$ and $\bar{c}_\perp$ from the constant c must be within the experimental errors.

The Lorentz-type transformations. Equations (7.2.12) show that the Robertson transformations take the same form as the usual Lorentz transformations. However, this does not physically implies the equivalence between the Robertson and Lorentz transformations. The reason is that the Lorentz-like coordinates $(\bar{x}_0, \bar{y}_0, \bar{z}_0, \bar{t}_0, \cdots)$ do not directly represent the physical coordinates. Robertson's test theory and Einstein's theory would give different predictions to experiments due to the difference of the two-way velocities of light.

An example: the time dilation. We now consider how to compare Eq. (7.2.44b) with experiments. The time interval $\Delta \bar{\tau}'$ is a proper time interval of a moving clock, and hence can be directly identified with an experimental measurement. However, $\Delta \bar{t}$ and $\bar{v}$ are defined by Robertson's but not Einstein's simultaneity, and therefore cannot be directly identified with experimental measurements. In fact, the relation between Robertson's time interval $\Delta \bar{t}$ and the Einstein's time interval Δt is similar to Eq. (1.3.22), i.e.,

$$\Delta \bar{t} = \Delta t + \Delta x \left(\frac{1}{\bar{c}_\parallel} - \frac{1}{c} \right) = \Delta t \left[1 + v \left(\frac{1}{\bar{c}_\parallel} - \frac{1}{c} \right) \right]. \tag{7.2.45}$$

Correspondingly, the Robertson velocity $\bar{v}$ is related to the Einstein velocity v through the equation:

$$\bar{v} = v \left[1 + v \left(\frac{1}{\bar{c}_\parallel} - \frac{1}{c}\right)\right]^{-1}. \tag{7.2.46}$$

These show that we should firstly identify the Einstein velocity v and Einstein's time interval Δt of Eqs. (7.2.45) and (7.2.46) with data given by experiments to obtain values of $\bar{v}$ and $\Delta \bar{t}$, and then put them in Robertson's time dilation formula (7.2.44b). In this way we can have a limit on the parameter d and the differences $(\bar{c}_\parallel - c)$ and $(\bar{c}_\perp - c)$.

Doppler effect. Derivation of Doppler frequency shift is similar to that in Sec. 6.5.5. While Fig. 8 is valid in the present case, the replacements $dr \to d\bar{r}, dt \to d\bar{t}, d\tau \to d\bar{\tau}, r_{A(B)} \to \bar{r}_{A(B)}$ and $v \to d\bar{v}$ should be made. Equations (6.5.28b) and (6.5.29) are also valid for this case:

$$d\bar{\tau} = d\bar{t} + \frac{\bar{r}_B}{c(\theta + d\theta)} - \frac{\bar{r}_A}{c(\theta)}, \tag{7.2.47}$$

$$\bar{r}_B = \bar{r}_A - d\bar{r}\cos\theta, \tag{7.2.48}$$

where

$$d\bar{r} = \bar{v}d\bar{t}, \quad d\theta = \frac{d\bar{r}.}{\bar{r}_A}. \tag{7.2.49}$$

Here the physical meaning of the quantities are the same as the original quantities in Fig. 8.

The velocity of light, $c(\theta)$, is now defined by Eq. (7.2.7):

$$\frac{1}{c(\theta)} = \frac{\sqrt{\bar{c}_\parallel^2 + (\bar{c}_\perp^2 - \bar{c}_\parallel^2)\cos^2\theta}}{\bar{c}_\parallel\bar{c}_\perp} \tag{7.2.50}$$

and

$$\frac{1}{c(\theta + d\theta)} = \frac{1}{c(\theta)} + d\theta\frac{d}{d\theta}\left(\frac{1}{c(\theta)}\right) = \frac{1}{c(\theta)} - d\theta\frac{(\bar{c}_\perp^2 - \bar{c}_\parallel^2)\cos\theta\sin\theta}{\bar{c}_\parallel\bar{c}_\perp\sqrt{\bar{c}_\parallel^2 + (\bar{c}_\perp^2 - \bar{c}_\parallel^2)\cos^2\theta}}. \tag{7.2.51}$$

By substituting Eqs. (7.2.48)–(7.2.51) into (7.2.47) we have

$$d\bar{\tau} = d\bar{t}\left[1 - \bar{v}\cos\theta\frac{\bar{c}_\parallel^2 + (\bar{c}_\perp^2 - \bar{c}_\parallel^2)(\cos^2\theta + \sin\theta)}{\bar{c}_\parallel\bar{c}_\perp\sqrt{\bar{c}_\parallel^2 + (\bar{c}_\perp^2 - \bar{c}_\parallel^2)\cos^2\theta}}\right]. \tag{7.2.52}$$

The frequency ω measured by the observer is defined by

$$\omega = \frac{dN}{d\bar{\tau}} = \frac{dN}{d\bar{\tau}'}\frac{d\bar{\tau}'}{d\bar{\tau}} = \omega'\frac{d\bar{\tau}'}{d\bar{\tau}}, \tag{7.2.53}$$

where ω' represents the proper frequency of source. By putting Eq. (7.2.52) into Eq. (7.2.53) we get

$$\omega = \omega' \frac{d\bar{\tau}'}{dt} \left[1 - \bar{v} \cos\theta \frac{\bar{c}_\parallel^2 + (\bar{c}_\perp^2 - \bar{c}_\parallel^2)(\cos^2\theta + \sin\theta)}{\bar{c}_\parallel \bar{c}_\perp \sqrt{\bar{c}_\parallel^2 + (\bar{c}_\perp^2 - \bar{c}_\parallel^2)\cos^2\theta}} \right]^{-1}, \qquad (7.2.54)$$

where $d\bar{\tau}'/dt$ is the time dilation effect given by Eq. (7.2.44b). Thus Eq. (7.2.54) becomes

$$\omega = \omega' \left(\frac{d'}{d}\right)\left(\frac{\bar{c}_\perp}{\bar{c}'_\perp}\right) \sqrt{1 - \frac{\bar{v}^2}{\bar{c}_\parallel^2}} \left[1 - \bar{v} \cos\theta \frac{\bar{c}_\parallel^2 + (\bar{c}_\perp^2 - \bar{c}_\parallel^2)(\cos^2\theta + \sin\theta)}{\bar{c}_\parallel \bar{c}_\perp \sqrt{\bar{c}_\parallel^2 + (\bar{c}_\perp^2 - \bar{c}_\parallel^2)\cos^2\theta}} \right]^{-1}, \qquad (7.2.55)$$

where the subscripts "$\parallel$" and "$\perp$" represent "parallel" and "perpendicular" to the direction of the velocity $\mathbf{v}$ of source, the primed and unprimed quantities are respectively defined in the frame $\bar{F}'$ connected rigidly with the source and the frame $\bar{F}$ in where the observer is at rest, and θ is the angle between the directions of $\mathbf{v}$ and light ray as seen from $\bar{F}$. Equation (7.2.55) is the *Doppler effect* formula in Robertson's test theory.

The *transverse Doppler effect* can be obtained by putting $\theta = \pi/2$ into Eq. (7.2.55),

$$\omega = \omega' \left(\frac{d'}{d}\right)\left(\frac{\bar{c}_\perp}{\bar{c}'_\perp}\right) \sqrt{1 - \frac{\bar{v}^2}{\bar{c}_\parallel^2}}, \qquad (7.2.56)$$

which is of second-order in $\bar{v}$.

The longitudinal Doppler effect is obtained by putting $\theta = 0$ into Eq. (7.2.55),

$$\omega = \omega' \left(\frac{d'}{d}\right)\left(\frac{\bar{c}_\perp}{\bar{c}'_\perp}\right) \sqrt{1 - \frac{\bar{v}^2}{\bar{c}_\parallel^2}} \left(1 - \frac{\bar{v}}{\bar{c}_\parallel}\right). \qquad (7.2.57)$$

Of course, to the first-order approximation the longitudinal Doppler effect is the same as that in Einstein's theory and the classical theory.

7.3. The Mansouri–Sexl (MS) Transformation

To generalize Robertson's theory, let us recall Edwards' theory that is a generalization of Einstein's theory. Edwards' theory is based on the Edwards inertial frames in where the one-way velocity of light is assumed to be

$$c_r = \frac{c}{1 - q_r}, \qquad c_{-r} = \frac{c}{1 + q_r}, \qquad \bar{c} = c. \qquad (7.3.1)$$

Similarly, one can generalize Robertson's postulate concerning the velocity of light, Eqs. (7.2.8), to the following (see subsection 1.3.6):

$$c_r = \frac{\bar{c}_r}{1 - q_r}, \qquad c_{-r} = \frac{\bar{c}_r}{1 + q_r}, \qquad -1 \le q_r \le +1, \qquad (7.3.2a)$$

where

$$\bar{c}_r = \frac{\bar{c}_{\parallel}\bar{c}_{\perp}}{\sqrt{\bar{c}_{\parallel}^2 + (\bar{c}_{\perp}^2 - \bar{c}_{\parallel}^2)\cos^2\alpha}}. \tag{7.3.2b}$$

In subsection 1.3.6, we have defined the MS inertial frame $F^*(x^*y^*z^*t^*)$ based on the postulate Eqs. (7.3.2). Then, the coordinate transformations between the Robertson frame $\bar{F}(\bar{x}\bar{y}\bar{z}\bar{t})$ and the MS frame $F^*(x^*y^*z^*t^*)$ are given by Eqs. (1.3.28), i.e.,

$$\bar{x} = x^*, \quad \bar{y} = y^*, \quad \bar{z} = z^*. \tag{7.3.3a}$$

$$\bar{t} = t^* + q_r\frac{r^*}{\bar{c}_r}, \tag{7.3.3b}$$

where

$$\frac{q_r}{\bar{c}_r} = \frac{q_x}{\bar{c}_x}\cos\alpha + \frac{q_y}{\bar{c}_y}\cos\beta + \frac{q_z}{\bar{c}_z}\cos\gamma = \frac{q}{\bar{c}_x}\cos\alpha. \tag{7.3.3c}$$

For simplicity, we consider here the special case for the directional parameter: $q_y = q_z = 0$ and $q \equiv q_x \neq 0$.

Similarly, the relations among $\bar{F}'(\bar{x}'\bar{y}'\bar{z}'\bar{t}')$ and $F^{*'}(x^{*'}y^{*'}z^{*'}t^{*'})$ are

$$\bar{x}' = x^{*'}, \quad \bar{y}' = y^{*'}, \quad \bar{z}' = z^{*'}. \tag{7.3.4a}$$

$$\bar{t}' = t^{*'} + q'_r\frac{r^{*'}}{\bar{c}'_r}, \tag{7.3.4b}$$

where we also assume $q'_y = q'_z = 0$ and $q' \equiv q'_x \neq 0$ for simplicity and hence

$$\frac{q'_r}{\bar{c}'_r} = \frac{q'}{\bar{c}'_x}\cos\alpha'. \tag{7.3.4c}$$

Relations between the MS velocities and Robertson velocities can be obtained from Eqs (7.3.3) and (7.3.4),

$$u^*_x = \frac{\bar{u}_x}{1 - q\bar{u}_x/\bar{c}_{\parallel}}, \quad or \quad \bar{u}_x = \frac{u^*_x}{1 + qu^*_x/\bar{c}_{\parallel}}, \tag{7.3.5a}$$

$$u^*_y = \bar{u}_y, \quad u^*_z = \bar{u}_z. \tag{7.3.5b}$$

In particular for the relative velocity we have

$$\bar{v} = \frac{v^*}{1 + qv^*/\bar{c}_{\parallel}}. \tag{7.3.6}$$

Using Eqs. (7.3.3), (7.3.4) and (7.3.5a) to transform the Robertson coordinates and Robertson velocity in Eqs. (7.2.10), (7.2.13) and (7.2.36) into the MS coordinates and MS velocity, and then putting the results into the general Robertson transformations (7.2.12), one can arrive at the general MS transformations. However, for simplicity, we consider here the special MS transformations between the MS frames F^* and the Einstein frame F.

By use of Eqs. (7.3.3) for changing the Robertson coordinates $\bar{r}$ and $\bar{t}$ in Eqs. (7.2.9) to the MS coordinates $\mathbf{r}^*$ and t^*, we can obtain the generalized transformations among the MS frame F^* and the Einstein frame F as follows:

$$x^* = (d)\frac{\bar{c}_\|}{\bar{c}_\perp}\frac{1}{\sqrt{1-(v^2/c^2)}}(x-vt), \tag{7.3.7a}$$

$$y^* = (d)y, \tag{7.3.7b}$$

$$z^* = (d)z, \tag{7.3.7c}$$

$$t^* = (d)\frac{c}{\bar{c}_\perp}\left\{\frac{1}{\sqrt{1-(v^2/c^2)}}\left[\left(1+\frac{v}{c}q\right)t-\left(\frac{v}{c}+q\right)\frac{x}{c}\right]\right\}. \tag{7.3.7d}$$

Recall the relation between Robertson and Lorentz transformations, we introduce the following *Edwards-like coordinates*:

$$\tilde{x}^* = d^{-1}\frac{\bar{c}_\perp}{\bar{c}_\|}x^*, \tag{7.3.8a}$$

$$\tilde{y}^* = d^{-1}y^*, \tag{7.3.8b}$$

$$\tilde{z}^* = d^{-1}z^*, \tag{7.3.8c}$$

$$\tilde{t}^* = d^{-1}\frac{\bar{c}_\perp}{c}t^*. \tag{7.3.8d}$$

Then the MS transformations (7.3.7) become the Edwards-like transformations:

$$\tilde{x}^* = \frac{1}{\sqrt{1-(v^2/c^2)}}(x-vt), \tag{7.3.9a}$$

$$\tilde{y}^* = y, \tag{7.3.9b}$$

$$\tilde{z}^* = z. \tag{7.3.9c}$$

$$\tilde{t}^* = \frac{1}{\sqrt{1-v^2/c^2}}\left[\left(1+\frac{v}{c}q\right)t-\left(\frac{v}{c}+q\right)\frac{x}{c}\right]. \tag{7.3.9d}$$

We see that the MS transformations (7.3.9) have the same forms as the Edwards transformations (6.3.7). This is the reason for referring the coordinates on the left-hand side of (7.3.8) to the Edwards-like coordinates.

These equations are equivalent to the original MS transformations. In fact, introducing the parameters $\epsilon = (\epsilon_x, \epsilon_y, \epsilon_z)$ defined by

$$\epsilon_x = -\frac{1}{\bar{c}_\|}\left(\frac{v}{c}+q\right), \tag{7.3.10a}$$

$$\epsilon_y = 0, \qquad \epsilon_z = 0, \tag{7.3.10b}$$

and a, b given by Eqs. (7.2.6), equations (7.3.7) then become

$$t^* = at + \epsilon \cdot \mathbf{x}^*,$$

$$x^* = b(x - vt),$$

$$y^* = (d)y,$$

$$z^* = (d)z. \tag{7.3.11}$$

There are just the transformations proposed by Mansouri and Sexl (1977) [4]. It is shown that the new parameters ($\bar{c}_{\parallel}$, $\bar{c}_{\perp}$ and $\mathbf{q}$), but not the original parameters (a, b and ϵ), have physically explicit meaning as mentioned in the previous section and subsection 1.3.6: The two-way velocities of light, $\bar{c}_{\parallel}$, and $\bar{c}_{\perp}$, can be determined by experiments, while the directional parameter, $\mathbf{q}$, has no any physical effect.

From the above derivation we can see that the MS transformations are a generalization of the Robertson transformations, just as the Edwards transformations are a generalization of the Lorentz transformations. Therefore, this generalization is trivial in both physics and mathematics.

Let us now check that the transformations, Eqs. (7.3.7), satisfy MS' definition of simultaneity described in subsection 1.3.6. To do this, substituting Eqs. (7.3.7) into (7.2.4), the equation of motion for a light ray in the frame F, we get the corresponding equation in the MS frame F^* :

$$c_r^2 \left[\frac{1}{\bar{c}_r^2} q_r^2 - \frac{1}{\bar{c}_{\perp}^2} - \left(\frac{1}{\bar{c}_{\parallel}^2} - \frac{1}{\bar{c}_{\perp}^2} \right) \cos^2 \alpha \right] + c_r \left(\frac{1}{\bar{c}_r} q_r \right) + 1 = 0, \tag{7.3.12}$$

where $c_r = r^*/t^*$, $\cos \alpha = x^*/r^*$, $r^* = \sqrt{(x^*)^2 + (y^*)^2 + (z^*)^2}$, and $q_r/\bar{c}_r$ is given by Eq. (7.3.3c).

Solutions to Eq. (7.3.12) for c_r and c_{-r} are given by

$$c_r = \left[\sqrt{\frac{1}{\bar{c}_{\perp}^2} + \left(\frac{1}{\bar{c}_{\parallel}^2} - \frac{1}{\bar{c}_{\perp}^2} \right) \cos^2 \alpha} - \frac{q_r}{\bar{c}_r} \right]^{-1}, \tag{7.3.13a}$$

$$c_{-r} = \left[\sqrt{\frac{1}{\bar{c}_{\perp}^2} + \left(\frac{1}{\bar{c}_{\parallel}^2} - \frac{1}{\bar{c}_{\perp}^2} \right) \cos^2 \alpha} + \frac{q_r}{\bar{c}_r} \right]^{-1}. \tag{7.3.13b}$$

These equations can be rewritten as

$$c_r = \frac{\bar{c}_r}{1 - q_r}, \qquad c_{-r} = \frac{\bar{c}_r}{1 + q_r}, \tag{7.3.14a}$$

where

$$\bar{c}_r = \frac{\bar{c}_\| \bar{c}_\perp}{\sqrt{\bar{c}_\|^2 + (\bar{c}_\perp^2 - \bar{c}_\|^2)\cos^2\alpha}}. \tag{7.3.14b}$$

Equations (7.3.14) are just Eqs. (1.3.26).

The general MS transformations between any two MS frames F^* and $F^{*'}$ can be easily derived from Eqs. (7.3.7) or (7.3.9). In fact, when the Edwards-like coordinates are used, the general MS transformations between any two MS frames F^* and $F^{*'}$ can be expressed as the same forms as the general Edwards transformations, which are just the Edwards-like transformations,

$$\tilde{x}'^* = \tilde{\xi}(\tilde{x}^* - \tilde{v}^*\tilde{t}^*), \tag{7.3.15a}$$

$$\tilde{y}'^* = \tilde{y}^*, \tag{7.3.15b}$$

$$\tilde{z}'^* = \tilde{z}^*, \tag{7.3.15c}$$

$$\tilde{t}'^* = \tilde{\xi}\left\{\left[1 + (q + q')\frac{\tilde{v}^*}{c}\right]\tilde{t}^* - \left[(1 - q^2)\frac{\tilde{v}^*}{c} + (q' - q)\right]\frac{\tilde{x}^*}{c}\right\}, \tag{7.3.15d}$$

with

$$\tilde{\xi} \equiv \frac{1}{\sqrt{(1 + q\tilde{v}^*/c)^2 - (\tilde{v}^*/c)^2}}, \tag{7.3.15e}$$

where the primed Edwards-like coordinates $\tilde{x}'^*, \tilde{y}'^*, \tilde{z}'^*$, and $\tilde{t}'^*$ are defined by

$$\tilde{x}'^* = (d')^{-1}\frac{\bar{c}'_\perp}{\bar{c}'_\|}x^{*'}, \tag{7.3.16a}$$

$$\tilde{y}'^* = (d')^{-1}y^{*'}, \tag{7.3.16b}$$

$$\tilde{z}'^* = (d')^{-1}z^{*'}, \tag{7.3.16c}$$

$$\tilde{t}'^* = (d')^{-1}\frac{\bar{c}'_\perp}{c}t^{*'}, \tag{7.3.16d}$$

$$\frac{\tilde{v}^*}{c} = \frac{v^*}{\bar{c}_\|}, \tag{7.3.16e}$$

with v^* being the MS velocity of $F^{*'}$ relative to F^* as measured in F^* by MS clocks.

7.4. Comparison between the MS Transformations and Experiments

As mentioned in the previous sections, Robertson transformations are physically non-trivial because the two-way velocity of light can be measured by experiments. However, the difference between the MS and Robertson transformations is just due to the different values of the directional parameter **q**, just as it does between Edwards and Lorentz transformations. This implies that the MS and Robertson transformations are physically equivalent to each other. Thus, the MS transformations are physically trivial. Furthermore, the MS transformations as a generalization of Robertson transformations are similar to Edwards transformations as a generalization of Lorentz transformations, and therefore the MS transformations are also trivial in mathematics. In this section we shall give some examples to show the equivalence between the MS and Robertson transformations, although these are unnecessary and trivial.

As mentioned in Sec. 7.3, the general MS transformations (7.3.15), which are expressed in terms of the new quantities defined by Eqs. (7.3.8) and (7.3.16), have the same forms as the general Edwards' transformations (6.3.4). Therefore we can get all physical equations of MS' test theory directly from the corresponding ones of Edwards' theory by use of Eqs. (7.3.8) and (7.3.16).

7.4.1. *MS' Law of the Addition of Velocities*

For simplicity we consider here the special MS transformations (7.3.7) connecting the MS frame F^* with Einstein frame F. From Eq. (7.3.7) one arrives at the transformation equations for velocities

$$u_x^* = \frac{\bar{c}_\parallel}{c} \frac{u_x - v}{[1 + (v/c)q] - [(v/c) + q](u_x/c)}, \tag{7.4.1a}$$

$$u_y^* = \frac{\bar{c}_\perp}{c} \frac{u_y\sqrt{1 - (v^2/c^2)}}{[1 + (v/c)q] - [(v/c) + q](u_x/c)}, \tag{7.4.1b}$$

$$u_z^* = \frac{\bar{c}_\perp}{c} \frac{u_z\sqrt{1 - (v^2/c^2)}}{[1 + (v/c)q] - [(v/c) + q](u_x/c)}, \tag{7.4.1c}$$

where u^* is the MS velocity defined by the MS simultaneity as seen in the MS frame F^*, u and v are the Einstein velocities measured in the Einstein frame F. This is the *MS law of the addition of velocities*.

Of course, the difference between MS' addition law and Robertson's addition law follows just from the different definitions of simultaneity, just as it does between Edwards' and Einstein's laws of the addition of velocities. In fact, by using the relations (7.3.5), MS' law (7.4.1) reduces to Robertson's law.

Equations (7.4.1) give that the relative velocity of F^* to F is $u_x = v, u_y = u_z = 0$, while the relative velocity of F to F^* is

$$V_{MS}^* \equiv u_x^* = -\frac{\bar{c}_\parallel}{c} \frac{v}{1 + (v/c)q} = \frac{\bar{V}_R}{1 - (\bar{V}_R/\bar{c}_\parallel)q}, \tag{7.4.2a}$$

where $\bar{V}_R$ is the Robertson velocity defined by

$$\frac{\bar{V}_R}{\bar{c}_\parallel} = -\frac{v}{c}. \tag{7.4.2b}$$

Again $V_{MS}^* \neq -v$, i.e., the reciprocity rule of velocity is not valid. The relations (7.4.2) can be derived directly from the following relationship between MS' and Robertson's definitions of simultaneity (see subsection 1.3.7):

$$dt_{MS}^* = d\bar{t}_R - dx \left(\frac{1}{\bar{c}_\parallel} - \frac{1}{c_x} \right) = d\bar{t}_R \left(1 - \frac{q}{\bar{c}_\parallel} \frac{dx}{d\bar{t}_R} \right), \qquad (7.4.3a)$$

where Eq. (7.3.2) is used. By definition, therefore, we have from (7.4.3a)

$$V_{MS}^* \equiv \frac{dx}{dt_{MS}^*} = \frac{dx}{d\bar{t}_R} \left(1 - \frac{q}{\bar{c}_\parallel} \frac{dx}{d\bar{t}_R} \right)^{-1}. \qquad (7.4.3b)$$

Equation (7.4.3b) is just Eq. (7.4.2a) where $\bar{V}_R \equiv dx/d\bar{t}_R$.

In short, it is shown from the above that a velocity is defined by a coordinate time interval, and hence related to a definition of simultaneity. Therefore, before making use of experimental data in a test theory, one must consider whether the simultaneity in the experiments is the same as the one in the test theory.

7.4.2. Time Dilation

From the MS transformations (7.3.15) we can obtain the MS time dilation formula in terms of the Edwards-like coordinates,

$$\Delta\tilde{\tau}^{*\prime} = \Delta\tilde{t}^* \sqrt{\left[1 + q\frac{\tilde{v}^*}{c} \right]^2 - \frac{\tilde{v}^{*2}}{c^2}}, \qquad (7.4.4a)$$

where

$$\Delta\tilde{t}^* = d^{-1}\frac{\bar{c}_\perp}{c}\Delta t^*, \qquad (7.4.4b)$$

$$\frac{\tilde{v}^*}{c} = \frac{v^*}{\bar{c}_\parallel}. \qquad (7.4.4c)$$

Equation (7.4.4a) has the same form as Edwards' formula (6.5.10). Putting Eqs. (7.4.4b,c) into (7.4.4a), we have

$$\Delta\tilde{\tau}^{*\prime} = \frac{d'}{\bar{c}_\perp'}\frac{\bar{c}_\perp}{d}\Delta t^* \sqrt{\left(1 + q\frac{v^*}{\bar{c}_\parallel} \right)^2 - \frac{v^{*2}}{\bar{c}_\parallel^2}}, \qquad (7.4.5)$$

where $\Delta\tilde{\tau}^{*\prime}$ is a proper time interval of the moving clock, Δt^* is the corresponding MS coordinate time interval, and v^* is the MS velocity of the moving clock.

By using Eqs. (7.3.3b) and (7.3.5a) with $u_x^* = v^*$ and $\bar{u}_x = \bar{v}$, the MS formula (7.4.5) reduces to the Robertson formula (7.2.44b). Note that $\Delta\tilde{\tau}^{*\prime}$ in Eq. (7.4.5) is equal to $\Delta\tau'$ in (7.2.44b) because it is a proper time interval. The physical meaning of the relation between MS' formula and Robertson's formula is the same as that between the Edwards and Lorentz formulas as mentioned in subsection 6.5.3. This is

to say that the MS formula (7.4.5) is physically equivalent to the Robertson formula
(7.2.44b). The comparison among MS' and Robertson's formulas is similar to that
between Edwards' and Einstein's formulas.

7.4.3. The Römer experiment and Slow Transport of Clocks

Figure 7 is a sketch for a Römer-type experiment, where the replacements $\tilde{t}_A \to$
$t_A^*, \tilde{t}_B \to t_B^*$ and $\Delta\tilde{t} \to \Delta t^*$ should be made in MS' test theory because they depend
on the definition of simultaneity. Furthermore, the one-way velocities of light, $c(\theta)$
and $c(\pi - \theta)$, are now defined by Eqs. (7.3.13) or (7.3.14). Although we do not
need to change the proper times $\Delta\tilde{\tau}_E = \tilde{\tau}_2 - \tilde{\tau}_1$ and $\Delta\tilde{\tau}'_J = \tilde{\tau}'_B - \tilde{\tau}'_A$ as well as the
spatial coordinate $\tilde{r}$, however, in order for the consistency of the symbols we still
replace them by $\Delta\tau_E^* = \tau_2^* - \tau_1^*$, $\Delta\tau_J^{*'} = \tau_B^{*'} - \tau_A^{*'}$ and r^* respectively. Again we now
give the physical meaning of these symbols: The primed and unprimed quantities
are measured in the MS frames $F^{*'}$ and F^* respectively; The symbols "τ" and "t"
represent "proper" and "coordinate" times respectively; $\Delta\tau_J^{*'}$ is a proper time interval
of Jupiter in which two light signals are emitted, as seen in $F^{*'}$; Δt^* is a coordinate
time interval corresponding to $\Delta\tau_J^{*'}$, as seen in F^*; $\Delta\tau_E^*$ is a proper time interval of
the observer at the point E, in which the two light signals reach the Earth.

In MS' test theory the derivations for the corresponding formulas are similar to
those in Edwards' theory. So that we can use the procedures in subsection 6.5.4 to
get them. For example, corresponding to Eq. (6.5.15) we have

$$\Delta\tau_E^* = \Delta t^* + \frac{r^*}{c(\pi - \theta)} - \frac{r^*}{c(\theta)}, \tag{7.4.6}$$

with

$$c(\theta) = \left[\sqrt{\frac{1}{\bar{c}_\perp^2} + \left(\frac{1}{\bar{c}_\parallel^2} - \frac{1}{\bar{c}_\perp^2}\right)\cos^2\theta} - \frac{q}{\bar{c}_\parallel}\cos\theta\right]^{-1}, \tag{7.4.7}$$

where $\mathbf{q} = (q, 0, 0)$ is assumed for simplicity. By putting Eq. (7.4.7) into Eq. (7.4.6),
one arrives at

$$\Delta\tau_E^* = \Delta t^* + (2r^*\cos\theta)\frac{q}{c_\parallel}. \tag{7.4.8}$$

The geometry in Fig. 7 showing $2r^*\cos\theta = \Delta x^*$ and $v^* = \Delta x^*/\Delta t^*$, equation (7.4.8)
becomes

$$\Delta\tau_E^* = \Delta t^*\left(1 + \frac{\Delta x^*}{\Delta t^*}\frac{q}{c_\parallel}\right) = \Delta t^*\left(1 + q\frac{v^*}{c_\parallel}\right) \tag{7.4.9}$$

Note that Δt^* is a coordinate time interval in F^*, while the corresponding interval
$\Delta\tau_J^{*'}$ as timed by the Jupiter clock is a proper time interval. The observable quantities in the Römer-type experiment [37], which do not depend on the definition of
simultaneity, are the proper time intervals $\Delta\tau_E^*$ and $\Delta\tau_J^{*'}$. For this reason we must
transform the coordinate time interval Δt^* in Eq. (7.4.9) into the proper time interval

$\Delta\tau_J^{*'}$. This transformation is nothing but the MS time dilation formula (7.4.5) where $\Delta\tau^{*'} \to \Delta\tau_J^{*'}$. Therefore, eliminating Δt^* from Eqs. (7.4.9) and (7.4.5) we obtain the expected equation,

$$\Delta\tau_E^* = \Delta\tau_J^{*'} \frac{d}{\bar{c}_\perp} \frac{\bar{c}'_\perp}{d'} \left(1 + q\frac{v^*}{c_\|}\right) \left\{ \sqrt{\left(1 + q\frac{v^*}{\bar{c}_\|}\right)^2 - \frac{v^{*2}}{\bar{c}_\|^2}} \right\}^{-1} . \qquad (7.4.10)$$

Since to first-order in $v^*/\bar{c}_\|$ we have $d = d' = 1$ and $\bar{c}_\| = \bar{c}'_\| = c$, so that equation (7.4.10) becomes

$$\Delta\tau_E^* = \Delta\tau_J^{*'}, \quad or \quad \frac{\Delta\tau_E^* - \Delta\tau_J^{*'}}{t} = 0, \qquad (7.4.11)$$

where $t = r/c$ is the (average) time in which the light signals reach the Earth. This shows that there is no the first-order effect in the so-called first-order experiment. However this argument is not exact because Eq. (7.4.10) involves the parameter q which represents anisotropy of the one-way velocity of light. But the existence of q does not mean its observability. In fact, by use of the relation (7.3.5) to transform the MS velocity v^* into the Robertson velocity $\bar{v}$, equation (7.4.10) reduces to Robertson's time dilation equation, the parameter q being removed. This shows that the difference between MS' time dilation and Robertson's time dilation follows just from the difference between the MS velocity v^* and Robertson velocity $\bar{v}$. This case is similar to that in subsection 6.5.3. Therefore, the problem is either of the velocities v^* and $\bar{v}$ should be identified with the experimental data. If the Robertson simultaneity was used in a measurement, the datum should be identified with $\bar{v}$. This is to say that we should substitute the datum for $\bar{v}$ in Robertson's formula (7.2.44b) without the parameter q. Therefore, the measurement cannot be a test of the anisotropy of the one-way speed of light. On the other hand, if you would like to use MS' formula (7.4.10), you must first to substitute the datum still for $\bar{v}$ in the relation (7.3.5) so as to get a value of v^* and then put it in MS' time dilation formula (7.4.10). In this way you will certainly obtain the same result as that from Robertson's formula (7.2.44b). Thus, we say that MS' and Robertson's test theories are physically equivalent to each other.

The above analyses are similar to those in subsection 6.5.3, and hence are indeed trivial and unnecessary. This is due to the MS transformation as a generalization of Robertson transformation just like the Edwards transformation as a generalization of Lorentz transformation.

Let us now discuss the problem of slow transport of clocks. In Fig. 7 the paths AE and BE of light signals do not form a closed path. Thus, this kind of experiment is sometimes called the unidirectional experiment or the so-called first-order experiment because an absolute space–time theory would predict the existence of the first-order effect. However, according to the theories of special relativity involving the test theories, the key point is how to compare the two time coordinates t_A^* and t_B^* at which two events occur at the two separate points A and B, respectively. This deals with the definition of simultaneity because a value of the difference of two time coordinates depends upon a frame in where the difference is measured. The Römer-type experiment under consideration is a typical example in which the Jupiter system could be regarded as a slowly transporting clock. The key quantities in this example are the difference $\Delta t^* = t_B^* - t_A^*$ and the velocity v^* of Jupiter, which are all defined in the Earth frame and depend on the definition of simultaneity. To the first-order in v/c, the velocity of Jupiter can be regarded as being independent of simultaneity, while the coordinate time interval Δt^* depends on simultaneity. On the other hand, .Jupiter's proper time interval $\Delta \tau_J^{*'} = \tau_B^{*'} - \tau_A^{*'}$ is given by the same clock connected rigidly with Jupiter, which is a observable quantity in the sense that being independent of simultaneity. The first step in the theoretical calculations is the derivation of the relation between $\Delta \tau_E^*$ and Δt^*, as shown in Sec. 6.5 and the above. The second step, being a key step, is to use the relation between Δt^* and $\Delta \tau_J^{*'}$, i.e., a time dilation formula. Which formula must be used in the present step? The answer is obvious: a time dilation formula to be used must follow from the same theory [1]. For this reason we can say that the problem of slow transport of clocks is nothing but a problem of the time dilation. In other words if you correctly employ a time dilation formula coming from the same test theory, then the open path experiment is certainly equivalent to a closed path experiment. Inversely, if you use a formula not coming from the same test theory, then the formula itself is in contrast with the time dilation effect predicted by the present test theory. In the later case, you do not need firstly to use the formula in the present experiment while you should first to compare your time dilation formula with the time dilation experiments. Therefore as shown in Sec. 6.5 and the above, both the Römer-type experiment and slow transport of clocks are nothing but the problem of time dilation.

7.4.4. *Transverse Doppler Effect and Mössbauer Rotor Experiment*

Transverse Doppler effect. Derivation of Doppler effect formula in MS' test theory is similar to that in Sec. 7.2.3. However, a simple method to get Doppler effect is to transform the Robertson velocity $\bar{v}$ in Eq. (7.2.55) into the MS velocity v^* by using Eq. (7.3.6). In this way we obtain the *transverse Doppler effect* from Eq. (7.2.56),

$$\omega = \omega' \left(\frac{d'}{d}\right) \left(\frac{\bar{c}_\perp}{\bar{c}_\perp'}\right) \sqrt{\left(1 + q\frac{v^*}{\bar{c}_\parallel}\right)^2 - \frac{v^{*2}}{\bar{c}_\perp^2}\left(1 + q\frac{v^*}{\bar{c}_\parallel}\right)^{-1}}. \qquad (7.4.12)$$

---

[1]Mansouri and Sexl [39] claimed that the Römer experiment is a test of the first-order effect in v/c or of anisotropic one-way velocity of light. However in their paper the equation to be compared with the experiment comes from Eq. (7.4.8) of the book. Their incorrect step is to identify the data of Römer experiment with Δt^* (see [25] for detail). This is equivalent to the employment of $\Delta t^* = \Delta \tau_J^{*'}$. In other words they *did not distinguish* the coordinate time interval Δt^* from the proper time interval $\Delta \tau_J^{*'}$. Therefore contradiction is obvious: The formula (7.4.8) is a result just from MS' test theory while the equation $\Delta t^* = \Delta \tau_J^{*'}$ comes only from the absolute space–time theory.

Again the transverse Doppler formula here do still not involve the first-order term. In fact, to the first-order approximation, equation (7.4.12) becomes

$$\omega = \omega', \quad or \quad \nu = \nu',$$

i.e., there is still no the first-order effect in the transverse Doppler frequency shift in MS' test theory [2].

Explanation of Mössbauer rotor experiment. We now discuss such a problem: How to use the (MS) transverse Doppler formula (7.4.12) for explaining the Mössbauer rotor experiments introduced in Secs. 8.2.2 (II) and 9.3.3 (II)?

It must be noted that the velocity v^* in Eq. (7.4.12) is a MS velocity, which is defined by the MS simultaneity. For the case of the rotor experiment, v^* should be the instantaneous velocity of an end of a diameter of the rotor, while the value of the rotor velocity given in the experiment is only an average value of the rotor velocity, i.e., $\bar{v}$. In other words, the definition of the rotor velocity,

$$\bar{v} \equiv \frac{2\pi R}{T} = \text{cycles/min},\tag{7.4.13}$$

represents an average velocity. We have know that the relation between the instantaneous velocity v^* and the average velocity $\bar{v}$ is given by Eq. (7.3.6):

$$v^* = \frac{\bar{v}}{1 - q\bar{v}/\bar{c}_{\parallel}}.\tag{7.4.14}$$

This shows that we should identify the data of the rotor velocity given in the experiment with the average velocity $\bar{v}$ in Eq. (7.4.14) to get a value of v^* and then put it into Eq. (7.4.12). In this correct way, we would certainly come the same result from Eq. (7.4.12) as Robertson's formula (7.2.56), because putting (7.4.14) into (7.4.12) would lead to (7.2.56). So that we have the conclusion: The rotor experiments and one-way experiments, as introduced in Secs. 8.2.2 (II) and 9.3.3 (II), could not give a test of the so-called first-order effect; In other words, these experiments were not a test of the one-way velocity of light.

7.4.5. Anisotropic Four-Dimensional Space–Time

A method for constructing an invariant 4-interval with respect to the MS transformations from the 4-interval $d\bar{s}^2$ defined in Sec.7.2.3 is similar to that in 6.4. In other words, by use of the combine transformation of Q and $\bar{Q}$, we could obtain the 4-interval in MS' theory from that in the Minkowski space–time.

---

[2] Mansouri and Sexl (1977) [39] claimed that MS' test theory would predict the first-order effect in the transverse Doppler shift. Again they did not distinguish the coordinate time interval from the proper time interval, as mentioned in the previous section.

As done in Secs. 6.4 and 7.2.3, let us define the 4-vectors:

$$dX^* \equiv ((dX^*)^\mu) = \begin{pmatrix} \bar{c}_\| dt^* \\ dx^* \\ dy^* \\ dz^* \end{pmatrix}, \tag{7.4.15a}$$

and

$$d\tilde{X}^* \equiv \left((d\tilde{X}^*)^\mu\right) = \begin{pmatrix} cd\tilde{t}^* \\ d\tilde{x}^* \\ d\tilde{y}^* \\ d\tilde{z}^* \end{pmatrix}, \tag{7.4.15b}$$

where $(d\tilde{X}^*)^\mu$ are the Edwards-like coordinates defined by Eqs. (7.3.8). Then equations (7.3.3) and (7.3.8) can be expressed as

$$d\bar{X} = Q dX^*, \tag{7.4.16}$$

$$d\tilde{X}^* = \bar{Q} dX^*, \tag{7.4.17}$$

where the matrixes Q and $\bar{Q}$ are given by Eqs. (6.4.6) and (7.2.22).

Equations (7.3.9) show that the relation between the Edwards-like coordinate $d\tilde{X}^*$ and the Lorentz-like coordinate $d\bar{X}_0$ is similar to Eq. (6.4.6):

$$d\bar{X}_0 = Q d\tilde{X}^*. \tag{7.4.18}$$

In other words, if we use the Lorentz-like coordinate $d\bar{X}_0$, equations (7.3.9) then become a Lorentz-like form. Thus, we can transform the MS coordinates dX^* into the Lorentz-like coordinates $d\bar{X}_0$ through the two equivalent ways:

$$dX^* \longrightarrow d\tilde{X}^* \longrightarrow d\bar{X}_0, \tag{7.4.19}$$

and

$$dX^* \longrightarrow d\bar{X} \longrightarrow d\bar{X}_0, \tag{7.4.20}$$

or explicitly

$$d\bar{X}_0 = Q d\tilde{X}^* = Q\bar{Q} dX^*, \tag{7.4.21}$$

and

$$d\bar{X}_0 = \bar{Q} d\bar{X} = \bar{Q} Q dX^*. \tag{7.4.22}$$

Here the two matrixes Q and $\bar{Q}$ are commutable:

$$Q\bar{Q} = \bar{Q}Q = \frac{1}{d} \begin{pmatrix} \left(\bar{c}_\perp/\bar{c}_\|\right) & q\left(\bar{c}_\perp/\bar{c}_\|\right) & 0 & 0 \\ 0 & \left(\bar{c}_\perp/\bar{c}_\|\right) & 0 & 0 \\ 0 & 0 & 1 & 0 \\ 0 & 0 & 0 & 1 \end{pmatrix}. \tag{7.4.23a}$$

The inverse of $Q\bar{Q}$ is

$$(Q\bar{Q})^{-1} = \bar{Q}^{-1}Q^{-1} = (d) \begin{pmatrix} \left(\bar{c}_\perp/\bar{c}_\parallel\right)^{-1} & -q\left(\bar{c}_\perp/\bar{c}_\parallel\right)^{-1} & 0 & 0 \\ 0 & \left(\bar{c}_\perp/\bar{c}_\parallel\right)^{-1} & 0 & 0 \\ 0 & 0 & 1 & 0 \\ 0 & 0 & 0 & 1 \end{pmatrix}. \qquad (7.4.23b)$$

In the same way as done in Sec. 6.4, we can get the expected formulas as follows. The 4-interval is

$$\begin{aligned} d\bar{s}^2 &= d\bar{X}_0^\dagger \eta d\bar{X}_0 \\ &= (dX^*)^\dagger g^* dX^*, \end{aligned}$$

$$(7.4.24)$$

where the metric is defined by

$$g^* \equiv (g_{\mu\nu}^*) = (Q\bar{Q})^\dagger \eta (Q\bar{Q}), \qquad (7.4.25)$$

or explicitly

$$g^* = \frac{1}{d^2} \begin{pmatrix} \left(\bar{c}_\perp/\bar{c}_\parallel\right)^2 & q\left(\bar{c}_\perp/\bar{c}_\parallel\right)^2 & 0 & 0 \\ q\left(\bar{c}_\perp/\bar{c}_\parallel\right)^2 & (-1+q^2)\left(\bar{c}_\perp/\bar{c}_\parallel\right)^2 & 0 & 0 \\ 0 & 0 & -1 & 0 \\ 0 & 0 & 0 & -1 \end{pmatrix}. \qquad (7.4.26)$$

The inverse metric then is

$$(g^*)^{-1} = (g^{*\mu\nu}) = (d^2) \begin{pmatrix} (1-q^2)\left(\bar{c}_\perp/\bar{c}_\parallel\right)^{-2} & q\left(\bar{c}_\perp/\bar{c}_\parallel\right)^{-2} & 0 & 0 \\ q\left(\bar{c}_\perp/\bar{c}_\parallel\right)^{-2} & -\left(\bar{c}_\perp/\bar{c}_\parallel\right)^{-2} & 0 & 0 \\ 0 & 0 & -1 & 0 \\ 0 & 0 & 0 & -1 \end{pmatrix}. \qquad (7.4.27)$$

It is similar for the primed metric:

$$g^{*\prime} \equiv (g_{\mu\nu}^{*\prime}) = (Q'\bar{Q}')^\dagger \eta (Q'\bar{Q}'). \qquad (7.4.28)$$

The MS transformation matrix Λ^* can be expressed as

$$\Lambda^* = (Q'\bar{Q}')^{-1}\bar{\Lambda}_0(Q\bar{Q}), \qquad (7.4.29)$$

where $\bar{\Lambda}_0$ is given by Eq. (7.2.34). The 4-interval, Eq. (7.4.24), is invariant under the MS transformations, i.e.,

$$g^* = (\Lambda^*)^\dagger g^{*\prime} \Lambda^*. \qquad (7.4.30)$$

In particular, this yields the expressions for the metrics:

$$g^* = [\Lambda^*(q' = 0)]^\dagger \, \eta \, [\Lambda^*(q' = 0)] \,, \tag{7.4.31}$$

and

$$g^{*'} = \left[(\Lambda^*)^{-1}(q = 0)\right]^\dagger \eta \left[(\Lambda^*)^{-1}(q = 0)\right] . \tag{7.4.32}$$

Equation (7.4.25) can be written as

$$g^* = (Q\bar{Q})^\dagger \eta (Q\bar{Q}) = Q^\dagger \bar{g} Q, \tag{7.4.33}$$

where Eq. (7.2.24) and the commutability of Q and $\bar{Q}$ are used. It is obvious that the relation between g^* and $\bar{g}$, Eq. (7.4.33), is similar to the relation, Eq. (6.4.15a), between g and η. Equations (7.4.30) and (7.4.33) mean the invariance of the 4-interval:

$$d\bar{s}^2 = d\bar{X}^\dagger \bar{g} d\bar{X} = (dX^*)^\dagger g^* dX^* = \left(dX^{*'}\right)^\dagger g^{*'} dX^{*'}. \tag{7.4.34}$$

This is to say that the MS transformation is a generalization of Robertson transformation just as Edwards transformation is a generalization of Lorentz transformation. Thus, the MS and Robertson transformations are physically equivalent to each other just as the Edwards and Lorentz transformations do, as shown in Secs. 7.4.1–7.4.4.

7.4.6. Dynamics

Dynamics for Robertson's test theory has not yet been constructed. Such a dynamical theory may be constructed on the basis of the invariant 4-interval $d\bar{s}^2$ defined by (7.2.23). Owing to the presence of the $\bar{Q}$ transformation that can be determined by experiments, Robertson's test theory would be not equivalent to Einstein's theory in both kinematics and dynamics.

It has been shown in Secs. 7.4.1–7.4.4 that the MS transformations give us the same predictions as the Robertson transformations in kinematics. We have also proved in Secs. 7.3 and 7.4.5 that the MS transformation is a generalization of Robertson transformation just as Edwards transformation is a generalization of Lorentz transformation. Therefore, if the expected dynamics in MS' theory were based on the 4-dimensional space–time described in the previous subsection, then MS' theory would also be physically equivalent to Robertson's theory in dynamics.

7.5. Relationship among Lorentz and Generalized Transformations [25]

From the previous sections we know the following relations among the Lorentz transformations, Edwards transformations, Robertson transformations, and Mansouri–Sexl transformations:

$$
\begin{array}{ccc}
\textbf{Lorentz} & \longleftarrow \mathbf{q} = 0 \longleftarrow & \textbf{Edwards} \\
\uparrow & & \uparrow \\
\bar{c}_\parallel = \bar{c}_\perp = c & & \bar{c}_\parallel = \bar{c}_\perp = c \\
\uparrow & & \uparrow \\
\textbf{Robertson} & \longleftarrow \mathbf{q} = 0 \longleftarrow & \textbf{Mansouri–Sexl}
\end{array}
$$

It is shown that advantages of the new set of parameters $(\bar{c}_\parallel, \bar{c}_\perp, \mathbf{q}, d)$ are not only their explicit meaning physically, but also the separation of the parameters representing the two-way speed of light from the one representing the one-way speed of light: $\bar{c}_\perp \neq \bar{c}_\parallel$ represents the anisotropy of two-way speed of light; a non-zero value of q_r implies the anisotropy of one-way speed of light. Furthermore, the parameter d becomes a common (a "conformal") factor, and thus, is theoretically a trivial constant although it can be determined by experiments. Therefore, $\bar{c}_\parallel$ and $\bar{c}_\perp$ may be determined by measuring the two-way speed of light. However, the directional parameter $\mathbf{q}$ could not appear in any experiments where simultaneity is defined by a light signal (The problem of slow transport of clocks is equivalent to that of the definition of simultaneity by using a light signal if we apply the prediction of the test theory to the time dilation of the slow moving clock; A good argument about this is the analysis of Römer experiment). This implies that when the different methods of synchronization are taken into account, the Edwards transformations (6.3.4) are equivalent to Lorentz transformations as shown in section 6.5, while the MS transformations (7.3.7) are also physically equivalent to the Robertson transformations (7.2.9) as seen in section 7.5. So that we come to the following conclusions:

(i) The directional parameter $\mathbf{q}$ cannot be observed in any physical experiments. This is to say that its modules can be taken as any value in the range $(-1, +1)$, or that there are infinite definitions of simultaneity which are physically equivalent to each other, and Einstein simultaneity and Robertson simultaneity are the simplest choice (i.e., $q = 0$).

(ii) In other words, the Mansouri–Sexl transformations predict the same observable effects with Robertson transformations, just as Edwards transformations do with Lorentz transformations. Therefore both Edwards' theory and MS' theory are physically trivial theories. Therefore, Edwards' theory cannot be called a test theory of Einstein's special relativity; while MS' theory is not a new test theory since it is physically equivalent to Robertson's theory.

(iii) The difference between Robertson transformations and Lorentz transformations is simply due to the two-way velocity of light, which can be test by physical experiments. These two theories are not physically equivalent to each other. Thus, we say that Robertson's theory is a nontrivial test theory of special relativity.

(iv) Since the directional parameter **q** representing the anisotropy of the one-way speed of light is not observable while the parameters $\bar{c}_\perp$ and $\bar{c}_\parallel$ describing the anisotropy of the two-way speed of light can be measured in physical experiments, a test of the Mansouri–Sexl transformations is just a test of the anisotropy of the two-way speed of light (and a test of the parameter d), but not a test of the anisotropy of the one-way speed of light.

(v) Because $\bar{c}_\parallel$ and $\bar{c}_\perp$ might be, in general, functions of v^2, then to the first-order in v, the Robertson transformations are equivalent to the Lorentz transformations and at the same time the MS transformations are equivalent to the Edwards transformations.

(vi) It is needed to point out that the errors of such a claim by some authors that the so-called one-way experiments (see Sec. 8.2) have given a test of the anisotropy of the one-way speed are in the following: (a) the authors did not separate the anisotropy of the one-way speed from that of the two-way speed; or/and (b) the coordinate time was not be distinguished from the proper time; (c) in addition, these authors did not understand the velocities defined in the MS frames are all the MS velocities which cannot, in general, directly be compared with the experimental values due to different definitions of simultaneity although there is no difference between them for predictions to second-order in v/c.

7.6. Comparison of Different Conventions

As mentioned in Secs. 7.2 and 7.3, the original three parameters a, b, and ϵ in the MS transformations do not have physically obvious meanings, while the physical meaning of each of the new three parameters $\bar{c}_\parallel, \bar{c}_\perp$, and **q** is very clear. For the conveniences of the remainder of the book, we make here a summary of the relationships between the different conventions.

The relations among the new and old parameters are given by Eqs. (7.2.6) and (7.3.10). They are

$$\bar{c}_\parallel = \frac{cb}{a}\left(1 - \frac{v^2}{c^2}\right), \qquad (7.6.1a)$$

$$\bar{c}_\perp = \frac{cd}{a}\sqrt{1 - \frac{v^2}{c^2}}. \qquad (7.6.1b)$$

$$q \equiv q_\parallel = -\frac{v}{c} - \bar{c}_\parallel\epsilon_\parallel, \qquad q_\perp = \epsilon_\perp = 0. \qquad (7.6.1c)$$

In this book we choose the direction of the relative velocity **v** parallel to x-axis, so the subscript "$\parallel$" represents x-component while "$\perp$" denotes y- or z-component.

The inverse transformations are

$$\frac{a}{d} = \frac{c}{\bar{c}_\perp}\sqrt{1 - \frac{v^2}{c^2}}, \qquad (7.6.2a)$$

The inverse transformations are

$$\frac{a}{d} = \frac{c}{\bar{c}_\perp} \sqrt{1 - \frac{v^2}{c^2}}, \tag{7.6.2a}$$

$$\frac{b}{d} = \frac{\bar{c}_\perp}{\bar{c}_\perp} \frac{1}{\sqrt{1 - v^2/c^2}}, \tag{7.6.2b}$$

$$\epsilon_\| = -\frac{1}{\bar{c}_\|} \left(\frac{v}{c} + q \right), \qquad \epsilon_\perp = q_\perp = 0. \tag{7.6.2c}$$

The physical meanings of the new parameters $\bar{c}_\|, \bar{c}_\perp$, and $\mathbf{q}$ can be seen from the expressions of the velocities of light, i.e., equations (7.3.14)

$$c_r = \frac{\bar{c}_r}{1 - q_r}, \qquad c_{-r} = \frac{\bar{c}_r}{1 + q_r}, \tag{7.6.3a}$$

$$\bar{c}_r = \frac{\bar{c}_\| \bar{c}_\perp}{\sqrt{\bar{c}_\|^2 + (\bar{c}_\perp^2 - \bar{c}_\|^2) \cos^2 \theta}}, \tag{7.6.3b}$$

where q_r is given by Eq. (7.3.3c), i.e.,

$$\frac{q_r}{\bar{c}_r} = \frac{q}{\bar{c}_\|} \cos \theta, \tag{7.6.3c}$$

and θ represents the angle between the direction of light ray (i.e., the direction of $\mathbf{r}$) and that of the relative velocity $\mathbf{v}$. In particular, for $\theta = 0°$, $180°$, $90°$, or $270°$, we have

$$c_{\pm x} = \frac{\bar{c}_\|}{1 \mp q}, \qquad c_{\pm y} = c_{\pm z} = c_\perp. \tag{7.6.4}$$

Equation (7.6.3b) shows that the two-way velocity of light in a general inertial frame F^* is not isotropic except $c_\| = c_\perp$. Equation (7.6.3a) indicates that the one-way velocity of light depends on not only the directional parameter q but also the two-way velocity of light. In Sec. 1.3 we have shown that a non-zero value of the directional parameter q (a simultaneity factor) describes an anisotropy of the one-way velocity, which cannot be tested by experiments provided we could not find a new means of clock synchronization different from the usual method of the light signal synchronization or slow transport clocks. In contrast with the case of $q \neq 0$, however, that $c_\| \neq c_\perp$ describes the anisotropy of the two-way velocity of light even though the one-way velocity of light is also a function of the angle θ. Thus in an analysis of an experiment we must pay attention to whether the anisotropy of the one-way velocity or the two-way velocity of light has been tested.

Following Clifford M. Will [35], the old parameters a, b, d, and ϵ are expanded in powers of velocity as

$$a(v) = 1 + \left(\alpha - \frac{1}{2} \right) \frac{v^2}{c^2} + \left(\alpha_2 - \frac{1}{8} \right) \frac{v^4}{c^4} + \cdots, \tag{7.6.5a}$$

$$b(v) = 1 + \left(\beta + \frac{1}{2}\right)\frac{v^2}{c^2} + \left(\beta_2 + \frac{1}{8}\right)\frac{v^4}{c^4} + \cdots, \tag{7.6.5b}$$

$$d(v) = 1 + \delta\frac{v^2}{c^2} + \delta_2\frac{v^4}{c^4} + \cdots, \tag{7.6.5c}$$

$$\epsilon = (\varepsilon - 1)\frac{\mathbf{v}}{c^2}\left(1 + \varepsilon_2\frac{v^2}{c^2} + \cdots\right). \tag{7.6.5d}$$

If the new parameters $\bar{c}_{\parallel}, \bar{c}_{\perp}$, and q are expanded as

$$\frac{\bar{c}_{\parallel}}{c} = 1 + \bar{\alpha}\frac{v^2}{c^2} + \bar{\alpha}_2\frac{v^4}{c^4} + \cdots, \tag{7.6.6a}$$

$$\frac{\bar{c}_{\perp}}{c} = 1 + \bar{\beta}\frac{v^2}{c^2} + \bar{\beta}_2\frac{v^4}{c^4} + \cdots, \tag{7.6.6b}$$

$$q = q_1\frac{v}{c} + q_2\frac{v^3}{c^3} + \cdots, \tag{7.6.6c}$$

then by use of Eqs. (7.6.2) we obtain the relations among the coefficients on the right-hand sides of Eqs. (7.6.5) and (7.6.6):

$$\alpha = -\bar{\beta} + \delta, \tag{7.6.7a}$$

$$\beta = \bar{\alpha} - \bar{\beta} + \delta, \tag{7.6.7a}$$

$$\alpha_2 = (\bar{\beta} - \delta)\left(\frac{1}{2} + \bar{\beta}\right) + \delta_2 - \bar{\beta}_2, \tag{7.6.7c}$$

$$\beta_2 = (\bar{\alpha} - \bar{\beta})\left(\frac{1}{2} - \bar{\beta} + \delta\right) + (\bar{\alpha}_2 - \bar{\beta}_2) + \frac{1}{2} + \delta_2, \tag{7.6.7d}$$

$$\epsilon = -q_1, \tag{7.6.7e}$$

$$(\epsilon - 1)\epsilon_2 = -q_2 + (q_1 + 1)\bar{\alpha}. \tag{7.6.7f}$$

In short, the new parameters can be expressed as

$$\frac{\bar{c}_{\parallel}}{c} = 1 + (\beta - \alpha)\frac{v^2}{c^2} + \left(-\frac{1}{2}\alpha - \frac{1}{2}\beta + \alpha^2 - \alpha\beta - \alpha_2 + \beta_2\right)\frac{v^4}{c^4} + \cdots, \tag{7.6.8a}$$

$$\frac{\bar{c}_{\perp}}{c} = 1 + (\delta - \alpha)\frac{v^2}{c^2} + \left(-\frac{1}{2}\alpha + \alpha^2 - \alpha\delta - \alpha_2 + \delta_2\right)\frac{v^4}{c^4} + \cdots, \tag{7.6.8b}$$

$$q = -\varepsilon\frac{v}{c} - (\varepsilon - 1)(\beta - \alpha + \varepsilon_2)\frac{v^3}{c^3} + \cdots, \tag{7.6.8c}$$

and the expansion for the parameter d is not changed:

$$d(v) = 1 + \delta\frac{v^2}{c^2} + \delta_2\frac{v^4}{c^4} + \cdots, \tag{7.6.8c}$$

We want to emphasize here again that the new parameters, compared to the old ones, have the advantages as follows: (a) the new parameters $\bar{c}_{\parallel}, \bar{c}_{\perp}$, and q have physically obvious meanings; (b) the anisotropy of the one-way velocity of light is distinguished from that of the two-way velocity of light: a non-zero value of q indicates the anisotropy of the one-way velocity of light, while $\bar{c}_{\parallel} \neq \bar{c}_{\perp}$ represents the anisotropy of the two-way velocity of light.

Part III

Experimental Tests of Special Relativity

CHAPTER 8

THE TESTS OF EINSTEIN'S TWO POSTULATES

8.1. Introduction

Testing the postulate of the constancy of the velocity of light should involve, for instance: (1) the independence of the velocity of light upon the motion of the source, (2) the frequency-independence of the velocity of light, (3) the isotropy of the two-way velocity of light, and (4) the isotropy of the one-way velocity of light.

In Chap. 1, we have pointed out that a coordinate time interval must be related to the definition of simultaneity which is arbitrary as shown in Chap. 6 (in particular Sec. 6.5). In other wards a specific value of the coordinate time interval has no absolute meaning. Thus, talking about the specific value has also no absolute meaning. Similarly, values of all of such quantities related directly to the coordinate time interval, such as a velocity and an acceleration of a moving particle, have no absolute meaning too. Therefore, we say that these quantities are not directly observable. In particular, any possible isotropy of the one-way velocity of light, i.e., the above (4), cannot be tested in the present laboratories, as shown in Chap. 6. The problem concerning slow transport of clocks has been discussed in subsection 7.4.3.

A proper time interval is in contrast with the coordinate time interval, which does not depend upon the definition of simultaneity. Therefore, all of such laws not related to the definition of simultaneity, such as the above (1), (2), and (3), could be tested directly by experiments. In principle, testing these laws are easy. For instance, in order to test a possible relationship between the velocity of a monochromatic light wave and its frequency, an observer at a given location might measure a possible difference among the arrival times of two independent light waves with different frequencies emitted simultaneously by a source at another point. In this way one can test the independence of the one-way velocity of light upon the frequency or upon the motion of the source, while one cannot determine the value of the one-way velocity of light. Another possible experiment is to use a clock at a given location for measuring the difference between the time when the signal leaves and the time when it return back. This difference is a proper time interval, at which the light signal travels a closed path. Then, by definition we can get the two-way velocity of light. So the two-way velocity of light is a observable quantity. Furthermore, a possible dependence of the two-way velocity on the propagation direction could be tested by experiments.

In this chapter we shall introduce and analyze experiments concerning the tests of the isotropy of the two-way velocity of light, and the tests of the independence of the velocity of light upon the motion of the source. Experimental results concerning the frequency-independence of the velocity of light will be give in Chap. 12.

The tests of Einstein's principle of special relativity will be introduced in Sec. 8.5.

8.2. Tests of Directionality

Experiments of testing the constancy of the velocity of light can be classified into two types. In the first kind of experiments, the (two-way) velocities of light in different directions within a given frame were compared. In the second class of experiments, comparison between the (two-way) velocities of light in different frames was performed.

The first kind of experiments can be divided into two groups: the "*closed path of light*" and the "*one-way path of light*" experiments. The closed path experiments involve the *experiments of the Michelson–Morley type* and *other "closed path" interferometer experiments*.

The purpose of some of the "closed path of light " experiments was to search for a possible second-order effect of the "*ether wind*" (or "*ether drift*") or for the absolute motion of the earth relative to an assumed *absolute frame*. The negative results of the experiments led to the *contraction hypothesis* (see Sec.8.2.1). However, according to the theory of relativity, introducing ether is unnecessary at all. On the other hand, the key point in the ether theory is how to define simultaneity, because a value of the relative velocity of the earth to the absolute frame depends directly upon the definition of simultaneity, as pointed out in subsection 1.5.

The "one-way path of light" experiments observed the *transverse Doppler frequency shifts*. According to the *stationary ether theory* the transverse Doppler effect involves the first-order in the absolute velocity. So the purpose of some of the "one-way path of light " experiments was to search for a possible first-order effect of the "ether wind" . The negative results of the experiments led to the *time retardation postulate* [41–43]: the rate of a moving clock relative to an absolute frame will reduce by a factor $(1 - v^2/c^2)^{1/2}$, which makes the first-order effect disappear. In the theory of relativity the transverse Doppler frequency shift follows from the time dilation which does not involve the first-order effect. Sometimes the experimental result of the transverse Doppler frequency shift was regarded as a test of the isotropy of the one-way velocity of light or a test of the principle of relativity [58]. However the test theory of special relativity is equivalent physically to Einstein's theory, as shown in Sec. 6.5.

Even so we still apply the stationary ether theory to explanation of the experiments in order just to provide a relative accuracy of them.

In the experiments of testing a possible effect of the motion of light source on the velocity of light, the interferometers were at rest in the laboratories. The negative results show the constancy of the (two-way) velocity of light.

8.2.1. The Closed-Path Experiments

In the view of Edwards' theory of special relativity, the negative result of the closed-path experiments naturally follows from its postulate concerning the constancy of the two-way velocity of light, equation (6.3.2). In order to explain the closed-path experiments with a medium, let us generalize the constancy of the two-way velocity of light in vacuum, i.e., equation (6.3.2), to that in a medium [92]:

$$u = \frac{\bar{u}}{1 - (\bar{u}/c)\mathbf{q} \cdot \mathbf{e}}, \tag{8.2.1}$$

where $\bar{u}$ is the average velocity of light along a closed path (i.e., the two-way velocity of light) in a medium as seen in an Edwards frame $\tilde{F}$, in particular in vacuum $\bar{u} = c$ (note that the constant c here represents the two-way, but not the one-way, velocity of light in vacuum); $q = |\mathbf{q}|$ is a constant, values of which being limited by Eq. (6.2.3); $\mathbf{e}$ is a unit vector in the propagating direction of light; and u is the one-way velocity of light in the propagating direction $\mathbf{e}$ in the medium as measured in $\tilde{F}$.

The time of a light ray traversing a round trip (a closed path) is then given by the integral:

$$\Delta t = \oint \frac{dl}{u} = \oint \frac{dl}{\bar{u}} + \frac{1}{c} \oint \mathbf{q} \cdot \mathbf{e}dl, \tag{8.2.2a}$$

where the integral in the second term on the right-hand side can be written as

$$\oint \mathbf{q} \cdot \mathbf{e}dl = \oint \mathbf{q} \cdot \mathbf{dl} = \int\int \nabla \times \mathbf{q} \cdot \mathbf{ds} = 0. \tag{8.2.2b}$$

Here $\mathbf{e}dl = \mathbf{dl}$ by the definitions of $\mathbf{e}$ and dl, and $\nabla \times \mathbf{q} = 0$ because $\mathbf{q}$ is a constant vector independent of the space–time coordinates. Thus, the time traversing a closed path, equation (8.2.2a), becomes

$$\Delta t = \oint \frac{dl}{\bar{u}}. \tag{8.2.2c}$$

In particular, in a uniform, and isotropic medium, $\bar{u}$ along the closed path is a constant, we then have

$$\Delta t = \frac{1}{\bar{u}} \oint dl = \frac{\Delta L}{\bar{u}}. \tag{8.2.2d}$$

This is just indeed the definition of the average velocity of light along the closed path. In other words, the result shows that equation (8.2.1) satisfies the constancy of the two-way velocity of light in the medium.

(I) *The Michelson–Morley-Type Experiment*

A Michelson interferometer, in plane view, is sketched in Fig. 9. The whole apparatus is firmly mounted on a massive base and can be rotated about a vertical axis through any desired angle. Light from the source S is partly reflected and partly transmitted by the half-silvered mirror P. The transmitted ray 1 is reflected by the mirrors M_1 and P, and then enters the telescope T. The reflected ray 2, traveling toward and being reflected from the mirror M_2, passes through P, and then in T is brought into interference with the light ray 1. By suitable means, interference fringes are observed in T.

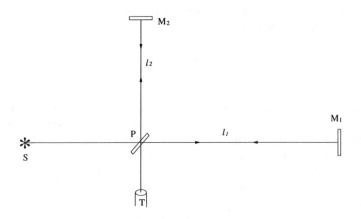

Fig. 9. Schematic diagram of Michelson interferometer.

We now calculate the travel-time difference of the two rays by use of the ether theory. The arrangement is assumed to move with the velocity $\mathbf{v}$ in the direction of the path PM_1 relative to the ether. According the ether theory the velocities of the ray 1 are $c - v$ and $c + v$ in the path PM_1 and M_1P, respectively. Let l_1 and l_2 represent the lengths of the paths PM_1 and PM_2 respectively. Then we obtain the traveling time of the ray 1 from P to M_1 and then back to P,

$$t_1 = \frac{l_1}{c - v} + \frac{l_1}{c + v} = \frac{2l_1 c}{c^2 - v^2}. \tag{8.2.3}$$

For calculating the traveling time of the second ray, we must consider the path actually traversed by it. Let $t_2/2$ represent the traveling time of the ray 2 from P to M_2 . Due to the motion of the arrangement with respect to the ether, the actual path is

$$l_2' = \sqrt{l_2^2 + (vt_2/2)^2}, \tag{8.2.4a}$$

where $vt_2/2$ represents the path passed through by the ether wind at the time $t_2/2$. So that the traveling time of the ray 2 from P to M_2 and then back to P is given by

$$t_2 = \frac{2l_2'}{c} = \frac{2}{c}\sqrt{l_2^2 + (vt_2/2)^2}$$

or explicitly

$$t_2 = \frac{2l_2}{\sqrt{c^2 - v^2}}. \tag{8.2.4b}$$

From Eqs. (8.2.3) and (8.2.4b) we obtain the travel-time difference of the two rays when they enter T,

$$\Delta t = t_1 - t_2 = \frac{2}{c} \left[\frac{l_1}{1 - v^2/c^2} - \frac{l_2}{\sqrt{1 - v^2/c^2}} \right]. \tag{8.2.5}$$

If now the entire apparatus be rotated through an angle of 90°, the two light paths l_1 and l_2 exchange their relationships to the direction of motion through the ether. After this rotation, the travel-time difference of the two rays is given by

$$\Delta t' = -\frac{2}{c} \left[\frac{l_2}{1 - v^2/c^2} - \frac{l_1}{\sqrt{1 - v^2/c^2}} \right]. \tag{8.2.6}$$

So a change of the travel-time difference after the rotation with respect to that before the rotation is then given by subtraction

$$\delta t = \Delta t - \Delta t' = \frac{2 (l_1 + l_2)}{c} \left[\frac{1}{1 - v^2/c^2} - \frac{1}{\sqrt{1 - v^2/c^2}} \right] \tag{8.2.7}$$

or, for small velocities $(v^2 \ll c^2)$,

$$\delta t \simeq \frac{(l_1 + l_2) v^2}{c} \frac{v^2}{c^2}. \tag{8.2.8}$$

This leads to the fringe shift

$$\Delta = \frac{c}{\lambda} \delta t = \frac{l_1 + l_2}{\lambda} \left(\frac{v^2}{c^2} \right), \tag{8.2.9}$$

where λ is the wavelength of light. This shows that after the apparatus is rotated through an angle of 90° an expected displacement of the interference fringes should be observed.

In 1881 A.A. Michelson [6] first performed this kind of experiment. No displacement on rotating the apparatus through 90° was observed. Considering the accuracy of the measurement the negative result implies that the velocity of an ether wind should be less than 21.2 km/sec if the ether theory is correct. Later, Michelson and Morley [7] performed a set of new measurements in 1887 in an improved experimental condition. The massive base of the apparatus floated in mercury and can thus, without disturbance, be rotated about a vertical axis through any desired angle. The length of the light path was $l_1 = l_2 = 11$ m, the light wavelength was $\lambda = 5.9 \times 10^{-7}$ m. If the velocity of the ether wind is regarded as $v = 30$ km/sec, the earth's orbital velocity, an expected displacement of interference fringes should then be $\delta \simeq 0.37$ according to Eq. (8.2.9). When they made the apparatus rotation, the largest one of the observed displacements of the fringes was less than 0.01. After a half year when the earth's orbital velocity had an opposite direction, they performed again a set of measurements and did still not observed an expected displacement. Considering the experimental accuracy the negative result of this experiment gives an upper limit of 4.7 km/sec on the velocity of the ether wind.

In order to explain the absence of any effect due to the motion of the ether in Michelson's experiment, Lorentz (1892) [44] and Fitzgerald (Lodge 1893 [45]) independently put forward the hypothesis that any rigid body moving through the ether with velocity $\mathbf{v}$ is contracted in its direction of motion, the relative contraction being equal to $(1 - v^2/c^2)^{1/2}$ (this is called *Lorentz–Fitzgerald contraction*). The length of the path PM_1 in Michelson's equal-path interferometer is then not l, but $l(1 - v^2/c^2)^{1/2}$, while the length of PM_2 is unchanged, since PM_2 at right angle to the direction of motion of the apparatus. For the time t_1 we then obtain, instead of (8.2.3),

$$t_1 = \frac{2l}{\sqrt{c^2 - v^2}} = t_2,$$

where t_2 is given by (8.2.4b) with $l_2 = l_1 = l$. In this case the phase difference Δ becomes zero, in agreement with Michelson's experiment.

Although the contraction hypothesis is to be regarded as an immediate forerunner of Einstein's theory of relativity, it should be emphasized that this concept, when standing by itself, contradicts the foundational principle of relativity. The Lorentz–Fitzgerald contraction is an absolute concept while the Lorentz contraction in Einstein's theory is a relative concept.

After A. Einstein developed his theory of special relativity, many experiments of the Michelson–Morley type in improved experimental conditions were carried out by some authors. For instance, Kennedy (1926) [46] as well as Illingwarth (1927) [47] independently performed measurements by means of a Michelson interferometer with non-equal lengths of arms (i.e., $l_1 \neq l_2$). Their measurements give negative results which lead to the upper limits of 5.1 km/sec and 2.3 km/sec on the velocity of the ether wind, respectively. In a repetition of the Michelson–Morley experiment carried out by Joos [48] in 1930, the sensitivity of his arrangement was extended to the point where an ether wind of velocity as small as 1.5 km/sec could have been observed. Again this observation gives a negative result that implies putting an upper limit of 1.5 km/s on the velocity of an ether wind.

A positive result for the measurement of the Michelson–Morley type was obtained by Miller [49] in 1933, which leads to an "ether wind" velocity of about 10 km/sec. However, Shankland *et al.* (1955) [50] analyzed this experiment in detail, and concluded that the observed fringe shift was produced by the statistical fluctuations and the temperature variation.

Another experiment was performed by Shamir and Fox [51] in 1969, in which the two arms of an interferometer were consisted of a transparent solid and the source was a He-Ne laser. Their arrangement could observe a fringe displacement of 10^{-5}. When

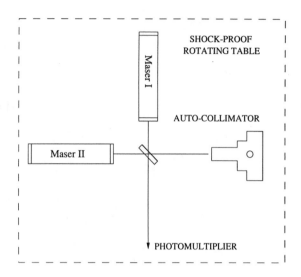

Fig. 10. Schematic diagram for recording the variations in beat frequency between two optical maser oscillators when rotated through 90° in space. Apparatus on the shock-proof rotating table is acoustically isolated from the remaining electronic and recording equipment.

the arrangement was rotated through 90°, no displacement of interference fringes was observed.

The above mentioned experiments employed the means of interference of light (the Michelson interferometer). In 1955, L. Essen [52] performed an experiment of the Michelson–Morley type by means of microwave resonator cavity as the arm of an interferometer, the negative result yielding an upper limit of 2.5 km/s on the velocity of an ether wind.

In 1964, Jaseja et al. [53] performed measurements by use of two He-Ne masers which can be regarded as equivalent to a Michelson–Morley experiment of improved precision. The experimental arrangement is shown schematically in Fig. 10. Two He-Ne masers were mounted with axes perpendicular on a rotating table carefully isolated from acoustical vibrations. An optical or infrared maser consisting of excited atoms between two parallel reflecting plates oscillates at a frequency given by

$$\nu \simeq \frac{Nc}{2L},$$

where N is an integer and L is the plate separation, or more precisely $2L/c$ is the travel-time for a round trip of the light between the two plates. When rotated the

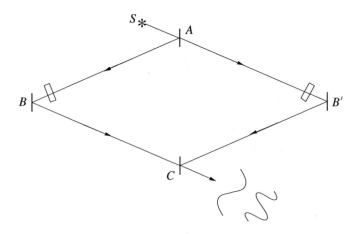

Fig. 11. Interferometer with frequency-doubling crystals.

maser in space, a change of the travel-time will produce a change in frequency. The experiment consisted basically in rotating the platform back and forth 90° and observing the change in frequency difference between the two masers.

Assume that the ether is streaming parallel to the axis of the maser 2 at velocity v. Then, the velocities of light in the masers 2 and 1 are, respectively, $c \pm v$ and $\sqrt{c^2 - v^2}$. Thus, the oscillation frequencies in the masers 2 and 1 are, respectively,

$$\nu_2 = \frac{Nc}{2L}\sqrt{1 - \frac{v^2}{c^2}}, \tag{8.2.10}$$

and

$$\nu_1 = \frac{Nc}{2L}\left(1 - \frac{v^2}{c^2}\right). \tag{8.2.11}$$

The frequency difference between the two masers are then

$$\nu_2 - \nu_1 \simeq \frac{1}{2}\left(\frac{Nc}{2L}\right)\frac{v^2}{c^2} = \frac{1}{2}\frac{v^2}{c^2}\nu. \tag{8.2.12}$$

When the apparatus is rotated through 90°, the relative change in frequency difference between the two masers should be

$$\frac{\Delta\nu}{\nu} = \frac{2(\nu_2 - \nu_1)}{\nu} \simeq \frac{v^2}{c^2}. \tag{8.2.13}$$

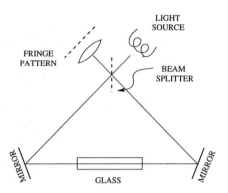

Fig. 12. Schematic diagram of triangular interferometer.

For an infrared maser, $\nu \simeq 3 \times 10^{14}$ Hz. Putting $v = 30$ km/s (the orbital velocity of the earth) into Eq. (8.2.13), we have the change in frequency difference $\Delta\nu \simeq 3 \times 10^6$ Hz. The measurements of this kind of change were completed by a photodetector, a discriminator, and a recorder. The negative result yields a limit of 0.95 km/s on the velocity of the ether drift.

In the above experiments of the Michelson–Morley type the paths of light are all closed paths. The negative results can be easily explained by use of the constancy of the two-way speed of light regardless of any assumptions concerning the one-way speed, or explicitly, equation (8.2.2d). Therefore, these experiments of the Michelson–Morley type are a test not of the one-way velocity of light, but just of the two-way velocity of light.

(II) Other Closed Light Path Experiments

An experiment differing from the Michelson–Morley type was completed by Silver-tooth [54] in 1972 by use of an interferometer of the form shown in Fig. 11. The source of light is a laser emitting at 1.06 μm. Frequency-doubling crystals are located at each of the path extremities B and B'. By this means, two sets of fringes result; one from the 1.06-μm wavelength, the other from the doubled 0.53-μm wavelength, which occurs only in the return portions of the paths. The azimuth scan was provided by the rotation of the earth rather than by rotation of the interferometer. The experiment observed the different motion of two sets of fringes produced by the motion of the earth. No fringe shift was observed in a rather extensive series of observations.

Any shift of a set of fringes is due to a change of the phase difference given by $\delta\phi = \omega_C \Delta t$ where $\Delta t = t_{ABC} - t_{AB'C} = (l_{AB}/c_{AB}) + (l_{BC}/c_{BC}) - (l_{AB'}/c_{AB'}) - (l_{B'C}/c_{B'C})$, and ω_C is the frequency of two light beams at the point C. For the original frequency,

$\omega_C = \omega$; while for the doubled frequency, $\omega_C = 2\omega$. It is easy to see that the postulate of the constancy of the two-way velocity of light, equation (8.2.1), would give the result of $\Delta t = $ constant independent of q_r. Thus, this experiment is also a test not of the one-way velocity of light, but just of the two-way velocity of light.

Another experiment was carried out by Trimmer, Baierlein, Faller, and Hill in 1973 [55] by use of a triangular interferometer shown in Fig. 12. Three arms of lengths 12 cm, 17 cm, and 12 cm form an isoscales triangle with interior angles $45°, 90°$, and $45°$. The glass is 12 cm long and is made of crown glass with a refractive index of 1.5. The interferometer was mounted on a two-inch thick piece of aluminum jig plate and enclosed in a vacuum chamber. The chamber rotates on a carefully-leveled granite surface plate. In order to keep the bearings from crushing the granite surface plate, a thin piece of shim stock is used to separate them. During the basic data-taking cycle of one minute, the average position of the fringe pattern is measured and then the apparatus is rotated through $60°$ to the next data-taking position. Data are taken continuously once a minute for several days at a time. No fringe shift was observed.

The null result is naturally explained by using Eq. (8.2.2c): The time traversing a closed path is given by

$$\Delta t = \oint \frac{dl}{\bar{u}} = \frac{L_0}{c} + \frac{L_g}{c/n}, \tag{8.2.14}$$

where n is the refractive index of glass, L_g is the length of the glass, and L_0 the light path in vacuum. The traveling time given by this equation is independent of both the directional parameter $\mathbf{q}$ and the rotation of the apparatus. This means that rotating the apparatus could not cause any displacement of the position of the fringes. So this experiment is a test not of the one-way velocity of light but only of the two-way velocity of light.

In order to test the isotropy of space, Brillet and Hall (1979) [40] performed a new measurement by use of a rotating laser. They stated the principle for this experiment as follows: A He-Ne laser (3.39 μm) wavelength is servostabilized so that its radiation satisfies optical standing-wave boundary conditions in a high stable, isolated Fabry–Perot interferometer; Because of the servo, length variations of this cavity—whether accidental or cosmic—appear as variations of the laser wavelength. To separate a potential cosmic cavity-length variation from simple drift, the whole system was continuously rotated about a rotation axis. A small portion of the laser beam was diverted up along the rotation axis to read out the cavity length via optical heterodyne with an "isolation laser" which is stabilized relative to a CH_4-stabilized reference laser. The beat frequency was shifted and counted. They analyzed the measurements by means of the metric transformation given by Robertson [3],

$$c^{-2}ds^2 = dt^2 - c^{-2}\left(dx^2 + dy^2 + dz^2\right), \tag{8.2.15a}$$

$$c^{-2}ds'^2 = (g_0 dt')^2 - c^{-2}\left[(g_1 dx')^2 + g_2^2\left(dy'^2 + dz'^2\right)\right]. \tag{8.2.15b}$$

The experiment gave a frequency shift limit of $\pm 2.5 \times 10^{-15}$, corresponding to $\pm 5 \times 10^{-15}$ in $g_2/g_1 - 1$ or a fractional length change of $\Delta l/l = (1.5 \pm 2.5) \times 10^{-15}$. (Joos' experiment [48] shows that $g_2/g_1 - 1 = (0 \pm 3) \times 10^{-11}$; The measurement by Jaseja *et al.* [53] yields that $g_2/g_1 - 1 = \pm 2 \times 10^{-11}$)

In the viewpoint of Robertson transformations, the quantity $g_2/g_1 - 1$ is related to our new parameters through Eq. (7.2.19), i.e.,

$$(g_2/g_1) - 1 = (\bar{c}_\parallel - \bar{c}_\perp)/\bar{c}_\perp. \qquad (8.2.15c)$$

This shows that these above experiments, e.g. Brillet–Hall's experiment, test nothing but the isotropy of the two-way velocity of light.

(IiI) *The Kennedy–Thorndike Experiment*

R.J. Kennedy and E.M. Thorndike (1932) [56] performed an important variation of the Michelson experiment, in which the difference in length of the two arms in the interferometer was kept large. They kept the interferometer be at rest in their laboratory, and observed a possible fringe shift in time. No annual and daily variations were observed. This excludes the possibility of the earth–velocity–dependence of the time needed by the light traveling through a closed path in the terrestrial laboratory. In other words, the experiment shows that the isotropy of the two-way velocity of light would not be changed by the motion of an observer.

Recently, Hils and Hall (1990) [91] carried out an improved Kennedy–Thorndike Experiment by searching for sidereal variations between the frequency of a laser locked to an I_2 reference line and a laser locked to the resonance frequency of a highly stable cavity. No variations were found at the level of 2×10^{-13}. This shows that the Lorentz transformation can be based on experimental facts at the 70-ppm level.

8.2.2. *The One-Way Experiments*

In the so-called *one-way experiments* performed before, light path was not closed. In one class of these experiments, Doppler frequency shift was observed. For instance, for transverse motion the absolute ether theory predicts the Doppler shift [57,58],

$$\nu_a^{ether} = \nu_0 \left(1 + \frac{\mathbf{v} \cdot \mathbf{u}}{c^2} \right), \qquad (8.2.16)$$

where higher-order terms are neglected, ν_0 and ν_a^{ether} are the frequencies of a ray as seen in the systems of source and observer, respectively, $\mathbf{u}$ is the relative velocity of the source to the observer, and $\mathbf{v}$ is the absolute velocity of the laboratory moving through the ether. This result shows that the transverse Doppler shift $\nu_a^{ether} - \nu_0$ is linear in $\mathbf{u}$ so that an experiment of this kind is called a first-order experiment, or a unidirectivity experiment.

In view of the theory of relativity, an experiment of this kind involves the measurement of transverse Doppler shift

$$\frac{\nu_a - \nu_0}{\nu_0} = \frac{u_a^2 - u_s^2}{2c^2}, \tag{8.2.17}$$

where u_a and u_s are, respectively, the velocities of the observer and the source as seen in laboratory frame, and higher-order terms are neglected. As mentioned in Sec. 6.5, Edwards' theory gives the same predictions as this equation. In other wards, no first-order effect would appears in this kind of experiment. Therefore, the so-called first-order experiment is indeed a test of transverse Doppler effect.

For the fact mentioned above, we shall introduce the experiments of this kind both in the present section and in Sec. 9.3.3.

(I) *The Two Masers Experiment*

The relative frequency stability of two beam-type maser oscillators was used by Cederholm, Bland, Havens, and Townes [59] in 1958 to test the dependence of the velocity of light on velocity of the frame of reference. The experiment involved comparison of the frequencies of two masers having their beams of NH_3 molecules traveling in opposite directions. The ether theory would give a change in frequency of a beam-type maser due to ether drift, assuming the molecules in the beam to have a velocity u with respect to the cavity through which they pass, and the cavity to have a velocity v with respect to the ether. The shift may be simply discussed by assuming that, if v is zero, radiation is emitted perpendicularly to the molecular velocity so that there is no Doppler shift. If the cavity and beam are then transported at a velocity v through the ether in a direction parallel to u, radiation must be emitted by the molecules slightly forward at an angle $\theta = \pi/2 - v/c$ with respect to u. The fractional change in frequency due to the Doppler effect is then $\epsilon = (u/c)\cos\theta$ or uv/c^2 due to motion through the ether, assuming that the proper molecular frequencies are unchanged by such motion. For a thermal molecular velocity of 0.6 km/sec and for the earth's orbital velocity of 30 km/sec, $\epsilon = 2 \times 10^{-10}$. The difference in frequency due to the ether drift between two masers with oppositely directed beams would be $2\epsilon\nu$, or about 10 cps for ν equal to 23870 Mc/sec, the NH_3 inversion frequency. Two masers oscillators with oppositely directed beams were mounted on a rack which could be rotated about a vertical axis. When the apparatus was rotated through 180°, an expected beat frequency change of 20 cps would be observed. However no such a change in beat frequency was recorded by them. This experiment give an upper limit of 0.03 km/sec on velocity of ether drift.

(II) *The Mössbauer Rotor Experiment* (see also Secs. 9.3.3 (II) and page 115)

Ruderfer (1960, 1961) [42,57] and Møller (1962) [58] pointed out a sensitive ether drift measurement may be performed using the Mössbauer effect. Such a measurement was carried out by Champeney, Isaak, and Khan in 1963 [60]. The arrangement is a development of that described by Champeney and Moon in 1961 [61]. Two arms screw into the central part of a rotor and sources and absorbers may be mounted

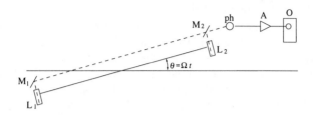

Fig. 13. Schematic diagram of two lasers experiment. L_1 and L_2 are two lasers. M_2 is a partially reflecting mirror. M_1 is a totally reflecting mirror, parallel to M_2. Ph is the photodiode. A and O are the amplifier and the oscilloscope.

at the tips resting against flanges, or at the center clamped in position between the two arms. The rotor spins in vacuum on a glass plate being accelerated initially by a rotating magnetic field. Two Xenon proportional counters are placed outside the vacuum vessel with their windows respectively North and South of the rotor, so that the apparatus is sensitive to a component of ether drift in an East-West direction. The measurements, interpreted on the classical assumption of an ether drift, place a limit on any steady drift past the earth, resolved parallel to the equatorial plane, of 1.6 ± 2.8 m/sec.

(III) *The Two Lasers Experiment*

In 1972, R. Cialdea [62] performed a new one-way path experiment by use of two equal completely independent lasers. The arrangement is shown in Fig. 13. Two lasers L_1 and L_2 are placed on an optical bench, which can rotate around a vertical axis normal to the plane of the Figure. Two light beams overlap at M_2 and reach a fast photodiode Ph, whose signal, amplified by A, is observed on a fast oscilloscope O. According to the ether theory, when the apparatus is rotated the beat frequency would vary periodically. But no this variation was observed, which is in agreement with the theory of special relativity. The negative result yields an upper limit of 0.0009 km/sec on the velocity of ether wind.

We have introduced above the one-way path experiments. In view of the classical ether theory, the experiments of this kind should involve first-order effects which are produced by the ether drift. However all these experiments gave a negative result.

However, in view of the theories of special relativity involving the test theories, which are in contrast with that of the ether theory, these experiments involve Doppler frequency shifts being of second-order effects from the time dilation effects (see Sec. 9.3.3), so that, in principle, they involve slow transport of clocks. Thus, the one-way path experiments, like the closed path experiments, could also not give a test of isotropy of the one-way velocity of light, as argued in Secs. 6.5 and 7.5.

(IV) *The TPA and JPL Experiments*

The TPA experiment was performed by Riis *et al.* (1988,1989) [30.31], in which they monitored the frequency of light emitted by atoms excited resonantly via two photon absorption (TPA) in an atomic beam as a function of the rotation of the Earth.

The JPL experiment was carried out by Krisher *et al.* (1990) [34], in that they monitored the time of flight of light signals along a fiberoptic link between two hydrogen maser clocks at the NASA-Jet Propulsion Laboratory (JPL) Deep Space Network, again as a function of the rotation of the Earth.

Clifford M. Will (1992) [35] gave limits on some of the parameters in the MS transformations by analyzing the two experiments. We shall here re-explain the physical meanings of the limits. Let us now recall the discussions in Ref. [35] as follows:

> We consider the following idealized model for this (the TPA) experiment. A laboratory moves with velocity **v** relative to the preferred frame F. A laser signal is emitted continuously from one end of a cavity and reflected from the other hand. The frequency of the laser is $\nu_L \equiv 1/t_0$, where t_0 is the interval between wave crests emitted, measured in the laboratory frame F^*. From stationarity or direct calculation, the frequency of the reflected beam is also ν_L. An atom moves with coordinate velocity $\mathbf{u} = u\mathbf{n}$, where $\mathbf{n}$ is a unit vector parallel to the axis of the cavity, pointing in the same direction as the emitted laser signal. The atom receives adjacent wave crests from the emitted laser beam and from the reflected laser beam.
> $\cdots$.

> If we define the frequency of the photon from the backward direction (i.e., moving in the same direction as the atom) as ν_+ and that of the photon from the forward direction as ν_-, then the actual measured quantities are

$$V \equiv \frac{\nu_- - \nu_+}{\nu_- + \nu_+}, \tag{4.2a}$$

$$\nu_L \equiv \sqrt{\nu_+ \nu_-}. \tag{4.2b}$$

Now substituting the low-velocity expansions, Eqs. (2.15) and (2.16), and assuming that $u \approx v$, we obtain

$$(\nu_+\nu_-)^{1/2}\nu_L^{-1} = 1 - \alpha u^2 - 2\alpha uv\cos\theta + \left(\alpha^2 - \frac{1}{2}\alpha - \alpha_2\right)u^4$$

$$+ \left(4\alpha^2 - 2\alpha\epsilon - 4\alpha_2\right)u^3 v\cos\theta + (2\alpha\delta - \alpha^2 - 2\alpha_2)u^2 v^2$$

$$+ [2\alpha(\beta - \delta - \epsilon) + 4\alpha^2 + \alpha - 4\alpha_2]u^2 v^2 \cos^2\theta$$

$$+ (2\alpha\beta - 4\alpha_2)uv^3 \cos\theta + O(6), \tag{4.5a}$$

$$V = u \left[1 + \epsilon u v \cos\theta + (\alpha - \delta)v^2 - (\beta - \delta)v^2 \cos^2\theta\right] + O(4), \qquad (4.5b)$$

where $\cos\theta \equiv \hat{v} \cdot \mathbf{n}$. Notice that the expression for ν_L in Eq. (4.5a) depends on the arbitrary synchronization parameter ϵ. This dependence is only apparent, however, because the coordinate velocity u is not directly measurable. Instead, V is the measured quantity through the accelerating voltage on the beam required to maintain resonant absorption. Using Eq. (4.5b) to express u in terms of V and substituting into Eq. (4.5a), and noting that in the atom's rest frame the resonance condition corresponds to $\nu_+\nu_- = \nu_1\nu_2$, we obtain, finally

$$\nu_L = (\nu_+\nu_-)^{1/2}\{1 + \alpha V^2 + 2\alpha V v \cos\theta + \left(\frac{1}{2}\alpha + \alpha_2\right)V^4$$

$$+4\alpha_2 V^3 v \cos\theta - \left(\alpha^2 - 2\alpha_2\right)V^2 v^2 - (\alpha - 4\alpha_2)V^2 v^2 \cos^2\theta$$

$$-2[\alpha(\alpha + \beta - \delta) - 2\alpha_2]V v^3 \cos\theta + 2\alpha(\beta - \delta)V v^3 \cos^3\theta + O(6)\} \quad (4.6)$$

The TPA experiment used a beam of neon atoms with kinetic energy per atom of about 120 keV, corresponding to a velocity of 3.5×10^{-3} of the speed of light. The laser frequency and the beam voltage (which was related to the velocity V) were monitored to look for a diurnal variation generated by the $\cos\theta$ term as the Earth rotated. A limit of 10^{-11} was put on such a variation, leading to the bound

$$|\alpha + 2\alpha_2 V^2 - [\alpha(\alpha + \beta - \delta) - 2\alpha_2]v^2| < 1.4 \times 10^{-6}. \qquad (4.7)$$

However, since $V^2 \approx v^2 \approx 10^{-6}$ the only useful bound is

$$|\alpha| < 1.4 \times 10^{-6}. \qquad (4.8)$$

For the JPL experiment, Clifford M. Will [35] considered the following idealized picture.

In a moving frame F^*, clock A is located at the origin of the frame; a traveling clock T moves slowly through F^*, passing points B and C, which are equidistant from the origin. In the JPL experiment, this motion is effected by the rotation of the Earth. As the traveling clock passes each of these two points, it receives a light signal from the clock A, and compares its own phase with that of the received signal.

$\cdots$

In the limit $v^2 \ll 1$, $\cdots$, the result (the phase difference at C) is

$$\frac{\Delta\phi}{\tilde{\phi}} = 2\alpha v(\cos\theta - \cos\theta_0) + (\delta - \beta)v^2(\cos^2\theta - \cos^2\theta_0) + O(v^3), \quad (3.14)$$

where $\tilde{\phi} \equiv 2\pi\nu L$, $\cos\theta \equiv \hat{\mathbf{v}} \cdot \mathbf{n}_C$, and $\cos\theta_0 \equiv \hat{\mathbf{v}} \cdot \mathbf{n}_B$.

In the JPL experiment, light signals were propagated simultaneously in both directions and phase comparisons were made at both ends of the fiberoptic link, yielding both $\Delta\phi(\theta)$ and $\Delta\phi(\theta + \pi)$. Then by summing and differencing the phase differences at both ends, it was possible to separate the v term in Eq. (3.14) from the v^2 term and from effects that do not change sign when the propagation direction is reversed, such as diurnal temperature effects in the fiberoptic link. The resulting limits are [34]

$$|\alpha| < 1.8 \times 10^{-4}, \qquad |\delta - \beta| < 2 \times 10^{-2}, \qquad (3.15)$$

where the projection of the propagation base line on the direction of $\mathbf{v}$ has been taken into account.

In the quoted equation (4.6), Clifford M. Will correctly expressed his result in terms of the measured quantity V. In fact, it is not necessary to use the MS transformations for the calculation. Instead, Robertson transformations would yield the same result.

We see that the results of the TPA and JPL experiments, just as the other one-way experiments, e.g., the Mössbeauer rotor experiments, are independent of the parameter ϵ, while they would be sensitive to the other parameters $\alpha, \beta, \delta, \cdots$. The problem is how to explain these limits given by the quoted equations (4.8) and (3.15). We have already shown in Secs. 7.3 and 7.6 that there might be two kinds of anisotropy, the anisotropy of the one-way speed of light and the anisotropy of the two-way speed of light. Thus, we must separate the one from the other. Only the new parameters $c_\perp, c_\parallel$ and q can used for the separation: a non-zero value of q indicates the anisotropy of the one-way velocity of light, while $\bar{c}_\parallel \neq \bar{c}_\perp$ represents the anisotropy of the two-way velocity of light. It is known from the relations between the original and new parameters, Eqs. (7.6.8), that the TPA and JPL experiments as well as the other one-way experiments would not determine the directional parameter q. Therefore, these so-called one-way experiments, just as the closed path experiments, are only the tests of the isotropy of the two-way speed of light.

8.3. The Experiments with Moving Sources of Light

This section will give an introduction to such experiments with moving light sources which involve binary systems, stars, the sun, extragalactic systems, moving media, and gamma ray sources. The light paths could be classified into the closed path type and the one-way path type.

Just in order to show a comparison among the experiments, a simple relationship between the velocity of light and the velocity of a moving source is assumed:

$$c' = c \pm kv, \qquad (8.3.1)$$

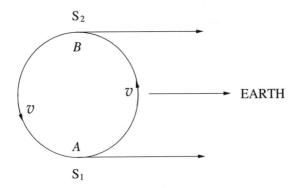

Fig. 14. Observations of a binary system.

where c is the velocity of a light ray from a stationary source, c' is the velocity of the ray as seen from a moving observer with the relative velocity v to the source, and k is a constant parameter to be determined by experiments, $k = 0$ corresponding to Einstein's theory, while $k = 1$ being a result given by the *"ballistic" hypothesis of light emission*.

The experiments with moving sources all show that the velocity of light is independent of motion of the sources. A negative result would give an upper limit on the parameter k. For instance, the observations of a binary system yield $k < 10^{-6}$.

8.3.1. The Astronomical Verification

(I) Observations of Binary Systems

The considerations of Comstock (1910) [63] and de Sitter (1913) [64] concerning the orbits of close binary stars are the oldest and best known astronomical evidences for the independence of the light velocity of the motion of the light source.

Since each of the twin stars moves with a high speed in its orbit, if the velocity of light did depend on the velocity of its source, then the light from the approaching star would reach the earth in a shorter time than that from the receding member of the doublet. Let us consider this effect. For simplicity we assume that twin stars S_1 and S_2 move in a circular orbit around their common center of gravity, as shown in Fig. 14. At the time t_1 the star S_1 reaches the point A, its tangential velocity being in a direction to the earth and a light ray emitted by the star being assumed to have a velocity $c + v$. An observer on the earth will receive this ray at the time

$$t_1' = t_1 + \frac{L}{c + v}, \tag{8.3.2}$$

where L is the distance between the binary and the earth. Let T represent the time when the star S_1 arrives at another point B. Another light ray emitted from the star at B will have a velocity $c - v$, and then reaches the earth at the time

$$t_1'' = t_1 + T + \frac{L}{c - v}. \tag{8.3.3}$$

One arrives at the time difference

$$T_1' = t_1'' - t_1' = T + \tau \tag{8.3.4a}$$

with

$$\tau = \frac{L}{c - v} - \frac{L}{c + v} = \frac{2vL}{c^2 - v^2} \simeq \frac{2vl}{c^2}. \tag{8.3.4b}$$

Here the difference T_1' is a half-period of the star S_1 as seen from the earth. We now consider the second star S_2. Let S_2 arrive at the point B at the time t_2, its tangential velocity being in a backward direction to the earth. A light ray emitted from S_2 at t_2 will arrive at the earth at the time

$$t_2' = t_2 + \frac{L}{c - v}. \tag{8.3.5}$$

Similarly, S_2 reaches the point A at the time $t_2 + T$. Another light ray emitted at such a time will be received by the earth at the time

$$t_2'' = t_2 + T + \frac{L}{c + v}. \tag{8.3.6}$$

Thus, the half-period of the second star S_2 as seen from the earth is given by

$$T_2' = t_2'' - t_2' = T - \tau, \tag{8.3.7}$$

where τ is defined by Eq. (8.3.4b). This results show that if the "ballistic" hypothesis for the velocity of light is correct the half-periods of the stars in a binary system will depend on the distance between the doublet and the earth. This would introduce a considerable spurious eccentricity into the orbits of the binary systems.

W. de Sitter (1913) [64] first considered this problem and pointed out that for many binary systems τ has the same order of magnitude with T. He came the following conclusions: If the velocity of light were not assumed constant, then for circular orbits of spectroscopic twin stars the time dependence of the Doppler effect would correspond to that of an eccentric orbit. Since the actual orbits have very small eccentricity, this leads one to conclude that the velocity of light is, to a large degree, independent of the velocity v of the twin star. From known data on β Aurigae, he calculated a k-value less than 0.002 by use of the expression (8.3.1). Later, Zurhellen's estimate (1914) [65] gave $k < 10^{-6}$.

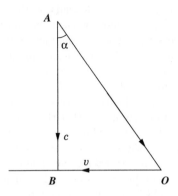

Fig. 15. Aberration of light.

(II) *Differential Aberrations of Stars and Nebulae*

Aberration of light may be described in Fig. 15. The point A is the location of a celestial body. An observer B is at rest with respect to the celestial body. Assume that another observer O moves with a velocity v relative to B. The spatial positions of B and O are assumed to coincide with each other at a given time when the same light ray from the celestial body is received by the two observers in the directions of $\overrightarrow{AB}$ and $\overrightarrow{AO}$, respectively. The directions of $\overrightarrow{AB}$ and $\overrightarrow{AO}$ are, respectively, called the "real" and the sight line directions of the light ray, the angle α between them being the aberration angle of light.

Observations of a celestial body on the earth would give two kinds of aberration, a daily aberration produced by the rotation of the earth and an annual aberration caused by the orbital motion of the earth. The daily aberration depends on a latitude at which the observer is located while the annual aberration varies with the season. To first-order approximation the relativistic aberration formula (2.10.8) give the same result with the classical theory:

$$\tan \alpha \simeq \frac{v}{c}. \tag{8.3.8}$$

Putting the orbital velocity of the earth $v \simeq 29.75$ km/sec into this equation, we get the maximum value of the annual aberration angle,

$$\alpha = 20'' \cdot 47. \tag{8.3.9}$$

This value is called the *aberration constant*. The observations of stars are all in agreement with this constant (in fact, it is the change $\Delta\alpha$ of the aberration angle α that is directly observed, during one year the maximum of the change $\Delta\alpha$ being 2α).

If the light velocity c in Eq. (8.3.8) depends on the source velocity v, then the aberration angle $\alpha = \tan^{-1}(v/c)$ could not yield the constant given by Eq. (8.3.9). A distant nebula might move fast enough to distinguish the possibility. For instance, the light from a nebula with a velocity of recession of 30 000 km/sec should, on the "ballistic" theory, be received at a velocity $0.9c$, and the aberration constant should accordingly be about 23" instead of $20'' \cdot 47$ [66]. So that comparing the aberration of a celestial body with a high velocity of recession and the aberration constant of a fixed celestial body could determine whether the velocity of light depends on the velocity of the light source. The annual aberrations of distant galaxies observed by G. Van Biesbroeck (1932) [67] are the same as those of the neighboring stars. In 1960, Heckmann's analysis [68] for four extragalactic nebulae showed that the measurements of differential annual aberration between stars and nebulae with large radial velocity gave negative results. The velocity of light is therefore independent of the velocity of the source.

8.3.2. The Laboratory Evidences

All of the past laboratory measurements show the constancy of the velocity of light from moving light sources, mirrors and sources of γ radiation, as well as from extraterrestrial sources.

(I) Laboratory Measurements of Moving Sources of Light

Q. Majorana (1919) [69] performed a moving source experiment by use of a Michelson interferometer with a glass-enclosed glowing mercury plasma which rotated about a fixed point at a high tangential velocity of 100 km/sec. R. Tomaschek (1924) [70] repeated the Michelson–Morley experiment by use of sunlight and starlight which were first reflected from heliostats into a telescope and then passed through a laboratory window to finally enter the interferometer. These experiments did not observe any effect of the motion of sources on the velocity of light. (A more careful examination of Majorana's result seems to indicate that the resolution used was not large enough to draw the conclusion due to the fact that only terms of the order v^2/c^2 are significant [90])

R.C. Tolman (1910) [71] performed a moving source experiment by use of a Lloyd mirror interferometer which consists of a slit (S) and a Lloyd mirror (M) as shown in Fig. 16. A part of light passed through the slip is reflected at grazing incidence from the mirror (M), while the other part of light directly reaches the screen. The two parts then form an interference pattern on the screen. Positions of the fringes depend on the time difference $\Delta t = t_{SAB} - t_{SB}$ where t_{SAB} and t_{SB} are the traveling times of light through the paths, respectively, SAB and SB. Assuming that the velocity of light from a moving source is $c \pm v$, and that the light velocity reflected from the

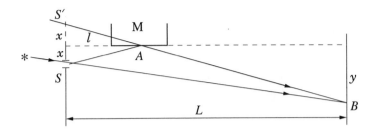

Fig. 16. Lloyd mirror interferometer. The screen plane is vertical with the Lloyd mirror (M) plane. The distances of the slit S to the screen and the mirror are, respectively, L and x. The distance of a point B on the screen to the mirror plane is y. The distance of a point A on the mirror to the screen is $(L - l)$.

mirror is changed to c, we then obtain

$$t_{SA} = \frac{\sqrt{x^2 + l^2}}{c(1 \pm \beta)}, \quad t_{AB} = \frac{\sqrt{y^2 + (L - l)^2}}{c}, \quad t_{SB} = \frac{\sqrt{(x - y)^2 + L^2}}{c(1 \pm \beta)}, \quad (8.3.10)$$

where $\beta = v/c$.

Let Δt_0 denote a time difference corresponding to $v = 0$ (the case of a stationary source). Noting the geometric condition $l = xL/(x + y)$, we have from equations (8.3.10)

$$\Delta t_\pm = t_{SA} + t_{AB} - t_{SB} \cong \pm \frac{L\beta}{c} \left(\frac{y}{x + y} \right) + \Delta t_0, \quad (8.3.11)$$

where the higher-order terms of (v/c), (x/L), and (y/L) are neglected because of smallness. The difference between Δt_+ and Δt_- is

$$\Delta t_+ - \Delta t_- = 2 \frac{L\beta}{c} \left(\frac{y}{x + y} \right),$$

and the corresponding phase difference is then

$$\Delta = \frac{c}{\lambda} (\Delta t_+ - \Delta t_-) = \frac{2\beta L}{\lambda} \left(\frac{y}{x + y} \right), \quad (8.3.12)$$

where λ is the wavelength of light. Equation (8.3.12) predicts a relative shift of two interference patterns corresponding to velocities $+v$ and $-v$ of light sources, if the "ballistic" hypothesis for the velocity of light is correct.

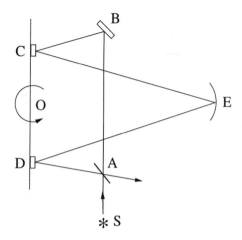

Fig. 17. Michelson's moving mirror experiment.

In Tolman's experiment, the image first of one limb, and then of the other limb of the Sun were focused by means of a lens upon a slip (S). In this way, Tolman did not find the predicted relative shift of two fringes in the interference pattern formed by the direct and reflected light from the slit illuminated by the image of each limb of the Sun. There are different explanations of this negative result. The main point is whether the slit would not alter the relative speed $c \pm v$ of light from a moving source to c from a stationary source. Furthermore the assumption that reflection at grazing incidence will change $c \pm v$ to c is still in question [72].

Other laboratory measurements of the velocity of sunlights from limbs of the Sun were carried out by Bonch–Bruewich (1960) [73] by means of a phase-modulation method. If the source velocity were added to the light velocity, then difference between traveling times of two rays, which were from two opposite limbs of the Sun equator, along a path of 2000 m should be 75×10^{-12} sec. The experimental result is indeed $(1.4 \pm 5.1) \times 10^{-12}$ sec, this corresponding to $k = 0.02 \pm 0.07$ in Eq. (8.3.1). This result shows that the velocity of light is independent of the motion of the limbs of the Sun.

(II) *The Moving Mirror Experiment*

A.A. Michelson (1913) [74] performed a moving reflection mirror experiment by means of an arrangement shown in Fig. 17. The distance OE was 608 cm. The distance between centers of the rotating mirrors D and C was calculated to be 26.5 cm on the basis of the known but unstated distance AB, so that AD and BC cannot be exactly deduced. They probably were about equal to DC or slightly more than half of DC so that the rotating mirrors could clear A and B. A light ray from the source S partially reflected at the beam splitter A was in turn reflected from the rotating mirror D. It was then reflected by the stationary concave mirror E, the moving mirror C, and the stationary mirror B to arrive at its starting point A where it was reflected into a telescope with a micrometer eyepiece. Michelson assumed that the velocities of the light rays reflected from the moving mirrors C and D were changed to $c \pm kv$, while the stationary concave mirror E could not alter the velocity of light. By also neglecting or regarding as negligible (compared to DEC) the light path increments CBA for the reflected ray and DA for the transmitted ray Michelson deduced a fringe shift Δ expressed as

$$\Delta = \frac{8vL}{c\lambda}\left(1 - \frac{k}{2}\right), \tag{8.3.13}$$

where v is the velocity of the moving mirrors, and $L = OE \simeq DEC/2$. The filtered wavelength λ was 6000 Å.

The observed fringe shift reduced to their equivalent value at a rotor speed of 1000 r.p.m. ranged from 3.1 to 4.1 with a weighted mean of 3.81 compared to the value of Δ of 3.76 computed with k equal to zero.

Another moving mirror experiment was carried out by Majorana (1918) [75] by use of a Michelson interferometer. A light ray reflected from a moving mirror enters a Michelson interferometer. The image of the source in the mirror has a equivalent velocity of 450 m/sec. Einstein's theory predicts a fringe shift of 0.71 due to Doppler effect of the moving mirror. Observed shift is between 0.7 and 0.8. This shows that the velocity of light is independent of the motion of the mirror.

The above two experiments were done in air. Beckmann and Madics (1965) [76] repeated the Tolman experiment by use of a Lloyd reflection mirror in vacuum (10^{-6} torr). They used two arrangements as shown in Fig. 18 (a,b). Their moving source was established by means of the reflection of light incident at 15° on a tiny mirror (M) mounted on a rotor at 12.8 cm from the axis of rotation. In the arrangement given in Fig. 18 (a), the reflected light ray passed through a slit S enters a Lloyd mirror A. In order to eliminate the doubt of a slit acting as a "new" source which might reduce the velocity of light to c, in the second arrangement as shown in Fig. 18 (b) the slit in front of the interferometer was removed. A light beam from a laser passed through a window (W) of the vacuum chamber was reflected by a mirror (SM) to a slit S. The light rays from the slit were focused by a lens (CL) to form an image of the slit in a space point, through where the moving mirror M passed and reflected the light to the Lloyd mirror A . The light rays from A will form an interference pattern on a photographic plate outside the vacuum chamber. When the tangential velocity of the mirror M was $+v$ one half of the plate was exposed, while when the velocity was $-v$ another half was exposed. In this way they took two interference patterns on a plate. Their measurements by means of the two arrangements shown in Fig. 18 (a,b) did not find a relative shift between the two sets of fringes. The wavelength of light

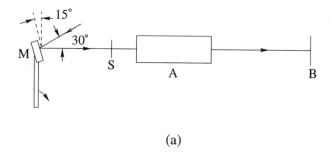

(a)

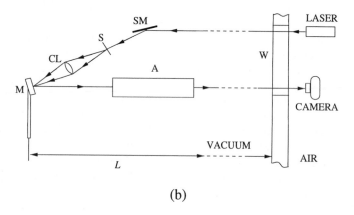

(b)

Fig. 18. Beckmann–Madics' experiment.

was $\lambda = 6328$ Å. The velocity of the mirror M was $v \simeq 1.52 \times 10^{-7}c$. $x/y \sim 10^{-4}$ could be neglected. $L = 2$ m and 4 m. Thus, the negative result would give an upper limit on the parameter in Eq. (8.3.1): $k \leq 0.05$.

(III) *The Moving Transparent Substance Experiment*

W. Kantor (1962) [77] reported that his moving glass experiment indicated the speed of light after passing through a moving piece of glass was a function of the velocity of the glass. The arrangement used by Kantor is shown in Fig. 19. A light beam was separated by the beam splitter A into two parts, which traversed a pentagonal path in opposite directions ($ABCDEA$ and $AEDCBA$) back to the beam splitter A to

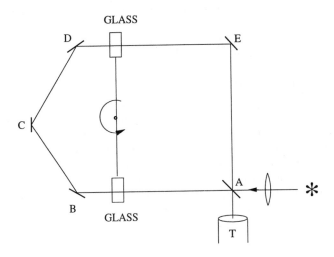

Fig. 19. Kantor's experiment.

produce vertical interference fringes which were seen with the aid of a telescope (T) focused almost for infinity. Two very thin ($l \simeq 0.127$ mm) transparent glass windows were mounted perpendicularly to the plane of a rotating disk at opposite ends of the same diameter. At one end of the disk diameter perpendicular to that on which the windows were mounted, a small magnet was embedded in the disk. A triggering signal was obtained by means of a small pickup coil mounted just below the disk magnet. This signal was used to produce a very short flash (15 μsec) of light from the xenon source S. The light flash occurred when the windows were in a position normal to each longitudinal light path of the interferometer, as shown in Fig. 19.

The windows were approximately four-tenths of the height of the interferometer optics, so that the lower part of the light beam went through the windows while the upper six-tenths passed above them. The optical path length through the windows was not quite the same as the geometrical path length above the windows due to the tiny but finite thickness of the windows. This slight difference produced no noticeable difference in the fringe separations of the fringe pattern seen in the whole field of view through the windows and above them when the disk was stationary. The disk rotation being only counterclockwise, the light beam transmitted by the beam splitter passed through the windows in opposition to their motion, while the reflected light beam went around the interferometer in the opposition direction and through the windows concurrent with their motion. The velocity of light in the glass is $(c/n) \pm fv$. Assume that the velocity of light passed through the moving window be $c \pm kv$ and then after reflection by the stationary mirror its velocity be reduced back to c. We can obtain the light path difference between the two light rays in opposite directions,

$$\Delta = c\frac{\Delta t}{\lambda} = \frac{2k\beta}{\lambda}(L-l) + \frac{4\beta nl}{\lambda} - \frac{4\beta ln^2}{\lambda}(1-f), \tag{8.3.14}$$

where $\beta = v/c$ with v being the velocity of the glass windows in the direction of light ray, and a light path difference corresponding to the case of the windows being at rest are neglected due to the small thickness of the windows. For Kantor's experiment, $\beta \simeq 1.56 \times 10^{-7}$, the refractive index of the glass $n \simeq 1.5$, the thickness $l \simeq 1.27 \times 10^{-2}$ cm, the length of each of the arms $L \simeq 118$ cm, and the wavelength $\lambda = 5000$Å.

According to Einstein's theory of special relativity, $k = 0$ and $f = 1 - n^{-2} - \lambda n^{-2}dn/d\lambda$ (Laub coefficient) with $dn/d\lambda \sim 10^{-5}$ for the glass. So that equation (8.3.14) would predict a negligible fringe shift of $\Delta \sim 10^{-5}$. This implies that no fringe shift could be observed by this experiment.

On the other hand, if $k \neq 0$ and noting $L \gg l$, equation (8.3.14) then is approximately

$$\Delta \simeq \frac{2k\beta nL}{\lambda}. \tag{8.3.15}$$

Putting $k = 1$ and the above conditions of Kantor's experiment into Eq. (8.3.15), one arrives at a maximum fringe shift of $\Delta \simeq 0.74$.

Kantor [77] reported that his experiment observed a relative fringe shift of $\Delta \simeq 0.5$. After that time some similar experiments were repeated. However all of the measurements gave the same negative result. For instance, Babcock and Bergman (1964) [78] repeated this kind of measurements. They used an interferometer like Kantor's (see Fig. 19). There were three principal differences in the equipment: (1) The optical path in the interferometer was longer; (2) a reversible motor was used; (3) the interferometer was enclosed in a vacuum chamber. For this reason the experiment of Kantor's type was performed with approximately four times the sensitivity, i.e., could detect a shift of about 0.02. The glass plate was thin ($l = 0.34$ cm), Einstein's theory then predicting a negligible shift of 0.0036 fringes. On the other hand, if equation (8.3.1) is assumed, the fringe pattern seen when the arm was rotating would have been shifted with respect to that seen when the arm was stationary by Δ fringes given in Eq. (8.3.15). In this experiment, observations were taken with the arm first rotating one way, then the other, so that the total shift would have been twice that given by Eq. (8.3.15), i.e.,

$$\Delta \simeq \frac{4k\beta nL}{\lambda}. \tag{8.3.16}$$

The average rotation rate was 45 rps and β was 1.25×10^{-7}. L (the length of the path AB, see Fig. 19) was 276 cm, and λ was 4.74×10^{-5} cm. On the assumption that $k = 1$, equation (8.3.16) predicts 2.9 fringes. The shift found in this experiment was less than 0.02 fringes. This leads to that $k \leq 0.01$.

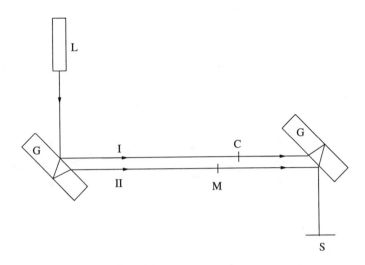

Fig. 20. Experimental arrangement.

Beckmann and Mandies (1964) [79] also repeated the experiment of Kantor's type by means of a similar arrangement. Their measurement gave a negative result that leads to the limit $k \leq 0.1$.

James and Sternberg (1963) [80] measured a possible effect of a glass plate moving vertical to a light path. No transverse shift of the light path was observed. This experiment yields an upper limit for the parameter k of 0.025.

Rotz (1963) [81] measured the difference between the velocities of three beams of light in vacuum, where one of three beams had passed through a piece of glass, which might be moved at a constant speed or kept stationary as desired. They used a three beam interferometer. The central beam passed through a piece of glass. The apparatus was arranged by mounting the three slits forming the interferometer beams on the periphery of a wheel and cementing a piece of glass over the central slit. The wheel, which can be made to rotate at a uniform velocity, also carried a self-shuttering arrangement which allowed illumination of the slit only when the glass was moving parallel to the beams, or was at rest in the proper position. The entire apparatus was enclosed in a vacuum chamber. The light beams leaving the glass and slits traveled a known distance in vacuum before reaching a flat ($\lambda/4$) glass window in the wall of the chamber. The diffraction pattern of the slits was formed by an objective lens and observed through a microscope mounted on a graduated optical bench in the usual manner. The light source was a CW gas laser providing monochromatic light at 6328 Å and operating in a single mode.

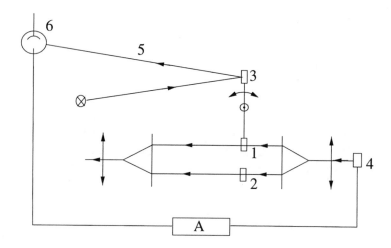

Fig. 21. Diagram of Zahejsky–Kolesnikov's experimental arrangement.

If it is assumed that light propagates in the glass with a velocity $(c/n) + fv$ and with a velocity $c + kv$ upon leaving the glass, then, by a similar derivation of (8.3.14), we obtain the phase difference between the central and side beams due to the motion of the glass as follows:

$$\Delta\varphi = \frac{\beta}{\lambda}\{k(l - L) + l\left[n^2(1 - f) - n\right]\}, \qquad (8.3.17)$$

where $\beta = v/c$ (v = velocity of glass), n is the index of refraction of the glass, l is the thickness of the glass, L is the distance from where light enters the glass on the central slit to where it enters the next optical element, and f is the velocity addition coefficient for glass [according to Einstein's special relativity, it is given by Eq. (4.4.11)].

In this experiment, $\beta = 2 \times 10^{-7}$, $n = 1.5$, $l = 1.5$ mm, $L = 0.75$ m and 1.48 m, so that Eq. (8.3.17) predicts that $\Delta\varphi = (k\lambda/6)$ and $(k\lambda/2)$ corresponding to the two values of L. If the value of k is on the order of 0.1 to 1.0, then it was easy to observe a fringe shift. The experimental negative result leads to that $k \leq 0.1$.

Waddoups, Edwards, and Merrill (1965) [82] completed an experiment with a moving mica window by use of an interferometer shown in Fig. 20. The beam from the gas-laser light source L is split by the glass plate G. Half of the lower beam II passes through the mica window M which is mounted on the rotating table. The other half of the beam passes through the stationary mica window C. The beams are recombined by the second glass plate, and the resulting interference pattern is observed and photographed at the screen S. No fringe shift was found. The estimated sensitivity of this method is 1/20 fringes, so that the negative result leads to that $k \leq 0.14$.

Zahejsky and Kolesnikov (1966) [83] test any possible effect of a moving glass window on the velocity of light by means of the arrangement shown in Fig. 21.

A xenon discharge tube 4 produced short flashes of white light which were collimated and split into two parallel light beams in the interferometer. One of these beams passed through the thin glass window 1 mounted on the rotating arm, while the other passed through a stationary window 2. To ensure that the light beam passed perpendicularly through the rotating window, a mirror 3 was placed at the other end of the rotating arm to reflect the triggering light beam 5 the photo-cell 6. Electric impulses from the photo-cell were transmitted to the amplifier and stroboscope A, which provided impulses for the discharged tube 4.

The sensitivity of apparatus was such that a fringe shift of 0.1 fringes would have been observed, but no fringe shift was observed. The experiment was made in air and turbulence of the air did not influence the fringe pattern. When streams of warm air were produced in the laboratory, the fringes became unstable and shifted irregularly. For this reason, Zahejsky and V. Kolesnikov claimed that it is possible that the shift in Kantor's experiment [77] and the disturbance of the fringe pattern in air in Babcock–Bergman's experiment [78] were caused by the temperature gradient in the air [1].

(IV) *Moving Sources of γ Radiation*

We have introduced the laboratory tests and astronomical evidences of the independence of the light velocity with respect to the motion of sources. In the laboratory tests all of the moving source have a low velocity. In the astronomical evidences although the extragalaxies would have higher velocity up to $0.1c$, however, the point at issue is the explanation of the red-shift given by Doppler effect. In addition, the astronomical evidences were questioned by some authors on the assumption of the extinction theorem. For instance, J.G. Fox (1962) [85] gave his argument as follows. The extinction theorem of dispersion theory shows that an incident light wave is extinguished at the surface of a dielectric. This may mean that information about the velocity of light from a moving source would be lost if the light passed through intervening transparent, stationary material before it was measured. All past laboratory measurements to verify the constancy of the velocity of light from moving light sources and mirrors and from extraterrestrial sources were made only after the light had passed through stationary material. Double stars, especially close binary pairs, are surrounded by a common envelope of gas which may contain enough matter to extinguish the direct light from the stars. Thus, de Sitter's proof of the constancy of

---

[1]Kantor (1972) [84] argued that his measurements were unambiguous, and that the subsequently published negative results were the consequence of a failure to properly adjust and/or maintain adjustment of the interferometer at its best condition for the detection of a fringe shift.

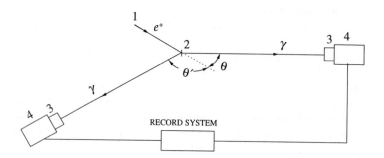

Fig. 22. Sadeh's experimental arrangement. (1) Cu⁶⁴ source; (2) Perspex; (3) NaI(TI) crystals; and (4) 56-AVP photomultiplier. $\theta = 20°$. $\theta' = 135°$.

the velocity of light may not be conclusive.

We now here introduce another kind of experiments where moving sources of γ radiation were used. The γ ray sources (e.g. π^0 mesons) move at high velocities comparable to c. On the other hand, γ rays have higher frequencies than visible light, and hence their extinction distances are longer. For instance, the extinction distances for γ ray of 4 MeV and 6 GeV are about 320 cm and 5 km of air, respectively [89], while the extinction distances for visible light is about 10^{-2} cm. This shows that the extinction theorem does not apply to γ rays. Therefore, a measurement of the γ-ray velocity would provide a strong test for the constancy of the velocity of light.

The first experiment for measuring the velocity of γ rays from a moving source is that by Luckey and Weil (see Ref. [85]), performed for another purpose. In this experiment, 310-MeV electrons were directed at a thin target, and 170-MeV γ rays were detected in coincidence with the degraded electrons of 140-MeV energy. The γ-ray velocity measured was 2.97×10^{10} cm/sec. (It must be emphasized that *this specific value of the γ-ray velocity has no absolute meaning because the clock synchronization in the experiment was made by means of electromagnetic signals. In fact, this value is nothing but the velocity of the signals taken by the authors. In other words, if the authors took another value for the velocity of the signals, then the same another value of the γ-ray velocity would be obtained by them.*)

It is not completely clear to know whether the velocity of the γ-ray source was that of the electrons. J.G. Fox [85] argued that the velocity of the source should be taken to be that of the center of mass of the radiating electron–nucleus system, e.g., free electrons cannot radiate; the high γ-ray energy and hence the large energy loss of the electrons in this experiment indicates a strongly coupled electron–nucleus system. The velocity of the center of mass might be less than the fractional error in the experiment. Then this experiment is inconclusive.

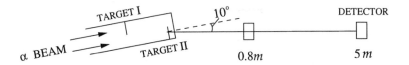

Fig. 23. Diagram for Alväger–Nilsson–Kjellman's experiment.

To overcome this doubt, Sadeh (1963) [86] measured the velocity of γ rays from the annihilation in flight of positrons. In the annihilation the center-of-mass system of the positron and electron moves with a velocity close to $c/2$, and two γ rays are emitted. In the case of annihilation in flight, the angle between the two emitted γ rays is smaller than 180° and depends on the energy of the positron. If the velocity of the γ ray adds on to the velocity of the center of mass, then the γ ray traveling with a component of motion in the direction of the positron flight will have a velocity greater than c, while the other γ ray having a component in the opposite direction will have a velocity smaller than c. If it is found that the two γ rays reach the counters at the same time for equal distances between the counters and the point of annihilation, this would prove that even for a moving source the two γ rays travel with the same velocity.

Sadeh used a 20-curie Cu^{64} positron source produced in the reactor. The source, in the shape of a disc 0.1 mm thick and 2 cm in diameter, was placed at one end of a lead collimator, and at the other end was placed a 1-mm thick layer of Perspex in which the annihilation took place, as seen in Fig. 22. Two detectors, the (3) and (4) in Fig. 22, were placed 60 cm from the Perspex, at the angles of 20° and 135° to the motion of the positrons. A time-to-amplitude converter converted the time differences to pulse heights. The multichannel analyzer was time-calibrated by measuring the annihilation at rest taking different distances between the source and the crystals. When the calibrated analyzer is used to measure the time difference, if the velocity of the γ rays from annihilation in flight were constant, the peak would occur in the same place as the peak occurring in the calibration. This experiment shows that the velocity of the γ rays is a constant ($\pm 10\%$) independent of the velocity of the source.

Alväger, Nilsson, and Kjellman (1963, 1964) [87] compared the velocity of γ rays from the moving sources with that from the source at rest by means of a time-of–flight technique. The moving γ-ray source was C^{12*} that was excited in the reaction $C^{12}(\alpha, \alpha')C^{12*}$ and the γ-ray source at rest was O^{16*} from the reaction $O^{16}(\alpha, \alpha')O^{16*}$. Half the 14-MeV α-beam intensity was used for the C^{12} target and half for the O^{16} target. As the half-life of the 4.43-MeV level in C^{12} is 6.5×10^{-14} sec, the γ-ray is emitted before the nucleus has been stopped. The half-life of the 6.13-MeV level in O^{16}, on the other hand, is 1.2×10^{-11} sec, so that the nucleus is at rest when the γ-ray

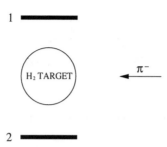

Fig. 24. Schematic diagram of the Fillippas–Fox experiment arrangement.

is emitted. Doppler shift measurements with a sodium iodide crystal showed that the mean forward velocity of the recoiling C^{12} nucleus was $(1.8 \pm 0.2)10^{-2}c$. In the case of O^{16} the Doppler shift measurements confirmed that the nucleus was stopped before emission of the 6.13-MeV γ-ray.

The arrangement is shown in Fig. 23. A large detector was placed at a mean distance of 5 m from the target arrangement and at an angle of 10° with respect to the α-beam. A small detector, placed at 0.8 m from the target arrangement, was used in parallel with the large detector to give reference peaks in the time spectra. The distance between the target I and target II is 30cm. In experiment, targets C^{12} and O^{16} were first placed at the positions I and II, respectively, and then they were interchanged. Half the beam intensity was used for the C^{12} target and half for the O^{16} target. The effective mean forward velocities of "the moving source" (C^{12*}) was $v = (1.75 \pm 0.1) \times 10^{-2}c$ and the effective mean forward velocities of "the source at rest" (O^{16*}) was $v_0 = 0.3 \times 10^{-2}c$. The difference between these velocities was $v - v_0 = (1.5 \pm 0.1) \times 10^{-2}c$, while the light path difference was 4.2 m, the distance between the two positions of the detectors. If the source velocity were to be added to the velocity of light, the distance $(\Delta p)_{CO}$ between the γ-peaks obtained from the detector when the target position I contains C^{12} and the target position II contains O^{16} would be larger by $1.5 \times 10^{-2} \times 14 = 0.21$ ns than the distance $(\Delta p)_{OO}$ obtained with identical targets (a gamma quantum travels 4.2 m in 14 ns). The reverse target combination, on the other hand, would give a distance $(\Delta p)_{OC}$ that would be 0.21 ns smaller than $(\Delta p)_{OO}$, since in this case the C^{12} gammas would reduce the lead of the target I gammas from O^{16}. It is thus seen that the quantity $\Delta\Delta = (\Delta p)_{CO} - (\Delta p)_{OC}$ would be 0.42 ns larger when the detector is located at 5 m from the target II than it would be at the 0.8 m location. The result obtained from the measurements, being the difference between the mean $\Delta\Delta p$ at 5 m and that at 0.8 m, is 0.03±0.04 ns. This means that $k = 0.1 \pm 0.1$.

Fillippas and Fox (1964) [88] measured the relative speed of the two γ rays emitted

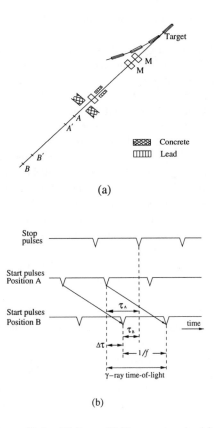

(a)

(b)

Fig. 25. The Alvager–Farley–Kjellman–Wallin experiment. (a) Arrangement; (b) Measurements of time.

forward and backward by a π^0 meson decaying in flight. Their arrangement is shown in Fig. 24. The experimental principle is similar to that in Sadeh's experiment. Neutral pions were produced through the reaction $\pi^- + p \to \pi^0 + n$ by stopping a beam of negative pions from the Carnegie Tech synchrocyclotron in the liquid hydrogen.

By comparing the observed aberration and Doppler shift with the predictions given by Einstein's theory, these neutral pions were known to have a velocity v of $0.20c$. Two γ-ray counters 1 and 2 were located on opposite sides of the H_2 target to measure the difference of the arrival times δt of the photon pairs from a π^0 decay for different detector distances. If the velocity of the two photons were $c \pm kv$, then for a detector target distance d we would have $\Delta t = \pm 2kvd/c^2$ to good approximation. However this time difference was not observed. The experiment with accuracy of 8%

proves that the velocity of γ rays is independent of the motion of source π^0. This gives an upper limit: $k \leq 0.4$.

It is reported by Alväger, Farley, Kjellman, and Wallin (1964, 1966) [89] that they measured the velocity of γ rays from the decay of π^0 mesons of energy > 6 GeV.

The arrangement is shown in Fig. 25 (a). π^0's were produced at the CERN Proton Synchrotron (PS) in an internal Be target bombarded by 19.2 GeV/c protons. The γ rays were observed at an angle of about 6° relative to the circulating protons. In order to achieve a high accuracy in the measurement of the γ ray velocity, the time–of–flight measurement was based on the radio frequency structure of the PS internal beam. The beam is locked in phase to the accelerating frequency so that it consists at the end of the acceleration process of narrow bunches of a few nsec half-width and separation $1/f$ ($\sim$105 nsec), where f is the radio frequency. During the target bombardment ($\sim$100 msec) the beam is slowly consumed by the target but the bunch structure can be conserved almost to the end of the irradiation if the radio frequency is maintained. Due to the very short lifetime of π^0 mesons ($\sim10^{-16}$ sec), the resulting γ-ray beam has the same radio frequency structure.

The principle for the time measurement is shown in Fig. 25 (b). As the counter assembly is moved back along the beam line, cable length being held constant, the start pulses will arrive successively later with respect to the stop pulses. When the detector has been moved a distance $L \approx c/f$ [$A \to B$ in Fig. 25 (a)], the start pulses will again show the same time relative to the stop pulses provided $c' = c$ (where c' represents the velocity of γ rays from moving π^0), and identical distributions will be obtained on the time analyzer in the two positions A and B. A non-zero value of $\Delta = \tau_A - \tau_B$ in Fig. 25 (b) will show that $c' \neq c$ (It is necessary, however, to stress that the anisotropy of the one-way velocity of light does not certainly mean a non-zero value of $\Delta = \tau_A - \tau_B$, as shown below.) The corresponding γ-ray flight time between A and B will then be

$$c' = L \left(\Delta\tau + \frac{1}{\nu} \right)^{-1}. \tag{8.3.18}$$

This experiment gave

$$\bar{\Delta} = 0.000 \pm 0.013 \text{ nsec.} \tag{8.3.19a}$$

By use of the experimental conditions, $L = 31.4503 \pm 0.0005$ m and $f = 9.53220 \pm 0.00005$ MHz, the velocity of the γ rays of energy ≥ 6 GeV is found to be

$$c' = (2.9979 \pm 0.0004) \times 10^{10} \text{cm/sec,} \tag{8.3.19b}$$

which is in agreement with the accepted value of the velocity of light from stationary sources.

Now we would like to prove that the velocity c' given by Eq. (8.3.19b) is not the one-way, but just the two-way, velocity of light.

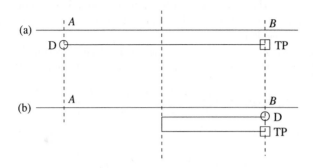

Fig. 26. Schematic diagram for explaining the Alvager–Farley–Kjellman–Wallin experiment.

The principle of this experiment can be shown in Fig. 26. A detector D is first located at the position A [see Fig. 26 (a)] and then moved to the position B [see Fig. 26 (b)]. An time analyzer TP is assumed to be fixed at the position B [see Fig. 26 (a,b)]. For simplicity we assume that the length of the electric cable between D and TP is equal to the distance L between A and B. Assume that the one-way velocity of light is given by Eq. (8.2.1) where the two-way velocity of light u and c as well as the parameter q are independent of the motion of source. Let t_A denote the time at which a γ-ray pulse from the decay of π^0 mesons was received by D and also at the same time an electric pulse was produced. The propagation time of the electric pulse from D to TP is given by

$$t_{AB} = \frac{L}{u} = \frac{L}{\bar{u}} - \frac{L}{c}q, \qquad (8.3.20)$$

where u is the one-way velocity of the electric pulse in the cable, which is given by Eq. (8.2.1), and $\bar{u}$ is the two-way speed of light in the cable. The arrival time of this pulse at TP is then

$$t'_A = t_A + t_{AB} = t_A + \frac{L}{\bar{u}} - \frac{L}{c}q. \qquad (8.3.21)$$

On the other hand, when the detector D is moved to the position B, the time, at which the γ ray arrives at D through the distance L in vacuum, is

$$t_B = t_A + \frac{L}{c'} = t_A + \frac{L}{c} - \frac{L}{c}q \qquad (8.3.22)$$

where c' is the one-way speed of γ ray in vacuum, which is given by Eq. (8.2.1) with $u = c'$ and $\bar{u} = c$. (The γ ray received by D at B is not the same γ ray received by D when it was located at A so that the time t_B given by Eq. (8.3.22) differs from an actual time by N/f where N is an integer number and f is the repeat frequency of the γ-ray pulses as seen in Fig. 25 (b). This difference, however, would not change the following final result.)

Furthermore this electric pulse produced by D at the position B at the time t_B will propagate along the cable of length L and reach the analyzer TP at the time

$$t'_B = t_B + \frac{1 - (\bar{u}/c)q}{\bar{u}}\frac{L}{2} + \frac{1 + (\bar{u}/c)q}{\bar{u}}\frac{L}{2} = t_B + \frac{L}{\bar{u}},$$

or, by use of Eq. (8.3.22),

$$t'_B = t_A + \frac{L}{\bar{u}} - \frac{L}{c}q + \frac{L}{c} = t'_A + \frac{L}{c}, \tag{8.3.23}$$

or

$$t'_B - t'_A = \frac{L}{c}. \tag{8.3.24}$$

Thus, by definition, the velocity of the γ ray is

$$c' = \frac{L}{t'_B - t'_A} = c. \tag{8.3.25}$$

Note that here c is defined in Eq. (8.2.1), which is the two-way velocity of light in vacuum. Then we conclude that this experiment determined only a value of the two-way (but not the one-way) velocity of the γ rays and the electric signals.

To determine a value of the parameter k in Eq. (8.3.1) by use of the experimental value given in Eq. (8.3.19), it is needed to have a value of the velocity of the π^0 meson. In this experiment, the value of this velocity was not directly observed, but was derived from Einstein's theory of special relativity. In view of the theory of relativity, in order to produce the γ ray of 6 GeV, the velocity of π^0 meson must be or at least be equal to $0.99975c$.

We have introduced above the experiments of moving γ-ray sources to test that the velocity of light is independent of the motion of the sources. In short, we want to clarify their physical meaning as follows.

In Sadeh's experiment [86] like the Fillippas–Fox experiment, the light paths (i.e., the γ-ray path and the detection apparatus circuit) formed a closed path. The detection apparatus was calibrated by means of a light source at rest. In other words, the difference between the two arrival times of two γ rays from a moving source to two separate detectors was measured by comparison with that in the case of a γ-ray source at rest. Thus, this experiment like other closed-path experiments is a test of the constancy only of the two-way velocity of light but not of the one-way velocity of light.

The Alväger–Nilsson–Kjellman experiment [87] involved the measurements of the time–of–flight for two γ rays arriving at the same detector (the small detector or the large detector). Thus, these measurements do not depend on any definition of simultaneity. This means that this experiment proved that the one-way velocity of light is independent of the motion of light source (Note that this does not mean the isotropy of the one-way velocity of light).

The Alväger–Farley–Kjellman–Wallin experiment gave the velocity of light from a moving source. By means of the constancy of the two-way velocity of light, i.e., equation (8.2.1), we have shown that the velocity of light measured in this experiment is the two-way velocity but not the one-way velocity. In other words, this experiment shows that the two-way velocity of light is independent of the motion of source.

8.4. Summary of the Tests of the Constancy of the Velocity of Light

In the previous sections we have introduced and analyzed the experiments for testing the constancy of the velocity of light. We have shown that the two-way velocity of light has been proved to be constant by means of the experiments and at the same time the one-way velocity of light has been proved to be independent of the motion of light source. It is needed to point out that the "independence" does not mean that either the one-way velocity c_r of light is equal to the two-way velocity c of light or the one-way velocity of light is isotropic.

We now make a summary of the results of these experiments in tables 8.1 and 8.2. It must be indicated that the velocities of light sources in table 8.1 are found by means of Einstein's theory of special relativity. There is no problem for small velocities compared to c, while it is not suitable for high velocities. However the values of the velocities of light sources given in table 8.2 are not essential to the conclusion, but just show the fact that the velocities of light sources could be compared with c.

Table 8.1. *Summary of Experiments on the Constancy of the Light Velocity*

Year	Experiment	Method used	Upper limit on "ether drift" (km/s)
1881	Michelson [6]	Optical technique	21.1
1887	Michelson and Morley [7]	Optical technique	4.7
1926	Kennedy [46]	Optical technique	5.1
1927	Illingworth [47]	Optical technique	2.3
1930	Joos [48]	Optical technique	1.5
1932	Kennedy and Thorndike [56]	Optical technique	null effect
1955	Essen [52]	Micro-waves	2.5
1964	Jaseja, Javan, Murray and Townes [53]	Two He-Ne masers	0.95
1972	Silvertooth [54]	Laser interferometer (frequency-double)	null effect
1973	Trimmer, Baierlein, Faller, and Hill [55]	Interferometer (with a glass piece)	null effect
1958	Cedarholm, Bland, Havens, and Townes [59]	Two masers (bit frequency)	0.03
1963	Champeney, Isaak, and Khan [60]	Rotated rotor (Mössbauer effect)	0.0016
1972	Cialdea [62]	Two lasers (bit frequency)	0.0009
1979	Brillet and Hall [40]	Laser resonator	0.03

Table 8.2. *Summary of Experiments on the Independence*
of the Light Speed of the Source Velocity

Year	Experiment	Moving source	Source speed	Limit on k
1913	de Sitter [64]	Double stars		$k \leq 0.002$
1914	Zurhellen [65]	Double stars		$k \leq 10^{-6}$
1960	Heckmann [68]	Stars & Nebulae	$\sim 10^{-1}c$	null effect
1910	Tolman [71]	Limbs of the Sun	$\sim 10^{-5}c$	null effect
1919	Majorana [69]	Mercury plasma	$\sim 10^{-7}c$	null effect
1924	Tomaschek [70]	Stars	$\sim 10^{-4}c$	null effect
1960	Bonch–Bruewich [73]	Limbs of the Sun	$\sim 10^{-5}c$	$k = 0.02 \pm 0.07$
1965	Beckmann and Mandies [76]	Rotating Reflection mirror	$1.52 \times 10^{-7}c$	$k \leq 0.05$
1962	Kantor [77]	Rotating glass-plate	$1.56 \times 10^{-7}c$	$k = 0.67$ (?)
1963	James and Sternberg [80]	Rotating glass-plate		$k \leq 0.025$
1963	Rotz [81]	Rotating glass-plate	$2 \times 10^{-7}c$	$k \leq 0.1$
1964	Babcock and Bergman [78]	Rotating glass-plate	$1.25 \times 10^{-7}c$	$k \leq 0.01$
1964	Beckmann and Mandies [79]	Rotating glass-plate		$k \leq 0.1$
1965	Waddoups, Edwards and Merrill [82]	Rotating mica-window		$k \leq 0.14$
1963	Sadeh [86]	e^+e^-	$0.5c$	$k \leq 0.3$
1963	Alväger, Nilsson and Kjellman [87]	C^{12}	$0.03c$	$k \leq 0.1$
1964	Fillippas and Fox [88]	π^0	$0.2c$	$k \leq 0.4$
1964	Alväger, Farley, Kjellman, and Wallin [89]	π^0	$0.99975c$	$k \leq 10^{-4}$

8.5. Tests of the Principle of Special Relativity

The principle of relativity implies that all inertial frames of reference are physically equivalent, or that there is no preferred inertial frame. Therefore, all the null results of experiments of searching for a preferred frame or ether drift, such as the experiments of Michelson–Morley type as shown in Sec. 8.2, are also evidences of the relativity principle.

The electromagnetism experiments of Trouton–Noble type [93–96] are usually regarded as the verifications of the relativity principle. According to the ether theory, when a plate condenser moves through the ether along the direction parallel to its plate planes, the charges on the plates will form two currents with the same magnitude but the opposite directions and a magnetic field parallel to the plates. Thus, there will be a torque force which would rotate the condenser. Trouton and Noble (1902, 1903) [93] carried out this type of experiments to observe the motion of the earth through the ether by use of plate condensers at rest in laboratory frame. No rotation of the plate condensers was observed. Later, Tomaschek (1926) [94] and Chase (1926) [95] repeated the experiments of Trouton–Noble type, and their null results proved the principle of relativity.

Another class of experiments is the tests of the isotropy of space [97–101]. Local Lorentz invariance requires that the local, nongravitational physics of a bound system of particles be independent of its velocity and orientation relative to any preferred frame. If the invariance were violated and such a frame existed, the energy levels of a bound system such as a nucleus could be shifted in a way that correlates the motion of the bound particles in each state with the preferred direction. Such a shift would lead to an orientation-dependent binding energy, i.e., an anisotropy of inertial mass. For example, Chupp *et al.* (1989) [101] test Lorentz invariance by searching for a time-dependent quadrupole splitting of Zeeman levels in ^{21}Ne. A component at twice the Earth's sidereal frequency would suggest a preferred direction which affects the local physics of the nucleus. Their experimental data produce an upper limit (1σ confidence level) of 2×10^{-21} eV on the Lorentz-invariance-violating contribution to the binding energy (see the References in Ref. [101] for the previous experiments). In addition, some experiments for testing Lorentz invariance at a small dimension have been reported in, e.g., Refs. [102–105].

CHAPTER 9

THE TESTS OF TIME DILATION

9.1. The Problem of Clock Paradox

The concept of time dilation was first introduced by Larmor in 1900 in «Aether and Matter» in where the Fitzgerald–Lorentz contraction is commented upon. According to Larmor, a clock moving with a velocity v through ether will go slower than a stationary clock by a factor of $\sqrt{1 - v^2/c^2}$. It is obvious that this dilation has an absolute meaning.

In contrast with Larmor's concept, the time dilation of special relativity seems to be a relative concept so that the so-called *clock paradox* or *twin paradox* has generated one of the longest standing controversies [106–113].

Let two clocks A and B be located at two separate positions in the laboratory frame F, while other two clocks A' and B' be at rest at two separate positions in a traveling frame F' which is moving with a velocity $\mathbf{v}$ relative to F, as shown in Fig. 27(a). We assume that the readings of the clocks A' and A are, respectively, $t_{A'}$ and t_A when A' leaves A [see Fig. 27 (a)], and that the readings of the clocks A' and B are, respectively, $t'_{A'}$ and t_B when the clock A with the reading t'_A meets the clock B' with the reading $t_{B'}$. The dilation formula (2.9.2) gives the relations among the readings:

$$\Delta \tau_{A'} = \Delta t_{BA} \sqrt{1 - \frac{v^2}{c^2}}, \tag{9.1.1a}$$

or

$$\Delta \tau_A = \Delta t_{B'A'} \sqrt{1 - \frac{v^2}{c^2}}, \tag{9.1.1b}$$

where

$$\Delta \tau_{A'} = t'_{A'} - t_{A'}, \qquad \Delta t_{BA} = t_B - t_A,$$

$$\Delta \tau_A = t'_A - t_A, \qquad \Delta t_{B'A'} = t_{B'} - t_{A'}. \tag{9.1.1c}$$

We know that $\Delta \tau_{A'}$ and $\Delta \tau_A$ are proper time intervals, while $\Delta t_{B'A'}$ and Δt_{BA} are coordinate time intervals.

The time dilation effects above show that the *one* clock A' is running slower than the difference between the readings of the *two* synchronized clocks A and B, or that the *one* clock A is running slower than the difference between the readings of the *two* synchronized clocks A' and B'. In short we can say that *a moving clock is running slower than two stationary and separate clocks*. In other words, the *proper time interval* $\Delta \tau_{A'}$ (or $\Delta \tau_A$) is smaller than the *coordinate time interval* Δt_{BA} (or $\Delta t_{B'A'}$). This statement is based on Einstein's synchronization between the two

175

separate clocks. We see that *there is no symmetry (paradox) in this statement,* and that this statement can, in principle, be tested by experiments.

A paradox occurs if you identify a proper time interval with its corresponding coordinate time interval, i.e.,

$$\Delta t_{BA} = \Delta \tau_A, \qquad \Delta t_{B'A'} = \Delta \tau_{A'}. \tag{9.1.2}$$

It will be clear that this identification is not correct. By definition, a coordinate time interval depends on the definition of simultaneity while a proper time interval does not. For instance, if the clocks A and B are synchronized by Edwards' definition of simultaneity, then the coordinate time interval between the two clocks $\Delta \tilde{t}_{BA}$ is different from Δt_{BA}. The relation between the two coordinate time intervals has been given by Eq. (1.3.23a), i.e.,

$$\Delta t_{BA} = \Delta \tilde{t}_{BA} + q\frac{l}{c}, \tag{9.1.3}$$

where l is the distance between A and B, q is a non-zero constant. In contrast with the coordinate time, the proper time would not be changed by the change of the definition of simultaneity, i.e.,

$$\Delta \tilde{\tau}_A = \Delta \tau_A. \tag{9.1.4}$$

On the other hand, following the identification (9.1.2) we should have

$$\Delta \tilde{t}_{BA} = \Delta \tilde{\tau}_A. \tag{9.1.5}$$

By using Eqs. (9.1.3)–(9.1.5), equation (9.1.2) leads to the error:

$$\Delta \tau_A = \Delta \tau_A + q\frac{l}{c}. \tag{9.1.6}$$

Therefore we see that the so-called clock paradox above follows from nothing but mixing the proper time with the coordinate time.

The clock paradox (the twin paradox) among two proper times can be stated as follows: An ideal clock which moves in the closed curve with respect to a clock at rest in the laboratory frame will indicate an elapsed proper time smaller than the one given by the stationary clock. The paradox arises because the difference in time between the two clocks is a velocity effect, and since special relativity is a reciprocal theory, the traveling clock can argue that it is the stationary clock that goes slower.

The proper time interval of the clock moving in the closed path has been given by Eq. (2.3.21):

$$\Delta \tau' = \oint dt \sqrt{1 - \frac{\mathbf{u}^2}{c^2}}, \tag{9.1.7}$$

where $\Delta \tau'$ is the reading difference of the traveling clock with a velocity $\mathbf{u}$ along the closed curve, and dt is a coordinate time interval as seen in the laboratory frame.

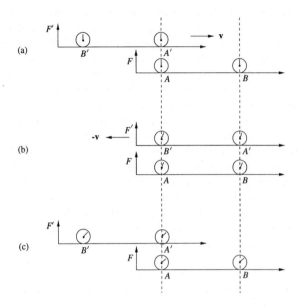

Fig. 27. Clocks moving with a velocity v relative to each other. (a) At a given time the clock A' leaves the clock A with a velocity $\mathbf{v}$ for another clock B; (b) At some later time the clock A' reaches and stops at the clock B, and at the same time moves backward with an inverse velocity $-\mathbf{v}$; (c) The clock A' then meets the clock A again.

For simplicity, we assume that the magnitude of $\mathbf{u}$ is constant in time, i.e., $u = |\mathbf{u}|$ =constant, so that the equation above reduces to

$$\Delta\tau' = \sqrt{1 - \frac{\mathbf{u}^2}{c^2}} \oint dt = \Delta\tau \sqrt{1 - \frac{\mathbf{u}^2}{c^2}}, \qquad (9.1.8)$$

where $\oint dt = \Delta\tau$ is the proper time interval of a stationary clock in the laboratory. Then we get a conclusion that it is the traveling clock that goes slower.

The calculation above is done in the laboratory frame. Of course, if a similar formula like Eq. (9.1.7) could be used by an observer moving together with the traveling clock, then the observer would obtain an inverse conclusion that it is the stationary clock that goes slower.

The key point for resolving the clock paradox above is to use correctly special relativity. It must be noticed that the dilation formula (9.1.7) is valid only in the laboratory frame which is assumed to be an inertial frame, but not in the frame of the accelerating clock. Thus, the above calculation made in the laboratory frame is correct while the one made in the frame of the traveling clock is wrong.

However, a real problem is how to apply correctly the special theory of relativity for the calculation in the frame of the traveling clock. Einstein stated that the resolution of the paradox lay outside the realm of special relativity [106]. One can resolve the paradox by using the general theory of relativity [107]. However, the principle of equivalence can be applied for an accelerating motion but not for a uniform motion.

Let us consider a particular case of rectilinear motion in Fig. 27: the clock A' comes back after it meets the clock B as shown in Fig. 27 (c). The motion of the traveling clock A' consists of two kinds of stages: the accelerating (decelerating) process (this process has been omitted in Fig 27) and the uniform motion. The principle of equivalence could be used to calculate the dilation effect of the accelerating process, but cannot be used to calculate that of the stage of the motion. To calculate the time dilation for the uniform motion, we must employ the equations of special relativity. Before doing this, we must first consider the difference between the laboratory frame and the rest frame of the traveling clock in the stage of the motion: The laboratory frame is one inertial frame, which has no any change although the traveling clock is moving forth and back. In contrast with the case of the laboratory frame, the rest frame of the traveling clock during its motion forward (when it has a velocity $+\mathbf{v}$) is not the same inertial frame with that during its motion back (when it has a velocity $-\mathbf{v}$). This means that we can simply use twice the time dilation formula (9.1.1a) in the laboratory frame as follows:

$$\Delta\tau_{A'} = \Delta t_{BA}\sqrt{1 - \frac{v^2}{c^2}} \qquad (9.1.9a)$$

when the traveling clock from A reaches B [see Fig. 27 (a)], and

$$\Delta\tau_{A'} = \Delta t_{AB}\sqrt{1 - \frac{v^2}{c^2}} \qquad (9.1.9b)$$

when it comes back to A. Here $\Delta t_{AB} = t'_A - t_B$ where t'_A is the reading of the stationary clock A when the clock A' comes back. So we have $\Delta t_{BA} + \Delta t_{AB} = t'_A - t_A = 2\Delta\tau_A$ which is the proper time interval of the stationary clock A. In this way we get the correct result: the traveling clock A' is running slower than the clock A, i.e.,

$$2\Delta\tau_{A'} = 2\Delta\tau_A\sqrt{1 - \frac{v^2}{c^2}}. \qquad (9.1.9c)$$

On the other hand, in the rest frames of the traveling clock A' we can write down the following similar formulas:

$$\Delta\tau_A = \Delta t_{B'A'}\sqrt{1 - \frac{v^2}{c^2}} \qquad (9.1.10a)$$

and

$$\Delta\tau_A = \Delta t_{A'B'}\sqrt{1 - \frac{v^2}{c^2}}, \qquad (9.1.10b)$$

where

$$\Delta t_{B'A'} = t_{B'} - t_{A'}, \qquad \Delta t_{A'B'} = t'_{A'} - t'_{B'}. \qquad (9.1.10c)$$

We see that if $t_{B'}$ were regarded as the same with $t'_{B'}$ the paradox would occur. However, $t_{B'}$ and $t'_{B'}$ in Eq. (9.1.10c) are the coordinate times in two different frames, one having the velocity $+\mathbf{v}$ relative to the laboratory while the other having the velocity $-\mathbf{v}$. Thus they are not equal to each other because of the simultaneity factor vx/c^2 in the Lorentz transformation. What is a relationship between them? A correct answer to the problem has been given by some authors [112–115]: If we account for the effects of acceleration by keeping track of the instantaneous inertial frame of the accelerating clock, calculations made by A' and A agree. The detailed calculation concerning the resolution of the twin paradox can be found in Ref. [113].

Although the clock paradox has been resolved by some physicists by using Lorentz transformations carefully, it is surprising that there are still some authors who question the reality of the time dilation effect. An example is the paradox of the transverse Doppler effect [116]:

> The experiments of transverse Doppler shift measure the frequency shift of light waves emitted by a light source in flight, when compared with the light frequency emitted by the same light source at rest. The observed frequency shift of the light waves emitted by a light source in straight-line flight is *interpreted* by Einstein as due to a rate slow down of clocks in flight. The experimental result is claimed to be evidence confirming the time-dilation according to Einstein's interpretation.
>
> Some physicists do not consider Einstein's the only explanation for that transverse Doppler shift. By the principle of relativity, an equivalent view is that the observer is in flight and the light source is at rest. One must the same frequency shift, irrespective of viewpoint. The same experimental result seems to them to mean that Einstein's interpretation entails alternately that clocks in flight speed up rather than slow down. Consequently, the interpretation of time-dilation in special relativity for them leads to contradiction.

This paradox follows from the error that the *"transverse"* is regarded as an absolute concept. In fact, a transverse Doppler effect as seen from the light source is no longer a transverse Doppler effect as seen from the observer due to aberration equation (2.10.8). In addition, we have known from the Doppler effect formula (2.10.7c) that if a light wave does not propagates in the transverse direction, i.e., $\theta \neq 90°$, contributions to Doppler shift will come from two sources which are the time-dilation [i.e., the denominator in a fraction of Eq. (2.10.7c)] and the classical Doppler effect [i.e., the numerator in a fraction of Eq. (2.10.7c)]. That a transverse shift as seen in a frame is no longer a transverse shift as seen from the other frame implies that a shift caused by time-dilation as seen in a frame is produced not only by time-dilation but also by directional effect as seen from the other frame. It is easy to prove this statement by making a calculation from the aberration and Doppler effect (see footnote on page 182).

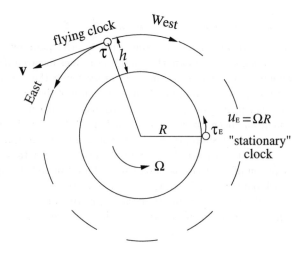

Fig. 28. Coordinate speeds in a nonrotating frame for a clock at rest on the earth's surface at the equator and an airborne clock circumnavigating the earth in the equatorial plane.

The prediction for the time-dilation effect of special relativity has been tested by many experiments. These experiments involve the flying atomic clock experiment, Doppler effect experiment, as well as measurements for the lifetime of mesons in flight, which will be introduced in the present chapter.

9.2. Around-the-World Atomic Clocks

During October 1971, Hafele and Keating [117] performed a flying atomic clock experiment in which four cesium beam clocks were flown on regularly scheduled commercial jet flights around the world twice, once eastward and once westward. They recorded directionally dependent time differences which are in good agreement with predictions of conventional relativity theory. This experiment is regarded as a resolution of the twin paradox with macroscopic clocks.

Figure 28 shows a schematic diagram of the flying clock experiment. It will be a sufficiently accurate approximation to assume that the earth is a spherically symmetric source of a gravitational field, and that the earth rotates on its axis once a day in an otherwise flat space. Let us consider a nonrotating frame F in the flat space. The velocity of a rest clock C_E on the surface of the earth is $u_E = \Omega R$, as seen in the frame F, where Ω is the angular velocity of the earth rotation and R is the radius of the earth. Assume the flying clock C_F follows a circular path around the earth at an altitude h above the surface and with a constant ground speed v, which is positive for eastward and negative for westward motion (see Fig. 28). The velocity of the flying clock with respect to the frame F can be found by use of Einstein's addition law of velocities:

$$u_F = \frac{(R+h)\Omega + v}{1 + (R+h)\Omega v/c^2} \simeq (R+h)\Omega + v, \qquad (9.2.1)$$

where we have neglected higher-order terms.

We now calculate the readings of the clock C_F and C_E by using the gravitational red-shift and the time-dilation of special relativity.

Owing to the motion of the clocks with respect to the nonrotating frame F being small, Einstein's time-dilation gives

$$d\tau_E = \sqrt{1 - \frac{\Omega^2 R^2}{c^2}}\,dt \simeq \left(1 - \frac{\Omega^2 R^2}{2c^2}\right)dt \qquad (9.2.2)$$

with only the lowest order term being retained due to $\Omega R \ll c$, and

$$d\tau_F = \sqrt{1 - \frac{u^2}{v^2}}\,dt \simeq \left(1 - \frac{v^2}{2c^2} - \frac{\Omega^2 R^2}{2c^2} - \frac{\Omega R v}{c^2}\right)dt. \qquad (9.2.3)$$

Here $d\tau_E$ and $d\tau_F$ are, respectively, the proper time intervals of C_E and C_F, and dt is a coordinate time interval in the frame F. By eliminating dt from Eqs. (9.2.2) and (9.2.3), we obtain

$$d\tau_F \simeq \left[1 - \frac{1}{2c^2}\left(v^2 + 2\Omega R v\right)\right]d\tau_E. \qquad (9.2.4a)$$

This is the time-dilation or kinematical effect due to the relative velocity between the two clocks C_E and C_F.

On the other hand, the gravitational field would give a similar time-dilation that is the so-called gravitational red-shift: To the lowest order approximation the difference between the two proper time intervals of C_E and C_F is proportional to the difference between corresponding gravitational potentials,

$$d\tau_F - d\tau_E = \left[\frac{GM}{c^2 R} - \frac{GM}{c^2 (R+h)}\right]d\tau_E \simeq \left(\frac{GM}{R^2}\right)\frac{h}{c^2}d\tau_E = \frac{gh}{c^2}d\tau_E, \qquad (9.2.4b)$$

where g is the measured surface value of the acceleration of gravity at the equator, and a higher order than h/R and an effect of the rotation of the earth on the potentials are neglected.

The total contributions produced by the velocity and potential effects are then from Eqs. (9.2.4a,b)

$$d\tau_F \simeq \left[1 + \frac{gh}{c^2} - \frac{1}{2c^2}\left(v^2 + 2\Omega R v\right)\right]d\tau_E. \qquad (9.2.5)$$

These predictions for equatorial circumnavigations will be modified somewhat for the actual flights because commercial around-the world jet flights do not follow equatorial paths, nor do they maintain constant altitude, ground speed, or latitude. So the expression (9.2.5) should be changed to [118]

$$\Delta\tau_F = \int \left[\frac{gh}{c^2} - \frac{1}{2c^2}(v^2 + 2v\Omega R \cos\varphi \cos\theta) \right] d\tau_E. \tag{9.2.6}$$

This expression [117,118] contains a slightly modified directionally dependent term, which for nonequatorial flights becomes proportional to both the eastward component of the ground speed, $v\cos\theta$, and the cosine of the latitude, $\cos\lambda$.

Table 9.1. *Observed results of the Hafele–Keating experiment*

Clock serial	$\Delta\tau$ (nsec)	
No.	Eastward	Westward
120	−57	277
361	−74	284
408	−55	266
447	−51	266
Mean ± S.D.	−59 ± 10	273 ± 7

Table 9.2. *Predicted time differences* (nsec)

	Direction	
Effect	Eastward	Westward
Gravitational	144±14	179±18
Kinematic	−184±18	96±10
Net	−40±23	275±21

Consider a particular case of $h = 0, \theta = 0$, and $\lambda = 0$. In this case, the total effect follows only from the prediction of special relativity, equation (9.2.4a), which shows an asymmetry in the time difference between the flying and ground clocks depending on the direction of the circumnavigation. This asymmetry should be easily understood in view of an observer in the nonrotating frame F: In the case of a flying clock westward having a velocity of $v = 2\Omega R$, the flying and ground clocks would have the same velocity with respect to the frame F and, therfore, have the same rates, i.e. they are dilated by the same amount compared to that of a stationary clock in F. This leads certainly to the result, $\Delta\tau = \tau_F - \tau_E = 0$. It is then obvious that for

$|v| < 2\Omega R$ we have $\Delta\tau > 0$ while for $|v| > 2\Omega R$ we have $\Delta\tau < 0$. In contrast with the claim in Ref. [119], a similar asymmetry would occur in case of rotating disk. In fact, if another γ-ray source S' is mounted on the rotating disk in Fig. 33 in such a way that its radius satisfies the condition $R_{S'} > R_a > R_S$, then the asymmetry will appear: the wavelength of a light ray from the new source S' would be displaced forward to the "red" direction as seen from the absorber, while a light ray from the old source S would be displaced forward to the "blue" direction.

The experimental results obtained by Hafele and Keating are shown in Tab. 9.1 which involves the observed relativistic time differences from application of the correlated rate-change method to the time intercomparison data for the flying ensemble. Table 9.2 gives predicted values. We see that the observations are in good agreement with the predictions.

9.3. Doppler Effect

In 1892, Doppler first pointed out that motion of a light source would produce an effect on the positions of its spectral lines. In 1868 Huggins firstly observed the Doppler shifts of the positions of spectral lines from distant stars. Later, Galizin and Belepolsky (1895) measured the Doppler frequency shifts of light waves reflected from a moving mirror; Stark (1906) observed the Doppler effect in hydrogen canal rays with high velocities. These measurements are in agreement with the Doppler effect predicted by classical theory or by first-order approximation of the relativistic Doppler effect formula.

In Sec. 2.10, we have derived Doppler frequency shift formula (2.10.20). It has been shown in this derivation that the denominator in the fraction of Eq. (2.10.20) has the same form as that in classical physics, which involves the first-order in v/c, while the numerator in the same fraction presents the time-dilation factor which is of the second-order in v/c. In particular, the transverse Doppler shift is caused only by the time-dilation effect. A measurement for the second-order Doppler effect (involving the transverse Doppler effect) could then provide a test for the time-dilation of special relativity.

9.3.1. The Hydrogen Canal Ray Experiment

The first suggestion as to a means by which the time-dilation effect might be observed experimentally came from Einstein and from Ritz about ninety years ago, namely that the newly discovered Doppler effect in canal rays involved velocities of the moving particles high enough to show the expected effect. This kind of experiment has been commonly imagined as performed by observing the canal rays at right angles to their direction of motion. However it would be extremely difficult to be sure that observation was made exactly at right angles to the direction of the rays, and very small deviations from this direction would introduce a first-order effect which

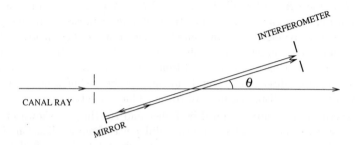

Fig. 29. Ives-Stilwell Schematic. The light was emitted by the canal rays involved
high speed hydrogen atoms, and also atoms at rest. It was viewed in a direction making
a small angle (6° in Otting's work, 7° in other) with the direction of the canal rays.
The light was observed directly and by reflection from a mirror on the line of sight.

will cover over the expected time-dilation effect. This will be clear by analyzing the
Doppler effect formula (2.10.7c) which can be expanded as a power series in v/c.
Omitting terms of order $(v/c)^3$ and higher, we have

$$\omega = \omega_0 + \omega_0 \left(\frac{v}{c} \cos\theta + \frac{v^2}{c^2} \cos^2\theta \right) - \omega_0 \frac{v^2}{2c^2}, \tag{9.3.1}$$

where the second and third terms are called the longitudinal and transverse shifts.
The formula above shows that at 90° the derivative of the longitudinal shift is a
maximum, so that the longitudinal shift for $\theta = 91°$ is of the same order as the
transverse shift, and the finite solid angle of the detector will cause a large broadening
of the lineshape at 90°.

These difficulties may be circumvented by measuring near 0 and 180°, where the
longitudinal shift changes slowly with angle. In this case the longitudinal shift is also
measured, allowing a determination of v/c.

Ives and Stilwell (1938) [120] observed the Doppler shift of light rays emitted at
near 0 and 180° by hydrogen canal rays. The experimental schematic diagram is given
in Fig. 29.

For observations of light waves emitted by a moving particle in the two opposite
directions, from Eq. (2.10.20) we know that Doppler effects are

$$\lambda = \frac{\lambda_0 \left(1 - (v/c) \cos\theta \right)}{\sqrt{1 - v^2/c^2}}, \tag{9.3.2a}$$

$$\lambda_r = \frac{\lambda_0 \left(1 + (v/c) \cos\theta \right)}{\sqrt{1 - v^2/c^2}}, \tag{9.3.2b}$$

where λ_r is the wavelength of the reflected light wave. Define $\lambda_\pm$ as

$$\lambda_+ = \frac{1}{2}(\lambda_r + \lambda) = \frac{\lambda_0}{\sqrt{1 - v^2/c^2}}. \tag{9.3.3}$$

$$\lambda_- = \frac{1}{2}(\lambda_r - \lambda) = \frac{\lambda_0(v/c)\cos\theta}{\sqrt{1 - v^2/c^2}}. \tag{9.3.4}$$

Equation (9.3.3) shows that the relation between λ_+ and λ_0 is the same as the transverse Doppler effect, and also shows that to the second-order approximation we have

$$\lambda_+ - \lambda_0 \simeq \frac{1}{2}\lambda_0\frac{v^2}{c^2}, \tag{9.3.5}$$

which was compared to the experimental results of the Ives–Stilwell experiment. Values of the three kinds of wavelengths λ_0, λ and λ_r can be given by experiment, and therefore the values of the left-hand side of the equation above are regarded as the observed values, while as theoretical predictions the right-hand side can be calculated by use of the following two methods. The first is an optical means, namely that the right-hand side of (9.3.5) is expressed in terms of the observed wavelengths λ_0, λ and λ_r:

$$\frac{1}{2}\lambda_0\frac{v^2}{c^2} \simeq \frac{(\lambda_r - \lambda)^2}{8\lambda_0\cos^2\theta}. \tag{9.3.6}$$

The second method is to calculate the right-hand side of (9.3.5) by use of electrodynamic quantities. Let V denote the voltage between two electrode plates, e be the charge on the hydrogen atom, and M be the mass of the particle observed. The equation of motion for the accelerated particle is given by

$$eV = \frac{1}{2}M_0v^2, \tag{9.3.7}$$

and then we have

$$\frac{1}{2}\lambda_0\frac{v^2}{c^2} = \lambda_0\frac{eV}{M_0c^2}. \tag{9.3.8}$$

Values for the left-hand side of Eq. (9.3.5) as observed values were given by the experiment, while the right-hand side as theoretical prediction is given by (9.3.6) or (9.3.8). Using Eq. (9.3.8) implies the recognition of the electrodynamic formula (9.3.7). Measurements performed by Ives and Stilwell (1938) [120] are in agreement with equations (9.3.5)–(9.3.8). However, their purpose was not to test special relativity because these equations could also be obtained in the contraction ether theory. Ives [121] had emphasized that he did not employ the constancy of the speed of light, but used the ether theory for the predictions. This experiment is, however, regarded as the first test in high accuracy of the second-order Doppler effect produced due to time-dilation.

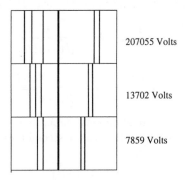

207055 Volts

13702 Volts

7859 Volts

Fig. 30. Spectrograms obtained for several voltages.

In a primary experiment, Ives and Stilwell tested the agreement between (9.3.4) and (9.3.7). The results are shown in Tab. 9.3, where v in the quantity "$\lambda_0 v/c$ Computed" in column 4 was evaluated by Eq. (9.3.7); the values of λ_- in the quantity "$\lambda_-/\cos 7°$" in column 5 were obtained from Eq. (9.3.4) by use of the observed values of λ_r and λ. Table 9.3 shows the agreement between the computed and observed values.

Table 9.3.

Plate	Voltage	Line	$\lambda_0 v/c$ Å Computed	Mean $\Delta\lambda(\text{Å})$ Observed $(\Delta\lambda/\cos 7°)$
169	6788	H_3	10.62	10.35
160	7780	H_2	14.04	14.02
163	9187	H_2	15.30	15.40
170	10574	H_2	16.34	16.49
165	11566	H_3	13.88	14.07
172	13560	H_2	18.50	18.67
172	13560	H_3	15.05	15.14
177	18350	H_2	21.55	21.37

The final results obtained by Ives and Stilwell are shown in Fig. 30 and Tab. 9.4. The figures indicate typical spectrograms obtained for several applied voltages. In each the center, undisplaced line is seen, accompanied at either side by two companions, which, by their separations from the center line by distances in the ratio $\sqrt{2}/\sqrt{3}$, are identified as H_2 and H_3, or double and triple atomic hydrogen. In the table, the values in column 4 were calculated from Eq. (9.3.8); the values in column 5 were obtained from Eq. (9.3.6); in column 6 the quantity $\Delta\lambda$ is defined by the left-hand side of Eq. (9.3.5), i.e., $\Delta\lambda = \lambda_+ - \lambda_0$. The table shows that the experimental values agree with the predictions.

Similar measurements with those in 1938 were carried out by Ives and Stilwell in 1941 [122], and the same results were obtained.

Table 9.4.

Plate	Voltage	Line	$\lambda_0 v^2/2c^2$ computed from voltage	$\lambda_0 v^2/2c^2$ computed from observed $\Delta\lambda$	$\Delta\lambda(\overset{\circ}{A})$ Observed
169	6788	H_3	0.0116	0.0109	0.011
160	7780	H_2	0.0203	0.0202	0.0185
163	9187	H_2	0.0238	0.0243	0.0225
170	10574	H_2	0.0275	0.0280	0.027
165	11566	H_3	0.0198	0.0203	0.0205
172	13560	H_2	0.0352	0.0360	0.0345
172	13560	H_3	0.0233	0.0237	0.0215
177	18350	H_2	0.0478	0.0469	0.047

In 1939, Otting [123] performed a similar experiment in which second-order Doppler shifts of the H_α lines emitted from hydrogen canal rays were observed. Kantor (1971) [124] pointed out that the rest wavelength λ_0 was not measured but was taken by Otting from the work of Curtis in 1914 who gave it as 6562.793 Å, and that if other values of λ_0 in the works of others are used, then the observation of Otting's can hardly be considered to accord with the Einstein–Doppler effect.

Mandelberg and Witten (1962) [125] repeated measurement of the H_α lines emitted from hydrogen canal rays. It is similar that he did not measure the rest wavelength but taken the same value of 6562.793 Å. In this way, his measurements show the agreement between Eqs. (9.3.5) and (9.3.6).

Recently, Hasselkamp, Mondry, and Scharmann (1979) [126] reported a new experiment which is similar to that of Ives and Stilwell. They measured the second-order Doppler shift of the $H\alpha$-line emitted by fast moving hydrogen atoms with velocities of 2.53×10^8 cm/s–9.28×10^8 cm/s. There is, however, one noticeable change: This is the first measurement of the second-order Doppler shift observing light emitted perpendicular to the direction of a linearly moving light source, while the experiments introduced above all used the longitudinal observation.

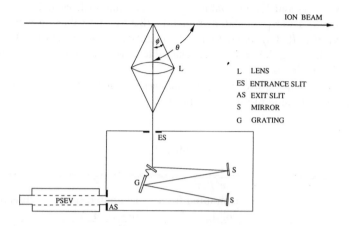

Fig. 31. Schematic diagram of Hasselkamp–Mondry–Scharmann's experiment; $\theta=$ angle of observation, $2\pi=$ opening angle of the optical system.

The experimental set-up is schematically given in Fig. 31. The radiation emitted by a small portion of the hydrogen beam is focused on the entrance slit of a monochromator, is then dispersed by a grating and is detected by a photomultiplier operating in the single photon counting mode. The observed value for the coefficient of the second-order Doppler shift is 0.52 ± 0.03 which compares well with the theoretical value of $1/2$.

9.3.2. Doppler Effect of γ-Rays from Capture Reactions

Olin *et al.* (1973) [127] observed a second-order Doppler shift of the O^{16} capture γ-rays of 10.09 ± 0.41 keV. The method used by them is similar to that by Ives and Stilwell, i.e., that the measurements of the longitudinal and transverse Doppler shifts were made at near 0 and 180°. In this experiment, the sources of γ rays were Ne^{20} in the capture reaction $O^{10} + He^4 \rightarrow Ne^{20}$. The capture reaction produces a well-collimated beam of Ne^{20} recoils with a very sharp velocity distribution. The measurements were made of γ-rays from the resonance for two recoil velocities. When a He^4 target was bombarded by 27.7-MeV O^{16} ions, the recoil velocity was $0.049c$. When an O^{16} target was bombarded by 6.93-MeV He^4 ions, the recoil velocity was $0.012c$. The high recoil velocity produced a transverse Doppler shift of 10 keV.

The relativistic Doppler effect equation (2.10.22) was used for comparison with observed shifts. Using the relation between energy and frequency $E = \hbar\omega$ for a photon, equation (2.10.22) becomes

$$E\left(\theta\right) = \frac{E_0\sqrt{1-\beta^2}}{1-\beta\cos\theta}, \tag{9.3.9}$$

where $\beta = v/c$. As mentioned in Sec. 2.10, the contraction factor $\sqrt{1-v^2/c^2}$ comes from time dilation, which will be tested by experiment. To do this, it will be replaced by an arbitrary function $F(\beta)$ of β, so that equation (9.3.9) is changed to

$$E\left(\theta\right) = \frac{E_0 F(\beta)}{1-\beta\cos\theta}, \tag{9.3.10}$$

where θ is the angle between the recoil velocity β and the direction of propagation of the γ ray. Equation (9.3.10) holds for general relativity, where F is also a function of the gravitational potential, and for classical physics, where $F = 1$.

From Eq. (9.3.10) one arrives at

$$\beta < \cos\theta >= \frac{\overline{E}\left(0°\right) - \overline{E}\left(180°\right)}{\overline{E}\left(0°\right) + \overline{E}\left(180°\right)}. \tag{9.3.11}$$

Here $\overline{E}$ is the mean energy of the γ rays absorbed in the counter, and $\cos\theta$ is averaged over the counter solid angle. $<\cos\theta>$ is known from the geometry of the gas cell to be 0.9963 ± 0.0010.

A convenient test of relativity can be made through the radio

$$R \equiv \frac{F(\beta)^{(O^{16}beam)}}{F(\beta)^{(He^4beam)}}. \tag{9.3.12}$$

In this experiment the observed value of the ratio is

$$R_{obs.} = \frac{\left(\frac{1}{\overline{E}(0°)} + \frac{1}{\overline{E}(180°)}\right)^{He^4beam}}{\left(\frac{1}{\overline{E}(0°)} + \frac{1}{\overline{E}(180°)}\right)^{O^{16}beam}}$$

$$= 1 - 0.001093 \pm 0.000038. \tag{9.3.13}$$

Prediction of special relativity for the ratio is given by

$$R_{theory} \equiv \frac{F\left(\beta_1\right)\left(O^{16}\right)}{F\left(\beta_2\right)\left(He^4\right)} = \sqrt{\frac{1-\beta_1^2}{1-\beta_2^2}} = 1 - 0.001114, \tag{9.3.14}$$

where β was determined nonrelativistically by Eq. (9.3.11): $\beta_1 = 0.0487$; $\beta_2 = 0.012$. The prediction R_{theory} is in agreement with the observed value of $R_{obs.}$.

In order to express their result more directly in terms of the measured quantities, Olin *et al.* analyzed their experiment in a less general manner. They calculated, but did not directly measure, the energy E_0 from Eq. (9.3.10). The quantity $E_0 F(\beta)$ for $\beta = 0.049$ recoils was calculated from

$$E_0 F(\beta) = 2 \left[E(0°)^{-1} + E(180°)^{-1} \right]^{-1}.$$

Then, the relativistic shift of the γ rays observed at $\beta = 0.0487$ is $E_0[1 - F(\beta)] = 10.09 \pm 0.41$ keV compared with a shift of 10.26 keV predicted by $F(\beta) = (1 - \beta^2)^{1/2}$. In time dilation measurements the relativistic correction is usually expressed as $\gamma - 1$, where $\gamma = 1/F(\beta)$. This experimental result may be expressed as $\gamma - 1 = 0.001165 \pm 0.000040$ at $\beta = 0.0487$.

9.3.3. The Doppler Effect Experiments Based on Mössbauer Effect

Rudolf L. Mössbauer (1958) demonstrated dramatically and yet accidentally the feasibility of observing gamma-ray resonance fluorescence by embedding the emitting and absorbing nuclei in a well-bound crystal lattice. He was investigating nuclear resonance scattering of the 129 keV gamma rays from Ir^{191}. For this transition at room temperature, there is an overlap of the absorption and emission lines yielding some resonance absorption. In order to determine the background in his experiment, Mössbauer cooled the system with the hope of reducing the Doppler broadening and eliminating the overlap of the emission and absorption lines. Contrary to his expectation, the resonance fluorescence increased considerably. With the help of a theory developed by Lamb (1939), this effect attributed to the fact that in solids the recoil momentum does not always produce a change in the vibrational state of the crystal lattice. Instead, for a fraction of the gamma transitions, the solid as a whole can take up the recoil momentum. Thus according to this theory, the emission and absorption spectra contain very strong lines of natural width superposed over a broad distribution reselling from the thermal motion of the atoms bound in a crystal lattice. Because of the extremely small recoil energy losses, these lines appear undisplaced at the resonance energy position. Mössbauer (1959) demonstrated the existence of these undisplaced resonance lines.

Immediately after the discovery of Mössbauer effect, several scientists all over the world began to examine the possibility of verifying with the help of the new discovery some of the predictions of the theory of relativity, such as the gravitational red shift, the time-dilation, and the role of acceleration.

(I) Temperature-Dependent Shifts of Recoil-Free γ Rays

Thermally excited vibrations of the nuclei in a crystal lattice would affect the energy of the recoil-free fraction of the γ rays. The value of any component of the nuclear velocity averages very nearly to zero over the nuclear lifetime, and thus the thermal motion cause little broadening through first-order Doppler effect. However, the aver-

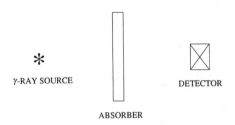

Fig. 32. Schematic for measurements of Mössbauer effect.

age of the square of the velocity has a finite value, which would cause the second-order Doppler shift of the γ rays to lower energy with increased temperature. From Eq. (9.3.3) we know $\nu_s \simeq \nu_0(1 - v_s^2/2c^2)$, where ν_0 is the frequency of the γ ray from a stationary nuclear; v_s is the average of the square of the velocity for the thermally excited vibrations, for instance for Fe^{57} at room temperature $v_s \sim 10^{-6}c$; and ν_s is the average frequency emitted by a thermal vibration nuclear. The second-order shift is then

$$\Delta\nu_s = \nu_s - \nu_0 \simeq -\frac{\nu_s^2}{c^2}. \tag{9.3.15}$$

It is known that v_s depends upon the temperature of the emitter. In the high-temperature classical limit (or at a temperature much larger than its Debye temperature), the average of the kinetic energy of the nuclei can be obtained from that of a perfect gas:

$$E = \frac{3}{2}kT_s \simeq \frac{1}{2}M_0\nu_s^2, \tag{9.3.16}$$

or

$$\nu_s^2 = \frac{3kT_s}{M_0}, \tag{9.3.17}$$

where T_s is the temperature of the emitter, k is the Boltzmann constant, and M_0 is the rest mass of the nucleus. From Eqs. (9.3.15) and (9.3.17) we get

$$\Delta\nu_s \simeq -\frac{3kT_s}{M_0c^2}. \tag{9.3.18a}$$

Similarly, for an absorber we have

$$\Delta\nu_a \simeq -\frac{3kT_a}{2M_0c^2}. \tag{9.3.18b}$$

In Fig. 32, if the temperatures of the emitter and absorber are not equal to each other, a frequency shift between the emission and absorption lines is then

$$\frac{\Delta\nu}{\nu_0} = \frac{\Delta\nu_s - \Delta\nu_a}{\nu_0} \simeq -\left(\frac{3k}{2M_0c^2}\right)\Delta T, \tag{9.3.19}$$

where $\Delta T = T_s - T_a$. For Fe^{57},

$$\frac{\Delta \nu}{\nu_0} = -\left(\frac{3k}{M_0 c^2}\right)\Delta T = -2.44 \times 10^{-15}\Delta T. \qquad (9.3.20)$$

This is the effect of Doppler shift on Mössbauer resonance absorption. In order to eliminate this effect so as to achieve a normal resonance absorption, we can make the emitter or absorber in motion. For instance, when $T_s > T_a$ and then $\Delta \nu_s < \Delta \nu_a$, we need to make the absorber (or the emitter) in motion with a small velocity in the direction far away from the emitter (or the absorber); while in case of $T_s < T_a$, we must make the absorber (or the emitter) in motion with a small velocity in the direction forward to the emitter (or the absorber).

In 1960, Josephson [128] predicted the temperature-dependent shift of γ rays emitted by a solid, equation (9.3.19). At the same time, Pound and Rebka (1960) [129] reported an experiment as a test of this shift. In their experiment, the temperature of Fe^{57} is room temperature (300K) while the Debye temperature is 467K, and therefore equation (9.3.20) is not valid for the case. At this lower temperature the shift should be 0.9 times of the value given by Eq. (9.3.20), namely,

$$\frac{\Delta \nu}{\nu_0} = -2.21 \times 10^{-15}\Delta T. \qquad (9.3.21)$$

Their experiment gave

$$\left(\frac{\Delta \nu}{\nu_0}\right)_{exp} = (-2.09 \pm 0.24) \times 10^{-15}\Delta T, \qquad (9.3.22)$$

which agrees with the prediction (9.3.21).

In 1961, Pound, Benedek , and Drever [130] tested the effect of hydrostatic compression on the energy of the 14.4-keV γ-ray from Fe^{57} in iron*. At the room temperature, the observed temperature-dependent shift is in agreement with the expected value in Eq. (9.3.20) to within an accuracy of 3% .

These kinds of experiments are usually regarded as a test of the time-dilation of special relativity, and as a resolution for the clock paradox. For instance, Sherwin (1960) [131] claimed that the experiments by Pound and Rebka, and by Hay *et al.* (see below) are shown to provide the first direct experimental verification of the time-keeping properties of accelerated clocks such as occur in the classical "clock paradox" of relativity.

(II) *The Experiments for Transverse Doppler Effect*

Hay, Schiffer, Cranshaw, and Egelstaff (1960) [132] first carried out a measurement of transverse Doppler shift by use of a rotating disk as shown in Fig. 33.

In Fig. 33, a Fe^{57} absorber and a Co^{57} source are placed at two positions of the same diameter on the disk, the distances of which from the center of the disk being,

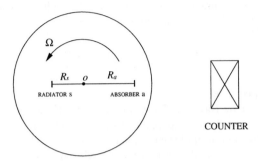

Fig. 33. Schematic diagram of rotating disk.

respectively, R_a and R_s. The gamma rays from the source pass through the absorber, and then are detected by a counter outside the disk. When the disk is at rest, the recoil-free 14.4-keV γ rays from the source will yield some resonance absorption in the absorber, so that the counting rate is minimal. When the disk is rotating, due to shift between the emission and absorption lines, the counting rate will increase. The expected shift can be calculated in two ways. According to the equivalence principle, one can treat the acceleration as an effective gravitational field and calculate the difference in potential between the source and absorber (the gravitational red shift). The other way is to treat the shift as a transverse Doppler effect following from the time dilation of special relativity, as done below.

We assume that the laboratory frame (the nonrotation frame) is an inertial frame, and that rotation motion does not affect the rate of clocks. Let Ω represent the angular velocity of the disk, and the linear velocities of the source and absorber are then, respectively, $v_s = \Omega R_s$ and $v_a = \Omega R_a$. From the transverse Doppler effect, Eq. (2.10.9), we have

$$\nu = \nu_s\sqrt{1 - \frac{v_s^2}{c^2}}, \qquad (9.3.23a)$$

$$\nu = \nu_a\sqrt{1 - \frac{v_a^2}{c^2}}, \qquad (9.3.23b)$$

where ν is the frequency of the γ ray as measured in the laboratory frame, ν_s and ν_a are the frequencies with respect to the source and absorber, respectively. By eliminating ν from Eqs. (9.3.23) we obtain the transverse Doppler shift between the source and absorber,

$$\frac{\nu_a - \nu_s}{\nu_s} = \sqrt{\frac{1 - v_s^2/c^2}{1 - v_a^2/c^2}} - 1 \simeq \frac{1}{2c^2}(v_a^2 - v_s^2) = \frac{\Omega^2}{2c^2}(R_a^2 - R_s^2). \qquad (9.3.24)$$

In order to analyze experimental results, it is usual to add a constant coefficient K to the right-hand side of the equation above:

$$\frac{\nu_a - \nu_s}{\nu_s} = K \frac{\Omega^2}{2c^2} \left(R_a^2 - R_s^2 \right), \tag{9.3.25}$$

where $K = 1$ for special relativity [1]

In the experiment by Hay, Schiffer, Cranshaw, and Egelstaff (1960) [132], a Co^{57} source of 14.4-keV γf-radiation was placed near the center of a rotating disk, and a resonant iron-57 absorber around the periphery of the disk. The resonance absorption was found to decrease as the disk was rotated, giving increase in transmission of up to 6 per cent. The corresponding frequency shift agreed, to within an accuracy of about 2 per cent, with that predicted by Eq. (9.3.24).

Champeney and Moon (1961) [61] performed a similar experiment in which the source and absorber of Fe^{57} were placed at opposite ends of a rotor (i.e., $R_s = R_a$ in Fig. 33) and had the velocity of $v_s = v_a \sim 8 \times 10^{-8} c - 5 \times 10^{-7} c$. The experiment shows that the Mössbauer absorption is unaffected by rotation, or is a test of the prediction of Eq. (9.3.24).

Champeney, *et al.* (1963) [60] repeated Champeney–Moon's experiment with some modifications, and yielded the agreement with time dilation prediction. This measurement gave an upper limit on the velocity of ether draft of 0.0016 km/sec (see Tab. 8.1).

Champeney, Isaak, and Khan (1963,1965) [134] reported some measurements in which Mössbauer sources and absorbers were attached to the center and tip (or vice versa) of a high speed rotor.

Kündig (1963) [135], by use of an ultracentrifuge rotor, observed the shift of 14.4-keV Mössbauer absorption line of Fe^{57} in a rotating system as a function of the angular velocity. An Fe^{57} absorber was placed at a radius of 9.3 cm from the axis of the rotor. A Co^{57} source was mounted on a piezoelectric transducer at the center of the rotor. By applying a triangularly varying voltage to the transducer, the source could be moved relative to the absorber. The measured transverse Doppler shift agrees within an experimental error of 1.1% with the predictions of special relativity.

---

[1]Dos Santos (1976) [133] claimed that his proposed experiment is able to distinguish between Einstein's and Lorentz's theories. In the proposed experiment a source and an absorber are, respectively, mounted on two rotors which give them a relative velocity. He claimed the existence of Doppler shift between the source and absorber: $\nu_a \neq \nu_s$. However, contrary to his expectation, the transverse Doppler shift of special relativity is independent of the direction of velocity. Equation (9.3.24) shows that $v_s = v_a$ will lead to $\nu_a = \nu_s$. In fact, if $\mathbf{v_a} = -\mathbf{v_s}$ and the direction of the gamma rays is perpendicular to the direction of the motion of the source and absorber seen from the laboratory frame, seen from the absorber the source moves with the velocity $v = 2v_s/(1 + v_s^2/c^2)$ and, due to the aberration, the angle between the directions of the gamma rays and the velocity of the source is no longer a right angle but an angle of $\theta = \cos^{-1} v_s/c$. So that putting these into Eq. (2.10.22) we still come the same result: $\Delta\nu = \nu_s - \nu_a = 0$. Thus we conclude that the proposal experiment by Dos Santos cannot distinguish Einstein's theory from Lorentz's electron theory.

An explanation of the rotor experiments by using Mansouri–Sexl's Doppler effect can be found on page 115.

9.3.4. Other Measurements of the Second-Order Doppler Effect

A first-generation laser test of the second-order Doppler effect was reported by Snyder and Hall in 1975 [136]. They used laser-saturation spectroscopy to transversely excite a beam of fast neon atoms which were free of first-order Doppler shifts. Their measurement agrees with the second-order Doppler shift of special relativity to within an accuracy of 0.5%.

A recent test of the second-order Doppler shift was carried out by Kaivola, Poulsen, Riis, and Lee (1985) [137] by use of two-photon absorption. In two-photon spectroscopy, the first-order Doppler shift is absent, and the second-order term becomes dominant. They measured the frequency difference between a two-photon transition in a fast neon-atom beam and in a cell. Two cw dye lasers were used, one stabilized to the fast-beam transition and the other to the cell. Their result is in excellent agreement with the second-order Doppler effect of special relativity. The experimental accuracy is 4×10^{-5}.

The combined measurement of the second-order Doppler shift and the gravitational red shift were carried out by Vessot and Levine (1979) [138a], and by Vessot et $al.$ (1980) [138b]. which is in agreement with the predictions on the 7×10^{-5} level.

9.4. Lifetime Dilation of Moving Mesons

The time dilation of special relativity implies that the mean lifetime of decay of radioactive particles in flight would increase by a factor of $\gamma = (1 - v^2/c^2)^{-1/2}$ compared to that of the particles at rest. Thus a measurement of the lifetime will provide a test of the prediction.

Let the number of radioactive particles at rest at an initial time $(t = t_0)$ in the laboratory frame be N_0. Then, at time t the number of remnant particles is denoted by N which is given by the decay law:

$$N = N_0 e^{-t/\tau_0}, \tag{9.4.1}$$

where τ_0 is called the proper lifetime.

For moving particles the time t in the equation above should be replaced by t/γ due to the time dilation effect, and thus the law of decay should be

$$N = N_0 e^{-t/\gamma\tau_0} = N_0 e^{-t/\tau}, \tag{9.4.2a}$$

where

$$\tau = \gamma\tau_0. \tag{9.4.2b}$$

Here τ is called the mean lifetime for the radioactive particles in flight. The equations above can be rewritten as

$$N = N_0 e^{-vt/v\tau} = N_0 e^{-x/\lambda}, \tag{9.4.3a}$$

$$\lambda = v\tau = \frac{v\tau_0}{\sqrt{1 - v^2/c^2}} = \frac{p\tau_0}{M_0}, \tag{9.4.3b}$$

where $x = vt$ is the path through which the particles pass during the time t, λ is the mean free path for the particles (when $x = \lambda, N/N_0 = e^{-1}$), M_0 is the rest mass of the particles, and p is the momentum.

There are two ways for testing Eq. (9.4.2) or (9.4.3). The first way is to determine the number of radioactive particles which have decayed in traversing a known flight path by measuring the numbers of remnant particles at two separate positions of flight path. The second way is to measure the relationship between the decay rate of radioactive particles and their velocity (or their momentum). In the performed experiments, the values of the proper lifetime τ_0 were measured by making the flying particles stop in the targets, and the observed values of the mean lifetime (τ) of the particles in flight were determined by means of the above two methods. The theoretical values of τ to be used for comparing with the observed values were calculated by use of Eq. (9.4.2b), where the values of the factor γ were calculated from the equation of special relativity $p = \gamma M_0 v$ (or $E = \gamma M_0 c^2$) by using observed values of the momentum p (or the energy E).

The first experiment of this type was carried out by Rossi and Hall in 1941 [139] by measuring μ-mesons in cosmic rays. Since that time, many measurements have been performed by use of mesons produced in the atmosphere and accelerators.

9.4.1. The μ-Mesons in the Cosmic Rays

μ mesons were first discovered in cosmic rays, the proper lifetime of which was measured to be $\tau_0 = 2.2 \times 10^{-6}$ sec. The μ-mesons in cosmic rays are generally produced in the atmosphere at high altitude ($\sim$10–20 km). A great part of them could reach sea level. This shows that the mean free path of the muons should be, at least, tens kilometers. If the decay lifetime τ of the flying muons were equal to the proper lifetime τ_0, then even though they moved with the light speed, their mean free path might mostly be $c\tau_0 = 660$ m. This shows that most of the muons could not reach sea level. There are two ways to resolve this problem. One way is to assume that the muons move faster than light. However, physicists never observed any particles with superluminal speed in the nature so that the velocity of the muons in cosmic rays are less than the light speed. Another way is to use the time dilation effect of special relativity: the lifetime of the flying muons increase by a factor of $\gamma = (1 - \beta^2)^{-1/2}$ compared to the proper lifetime, i.e., $\tau = \gamma\tau_0$. On the other hand, in view of an observer co-moving with the flying muons, although the lifetime of the muons is τ_0 the distance from the sea level contracts by a factor of γ compared to the tens of kilometers. The explanation given by the time-dilation effect is thus equivalent to that from the contraction effect.

9.4.2. Measurements of the Lifetime of the μ-Mesons in the Cosmic Rays

Rossi and Hall (1941) [139] first observed the momentum-dependence of the decay rate of the μ-mesons in cosmic rays, and yielded a result in agreement with a prediction by Eq. (9.4.3b).

Frisch and Smith (1963) [140] repeated, with substantial modifications, the first observation of the effect of the time dilation for radioactive particles moving with a high speed. The radioactive particles used in their experiment are μ-mesons which are produced high in the atmosphere and come shooting down toward the earth with a speed greater than $0.99c$. As they come down, some of them disintegrate in flight. The number arriving at medium altitude is, therefore, greater than the number surviving to reach sea level. At a medium altitude on top of Mt. Washington, they counted the μ-mesons. Then, they went down to sea level and counted the μ-mesons surviving to arrive at sea level. The difference of these numbers represents the number which decayed in flight. In addition, they slowed down and stopped a sample of μ-mesons and measured the proper lifetime. Their measurements yielded an observed time dilation factor $\gamma_{obs} = \tau/\tau_0 = 8.8 \pm 0.8$. By use of an observed value of the mean energy of the μ-mesons, they calculated from the equation of special relativity, $E = \gamma m_\mu c^2$, an expected value of the time dilation factor $\gamma_{theory} = 8.4 \pm 2.0$. These show the agreement between the observation and the prediction.

9.4.3. Measurements of the Lifetime of the Mesons Produced by Accelerators

Lederman *et al.* (1951) [141] observed the decay process $\pi^- \to \mu^- \nu^0$ for a beam of the flying negative pi-mesons with a velocity of $\sim 0.73c$ (corresponding to the dilation factor $\gamma \sim 1.5$) produced by an accelerator, and obtained the mean free path λ of (9.93 ± 1.10)m which implied a lifetime for the flying pion of $\tau = (4.55 \pm 0.52) \times 10^{-8}$ sec. By use of this observed value of τ, equation (9.4.2b) gives a proper lifetime τ_0 of $(2.92 \pm 0.32) \times 10^{-8}$ sec which agrees with the observed value of $\tau_0 = 2.60 \times 10^{-8}$ sec for pions at rest to $\sim 10\%$.

Durbin *et al.* (1952) [142] reported measurements of the lifetime for $\pi^\pm$-mesons in flight. Their observation showed that the decay of pions satisfies the law (9.4.3), where the observed value of the mean free path $\lambda = (8.5 \pm 0.6)$ m gives the lifetime for the flying pions $\tau = (3.8 \pm 0.3) \times 10^{-8}$ sec. This yields $\gamma_{exp} = \tau/\tau_0 = 1.5$. On the other hand, from the velocity of the pions $v \simeq 0.75c$ we get $\gamma_{theory} \simeq 1.5$. These show the agreement of the observation and prediction [143].

Greenberg *et al.* (1969) [144] measured the lifetime of the charged pions in flight, the principle result being $\tau = 26.02 \pm 0.04$ nsec. By use of the value they calculated the proper lifetime τ_0 from the equation $\tau = \gamma \tau_0$, where $\gamma \approx 2.4$, which agrees with the other measurements of τ_0 for pions at rest. Viewed another way, the observed change in lifetime, $\tau - \tau_0$, agrees with the predicted value, $(\gamma - 1)\tau_0$, to 0.4%.

Ayres *et al.* (1971) [145] compared the lifetimes of π^+ and π^- by measuring the fraction of surviving pions at several positions along a 1000-cm (one-half lifetime) decay path. The speed of the pions was $0.92c$, corresponding to a dilation factor of $\gamma \simeq$ 2.44. By inserting the observed value of the lifetime for pions in flight into the time dilation formula (9.4.2b), they obtained a value for the proper lifetime $\tau_0 = 26.02 \pm$ 0.04 nsec which was in good agreement with the result of independent measurements with pions at rest to 0.4%.

A series of measurements for the lifetime of muons in flight were carried out in the CERN muon storage ring. Farley *et al.* [146] reported a value of the muon lifetime in flight for an expantion factor $\gamma \simeq 12$, which agreed within 1% with the predicted value obtained by applying the above expantion factor and measured lifetime at rest to equation (9.4.2b). In 1977, they [147] reported the separate measurements for μ^+ and μ^-, with a γ factor of 29.33, which are an order of magnitude more precise and which show that the predictions of special relativity obtain even under accelerations as large as $10^{18}g$ and down to distances less than 10^{-15} cm.

The decay lifetimes for $K^\pm$-mesons in flight have also been measured. For example, a measurement for the lifetime of 1 GeV K-meson by Burrowes *et al.* (1959) [149] yielded a result in agreement with the time-dilation prediction to 5%.

Table 9.5. *Summary of Experiments on Time Dilation*

Year	Experiment	Method	"Clock" speed	Result
1938	Ives and Stilwell [120]	Moving hydrogen atoms longitudinal observation	$\leq 0.004c$	agrees with second-order Doppler shift of SR
1941	Ives and Stilwell [122]	Moving hydrogen atoms longitudinal observation	$\leq 0.006c$	See Tabs. 9.3 and 9.4
1939	Otting [123]	Moving hydrogen atoms longitudinal observation	$\sim 0.003c$	agrees with second-order Doppler shift of SR
1962	Mandelberg and Witten [125]	Moving hydrogen atoms longitudinal observation	$\leq 0.009c$	$(\lambda_0 \beta^2)_{obs} = 0.238 \pm 0.006$ $(\lambda_0 \beta^2)_{theory} = 0.238 \pm 0.0004$
1979	Hasselkamp, Mondry, and Scharmann [126]	Moving hydrogen atoms transverse observation	$8.4 \times 10^{-3}c$ $-3.1 \times 10^{-2}c$	agrees with second-order Doppler shift of SR

Table 9.5. *Summary of Experiments on Time Dilation (continued)*

Year	Experiment	Method	"Clock" speed	Result
1971	Hafele and Keating [117]	Atomic clocks in flight	$\leq 10^{-6}c$	Tests of time dilation and gravitational red shift
1973	Olin *et al.* [127]	Longitudinal observations for γ rays from Ne^{20} nuclei	$0.05c$	agrees with second-order Doppler shift to 3.5%
1975	Snyder and Hall [136]	Laser-saturation spectroscopy to excite neon atom beam	$\sim 10^{-3}c$	agrees with second-order Doppler shift to $\sim$ 0.5%
1985	Kaivola *et al.* [137]	Two-photon transition in fast neon atoms	$4 \times 10^{-3}c$	agrees with second-order Doppler shift to $\sim 4 \times 10^{-5}$
1960	Pound and Rebka [129]	Temperature-dependence of Mössbauer effect	$\sim 10^{-6}c$	agrees with second-order Doppler shift to $\sim$10%
1961	Pound *et al.*, [130]	Temperature-dependence of Mössbauer effect	$\sim 10^{-6}c$	agrees with second-order Doppler shift to $\sim$ 3%
1960	Hay *et al.*, [132]	Mössbauer isotopes were at the center and periphery of a rotor	$\sim 7 \times 10^{-7}c$	A test of transverse Doppler effect to $\sim$2%; $K = 1.001 \pm 0.062$ for 95% probability
1963	Kündig [135]	Mössbauer isotopes were at the center and periphery of a rotor	From $10^{-7}c$ to $10^{-6}c$	A test of transverse Doppler effect to $\sim$1.1% ; $K = 1.0065 \pm 0.011$
1961	Champeney and Moon [61]	Mössbauer isotopes were at opposite tips of a rotor	From $8 \times 10^{-8}c$ to $5 \times 10^{-7}c$	Null shift between source and absorber
1963	Champeney, Isaak, and Khan [60]	Mössbauer isotopes were at opposite tips of a rotor		Null shift between source and absorber

Table 9.5. *Summary of Experiments on Time Dilation (continued)*

Year	Experiment	Method	"Clock" speed	Result
1965	Champeney, Isaak, and Khan [134]	Mössbauer isotopes were placed at the center and tip of a rotor	$\sim 10^{-6}c$	$K = 1.021 \pm 0.019$
1941	Rossi and Hall [139]	Measurements of the momentum-dependence of the decay rate μ-mesons in cosmic rays	$\sim 0.97c$ $(\gamma \sim 4.1)$	A test of time dilation
1963	Frisch and Smith [140]	Measurements for the altitude-dependence of the number of μ-mesons in cosmic rays	$\sim 0.994c$	$\gamma_{theory} = 8.4 \pm 2.0$ $\gamma_{obs} = 8.8 \pm 0.8$
1972	Bailey *et al.* [146]	Measurement of the muon lifetime in flight	$0.998c$ $(\gamma \simeq 12)$	agrees within 1.1% with time dilation effect
1979	Bailey *et al.* [147,148]	Separate measurements for the lifetime of μ^+ and μ^- in flight	$0.9994c$ $(\gamma \simeq 29.3)$	agreed within $\sim 0.1\%$ with the time dilation effect
1951	Lederman *et al.* (1951) [141]	Measurement of the lifetime for π^--mesons in flight	$\sim 0.73c$ $(\gamma \simeq 1.5)$	agrees with time dilation to $\sim 10\%$
1952	Durbin *et al.* (1952) [142]	Measurement of the lifetime for $\pi^\pm$-mesons in flight	$\sim 0.75c$ $(\gamma \simeq 1.5)$	agrees with time dilation
1969	Greenberg *et al.* (1969) [144]	Measurement of the lifetime of the charged pions in flight	$0.91c$ $(\gamma \simeq 2.4)$	Observation agrees with time dilation to 0.4%
1971	Ayres *et al.* [145]	Measurements of the lifetime of $\pi^\pm$-mesons in flight	$0.92c$ $(\gamma \simeq 2.44)$	agrees with the time dilation to 0.4%

CHAPTER 10

THE ELECTROMAGNETISM EXPERIMENTS

10.1. Introduction

Long before Einstein's theory of special relativity was developed, electromagnetic phenomena for moving bodies had been investigated. In 1831 M. Faraday [150] first studied the so-called *"unipolar induction"*. He found that a stable electric current was produced in a stationary conducting wire connected with a rotating magnet. Although this discovery had been widely applied in engineering for production of electric generator (i.e., *unipolar machine*), many different explanations of the "unipolar induction" were in dispute for long time (see, e.g., Refs. [151,153,154]). In order to test Fresnel's ether dragging theory [156], in 1851, H.L. Fizeau [157] measured the velocity of light in a moving medium (a water current). However, Fresnel's ether theory cannot give any explanation to the dispersion phenomena [13]. Later, Rowland (1876), Röntgen (1888), as well as Eichenwald (1903) investigated the magnetic induction phenomenon in moving bodies (see, e.g., Refs. [151,153,155]). According to the theory of electrons developed by Lorentz (1892, 1895) [168,169], all of the electrodynamics phenomena are regarded as certain effects of moving electric charges. A charge moving within a magnetic field should be acted upon by a force directly proportional to its velocity. The interactions among electrons are achieved through the so-called ether. The ether is not dragged at all by moving bodies. Lorentz's theory of electrons can give explanations to the magnetic induction of moving bodies, as well as of Fizeau's flowing water experiment and other Fresnel drag experiments involving dispersion. However Lorentz's theory cannot give a complete explanation of the Michelson–Morley experiment [7], as well as of Faraday's unipolar induction phenomenon.

The electric and magnetic effects, which are related only to the relative velocities of moving bodies to observer, appear to have a certain symmetry. The classical electrodynamics cannot give an explanation of this kind of symmetry. To explain the symmetry is one of Einstein's motives for developing his theory of special relativity, as stressed in his first article [1]. Einstein gave the relativistic transformation of the electromagnetic fields in an empty space, but did not provide a general construction of electromagnetic field equations in moving media. Later, it is Minkowski (1908) [18] who proved that the electromagnetic field equations in moving media could be easily written, if once the covariance of the field equations with respect to Lorentz transformations was required. The performed measurements for electromagnetic induction were only of the first-order in v/c with v being the velocity of the moving body. Predictions of the Maxwell–Minkowski electrodynamics to the same order are in agreement with the electromagnetism experiments. The experiments have been, therefore, regarded as tests of special relativity. For this reason, some authors claimed that the experiments are tests of Einstein's simultaneity factor. However, this is not the case, as stressed in Chap. 6. In fact, Edwards transformations would physically give the same predictions.

Electromagnetic experiments, with the exception of Fresnel drag experiments, have not been repeated for long time, due to the difficulty of improving experimental accuracy. In this chapter we shall introduce the electromagnetic experiments, and at the same time compare them with the Maxwell–Minkowski electrodynamics.

10.2. Electromagnetic Induction of Moving Bodies

In the present section we shall introduce unipolar induction, Rowland's experiment (1876), Röntgen's experiment (1888), Eichenwald's experiment (1903), as well as the Wilson–Wilson experiment (1913).

10.2.1. Unipolar Induction

For motion of a magnet ($\mathbf{P}' = 0, \mathbf{M}' \neq 0$) the transformation (4.2.9a) in first-order approximation gives

$$\mathbf{P} = \frac{\mathbf{v}}{c} \times \mathbf{M}'. \tag{10.2.1}$$

This is the electrical polarization of a moving magnet. This effect has been known in the technical field for a long time under the name *unipolar induction* (this name was first given by Weber). The problem of the unipolar induction has been famous since the days of Arago and Faraday. The literature on this subject is voluminous and by no means free of contradictions.

The unipolar induction has been used for the production of the so-called "*unipolar generator*". In technology the unipolar generator often takes the form of a cylindrical iron body which rotates on its axis and is magnetized parallel to it (see Fig. 34). By means of two brush contacts, A (on the axis) and B (on the equator), an electric current can be taken in the conducting wire between A and B.

There were some different explanations of the unipolar generator (see, e.g., Refs. [153,154]). Faraday claimed that the magnetic lines of force do not rotate with the rotation of the magnet. The part of the magnet, the path BAC (see Fig. 34), then moves in the magnetic field. The motion of the magnet then produces an *electromotive force* (e.m.f.) which is the source of the current in the wire. Weber claimed an opposite view point: the magnetic lines of force rotate with the rotation of the magnet. Thus it is the wire that moves in the magnetic field, and then produces an e.m.f. which causes a current in the path $AVBCA$. This point of view was criticized by, e.g., Becker [151], who pointed out that this viewpoint cannot be consistent with any field theory. Although Lorentz's theory of electrons can give an explanation of the Fresnel drag effect, as mentioned in next section, but it cannot explain the unipolar induction.

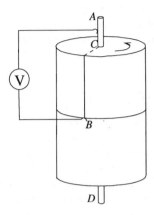

Fig. 34. Schematic diagram of the unipolar generator.

It is Einstein who gave an relativistic explanation of the unipolar induction by use of the transformation equations for electromagnetic field quantities [1]. We now calculate the electromotive force by using the Maxwell–Minkowski electrodynamics.

The inverse transformations of Eqs. (4.2.2) give the first-order approximation equations

$$\mathbf{E} = \mathbf{E}' - \frac{1}{c}\mathbf{v} \times \mathbf{B}' = -\frac{1}{c}\mathbf{v} \times \mathbf{B}', \tag{10.2.2a}$$

$$\mathbf{B} = \mathbf{B}' + \frac{1}{c}\mathbf{v} \times \mathbf{E}' = \mathbf{B}', \tag{10.2.2b}$$

where $\mathbf{E}' = 0$ and $\mathbf{B}'$ are the fields in the rest magnet. The equations are valid only in an inertial frame. However in the case that the angular velocity of the magnet is small, and hence, its effect on the electromagnetic fields can be neglected, equations (10.2.2) may be approximately valid for the slow rotation.

Equations (10.2.2) shows that in the stationary wire AVB, the electric field

$$\mathbf{E} = -\frac{1}{c}\mathbf{v} \times \mathbf{B}' = -\frac{1}{c}\mathbf{v} \times \mathbf{B} \tag{10.2.3}$$

would produce the current due to Ohm's law

$$\mathbf{J} = \sigma\mathbf{E}, \tag{10.2.4}$$

whose e.m.f. is

$$\varepsilon = -\int_{AVB} \frac{1}{c}(\mathbf{v} \times \mathbf{B}) \cdot d\mathbf{l} = \frac{1}{c}\int_{AVB} \mathbf{B} \times (\mathbf{r} \times \mathbf{\Omega}) \cdot dl, \tag{10.2.5}$$

where Ω is the angular velocity of the magnet.

It is needed to note that some authors (see, e.g. Ref. [151,152]) claimed that the unipolar induction is a direct consequence of Einstein's definition of simultaneity. However this is a misunderstanding. In fact Edwards' definition of simultaneity would give the same phenomenon.

10.2.2. Magnetic Effects of Moving Bodies

In 1838, before the electromagnetic field theory was developed by Maxwell, Faraday had pointed out that effects of a moving charged body is equivalent to that of an electric current. Later, Maxwell repeated this point of view: a moving electrified body is equivalent to an electric current. To decide the matter, an experiment was performed by Rowland in 1876. In Rowland's experiment the electrified body was a disk of ebonite, coated with gold leaf and capable of turning about a vertical axis between two fixed plates of glass, each gilt on one side. The gilt faces of the plates could be earthed, while the ebonite disk received electricity from a point placed near its edge; each coating of the disk thus formed a condenser with the plate nearest to it. An astatic needle was placed above the upper condenser plate, nearly over the edge of the disk; and when the disk was rotated a magnetic field was found to be produced. This implies the existence of an electric current produced by the moving electrified body. This kind of current corresponding to a moving charged body is named the *convection current* so as to distinguish from a *conduction current* flowing in a wire due to motion of electrons. Later, this kind of experiments were repeated under improved conditions by Rowland and Hutchinson (1889), H. Pender (1901, 1903), Einchenwald (1901), E.P. Adams (1901), as well as H. Pender and V. Crémieu (1903). These experiments show that a Rowland convection current is equivalent to an ordinary conduction current.

In 1888, Röntgen performed another kind of experiment which proved that a magnetic field exists around an uncharged dielectric moving at right angles to the electric lines of force of a constant electrostatic field. This implies there was a surface current on the surface of the dielectric. This kind of current produced by a moving dielectric is called the *current of dielectric convection* or the *Röntgen convection current*.

An experiment performed by Eichenwald in 1903 shows that a Röntgen surface current density is given by

$$J = \frac{v}{4\pi}|\mathbf{P}| = \frac{\varepsilon - 1}{4\pi}v|\mathbf{E}|, \qquad (10.2.6)$$

where $\mathbf{P}$ is the polarization vector of a dielectric, $\mathbf{E}$ is the electrostatic field, ε and v are the dielectric constant and the velocity of the moving dielectric, respectively.

In his another experiment, in which the dielectric disk between the two plates of a condenser was rotated together with the condenser, Eichenwald proved that the magnetic action was the sum of the Rowland and the Röntgen currents, and was independent of ε.

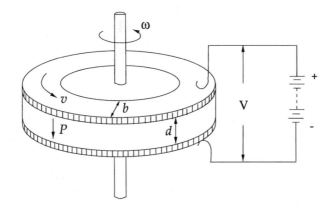

Fig. 35. Schematic diagram of an arrangement for investigating magnetic effects of moving bodies.

An arrangement for the above three kinds of experiments can be shown in Fig. 35, a dielectric disk of thickness d being placed between two metal plates of a condenser, and the two metal plates being connected to the terminals of a voltage source and thus becoming charged to a potential difference V. For Rowland's experiment, the two metal plates are rotated while the dielectric disk remains at rest. For Röntgen's experiment, the dielectric is rotated while the metal plates remain at rest. For Eichenwald's experiment, both the dielectric and metal plates are rotated together.

We now explain the three kinds of experiments by making use of the electrodynamics of moving media. A possible effect of rotation on the disk will be neglected. In Fig. 35 the polarization vector of the dielectric disk is given by $\mathbf{P}' = (\varepsilon - 1)\mathbf{E}/4\pi$, where $\mathbf{E} = V/d$ is the electric field inside the condenser. When the disk is rotated with an angular velocity $\mathbf{\Omega}$, equation (4.2.9b) shows that in the laboratory frame a magnetization vector will be given by $\mathbf{M} = -(\mathbf{v} \times \mathbf{P}')/c = -(\varepsilon - 1)\mathbf{v} \times \mathbf{E}/(4\pi c)$, where $\mathbf{v} = \mathbf{\Omega} \times \mathbf{r}$. It is obvious that the curl of $\mathbf{M}$ vanishes, there being no any current inside the dielectric disk. Because $\mathbf{M} \neq 0$ inside the disk while $\mathbf{M} = 0$ outside the disk, there must be a surface current on the surface of the disk given by $\varepsilon \mathbf{M}$, or explicitly

$$J = \frac{\varepsilon - 1}{4\pi} vE. \tag{10.2.7}$$

This is the same as the experimental result (10.2.6).

For Eichewald's experiment, the disk and metal plates are rotated together. The surface charge density on the surfaces of the condenser is $D/(4\pi) = \varepsilon E/(4\pi)$, and thus there will appear a surface current (density) of $vD/(4\pi) = \varepsilon vE/(4\pi)$. The total

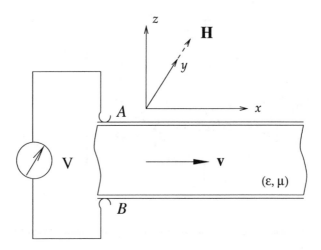

Fig. 36. Schematic diagram of the Wilson–Wilson experiment. A large condenser is filled with a medium (ε, μ). The condenser and medium is in motion at a velocity v in the direction of x-axis. The two plates of the condenser are connected to the terminals of a current meter. There is a uniform magnetic field **H** in the direction of y-axis. When the direction of **H** is inverted, A current passes through the current meter.

current is then given by

$$J = \frac{\varepsilon}{4\pi} vE - \frac{\varepsilon - 1}{4\pi} vE = \frac{1}{4\pi} vE. \qquad (10.2.8)$$

This shows that in case of the disk and condenser being rotated together the total magnetic effect is independent of ε, which was just proved by Eichenwald.

10.2.3. The Wilson–Wilson Experiment

In the previous subsections we have introduced the electric effect of moving magnet and the magnetic effects of moving dielectric. In the present subsection we will introduce the electromagnetic effect of a moving electromagnetic medium.

In order to test the Maxwell–Minkowski electrodynamics of moving media, Einstein and Laub (1908) proposed a possible experiment. In 1913, M. Wilson and H.A. Wilson performed a similar measurement. A schematic diagram of the Wilson–Wilson experiment is shown in Fig. 36.

For the case shown in Fig. 36, the magnetic field is static, i.e., $\partial \mathbf{B}/\partial t = 0$. In laboratory frame, Eq. (4.2.1a) gives $\nabla \times \mathbf{E} = 0$. This implies that the integration vanishes:

$$\oint \mathbf{E} \cdot d\mathbf{l} = \int\int (\nabla \times \mathbf{E}) \cdot \sigma = 0, \tag{10.2.9}$$

where the closed integration path is $AVBA$. Due to $\mathbf{E} = 0$ in outside the condenser, the above equation can be written as

$$\oint_{AVBA} \mathbf{E} \cdot d\mathbf{l} = \int_{AVB} \mathbf{E}_{ext} \cdot d\mathbf{l} + \int_{BA} \mathbf{E}_{int} \cdot d\mathbf{l} = \int_{BA} \mathbf{E}_{int} \cdot d\mathbf{l} = 0. \tag{10.2.10}$$

This means $(\mathbf{E}_{int})_x = 0$ inside the medium. By using the first-order approximation of the constitutive relations (4.2.5),

$$\mathbf{D} = \varepsilon\mathbf{E} + (\varepsilon\mu - 1)\frac{1}{c}\mathbf{v} \times \mathbf{H}, \tag{10.2.11}$$

we obtain

$$D_x = (\varepsilon\mu - 1)\frac{1}{c}vH_y. \tag{10.2.12}$$

Because $\mathbf{D}_{ext} = 0$ and ρ(the charge density)$= 0$ outside the condenser, the boundary condition of $\mathbf{D}$ leads to the following surface charge density on the surfaces of the condenser:

$$\sigma = -\frac{(D_{int})_z}{4\pi} = -\frac{\varepsilon\mu - 1}{4\pi}\frac{v}{c}H. \tag{10.2.13}$$

This shows that the condenser will be charged by the moving medium, and that $\sigma \to -\sigma$ when $\mathbf{H} \to -\mathbf{H}$. Thus a pulse of electric current will be produced in the path AVB, which can be measured by the current meter V, its magnitude being directly proportional to $(\varepsilon\mu - 1)$.

The arrangement of Wilson–Wilson's experiment (1913) consists of a hollow cylinder made by sealing-wax embedding steel spheres, the dielectric constant and relative permittivity in the cylinder being $\varepsilon = 6.0$ and $\mu = 3.0$. The inner and outer metal coatings of the cylinder form a condenser. In the experiment the cylinder condenser was rotated on its axis, and the direction of a magnetic field was parallel to the axis. This experiment gave an average value of $(\varepsilon\mu - 1) = 24$, while the theoretical value is 17.

10.3. The Fresnel Drag Effect

The first ether theory describing the propagation of light rays in moving media was developed by A.J. Fresnel in 1918 [156]. By analogy with the velocity of sound in elastic media, he assumed that the velocity of light in media are proportional to the square root of the density of an elastic ether, and then derived the formula for the velocity of light in an elastic medium without dispersion,

$$\mathbf{u} = \mathbf{u}' + f\mathbf{v}, \tag{10.3.1a}$$

with

$$f = 1 - \frac{1}{n^2} \tag{10.3.1b}$$

being the so-called Fresnel's drag coefficient, where $\mathbf{v}$ is the velocity of the medium with the index of refraction n relative to the ether and $\mathbf{u}'$ is the velocity of light in the medium in case of $\mathbf{v} = 0$. The second term on the right-hand side of Eq. (10.3.1a) can be regarded as a velocity of an ether wind, while the quantity $\mathbf{v} - f\mathbf{v} = \mathbf{v}/n^2$ is then the velocity of the medium relative to the ether wind. This implies that the ether is dragged along partly by the moving medium.

In order to test Fresnel's ether theory, in 1851, H. L. Fizeau [157] firstly measured the velocity of light in running water and obtained a result which was regarded as being in agreement with the above formulas (10.3.1). Later, He performed his second experiment in 1895 [158] and gave the same result.

In Sec. 4.4, we have derived the same formula from the relativistic electrodynamics in moving media. This formula can be obtained from the first-order approximation of the addition law. To see this, we assume that both the direction of the motion of the medium and that of the light ray are parallel with x-axis in F. In the frame F' connected with the medium, the velocity of the light is c/n. By using $u' = u'_x = c/n$, Einstein's law (2.4.3a) becomes

$$u = \frac{(c/n) + v}{1 + (v/c)/n} = \frac{c}{n} + \left(1 - \frac{1}{n^2}\right)v + O\left(\frac{v^2}{v^2}\right). \tag{10.3.2}$$

To the first-order in v/c, the addition law (10.3.2) reduces to Fresnel's formulas (10.3.1). Thus, Fizeau's experiment has been regarded as a test of Einstein's law of the addition of velocities or of the electrodynamics in moving media.

After 1905, some Fizeau-type experiments were performed. According to the relationship between the direction of the motion of the medium and that of the light ray, these "drag" experiments can be classified three kinds. The first kind is called the longitudinal drag experiment, such as the original Fizeau's experiments [157,158], Michelson–Morley experiment (1886) [159] and Zeeman's experiments [160–163], where the direction of the velocity of the medium is parallel to that of the light ray; The second is the transverse drag experiments, for example, Jones' (1971, 1975) experiments [164,165], where the direction of the velocity of the medium is perpendicular to that of the light ray; The third kind involves other experiments [166,167] where the angle between the directions of the velocities of the medium and light ray is Brewster's angle. All the three kinds of experiments involve the media with and without dispersion. The results given by the experiments without dispersion seem to be in agreement with Fresnel's formulas (10.4.1), while other experiments with dispersion cannot be explicitly explained by Fresnel's ether theory. The difficulty is closely connected with the fact that the index of refraction in this theory is constant. Lorentz's electron theory [168,169] was regarded as able to give a satisfactory explanation to the dispersion and drag experiments.

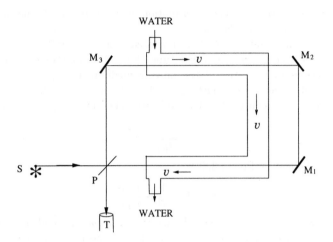

Fig. 37. Fizeau's experiment.

We want here, however, to emphasized that the definition of velocity depends closely upon the definition of simultaneity just as mentioned in Sec. 1.5. We know that the definitions of simultaneity in Fresnel's ether theory and Lorentz's electron theory are all different from Einstein's definition. We also know that Einstein's simultaneity is always used in laboratories. Thus the velocity v in both the two kinds of theories cannot be identified with data in experiments just as mentioned in subsection 6.5.3. Of course, the difference between the velocity v and Einstein velocity is of first-order. Therefore, to the first-order, Fresnel's formula and Lorentz's theory are in agreement with the experimental results.

In this section we shall introduce the Fresnel drag effect experiments.

10.3.1. Fizeau's and Zeeman's Experiments

The arrangement of Fizeau's experiment is shown in Fig. 37. A light ray from a source of light S is divided by a (weakly silver-coated) glass plate P, which is placed at an angle of $45°$ to the direction of propagation, into a transmitted part 1 and reflected part 2. The transmitted ray 1 reflected by the mirrors M_1, M_2, M_3, and traverses a rectangular path $PM_1M_2M_3P$; again a certain fraction of the ray passes the plate P and enters the telescope T. The reflected ray 2 traverses the same rectangle in the opposite direction. On its return to P it is partly reflected into T where it interferes with 1. Between M_3 (or P) and M_2 (or M_1) is inserted a tube filled with water, so that the light rays 1 and 2 are passing through water on the path PM_1, as well as on the path M_2M_3. As indicated in Fig. 37, the light ray 1 is traversing the water in a direction opposite to the direction of motion of the water, while the ray 2 has the same direction of motion as the water.

We now use the equations in Sec. 4.3 to calculate the shift of the interference fringes when the medium moves in the tube. Since the directions of the light rays and the directions of velocities of the moving medium in Fig. 37 are parallel or opposite parallel to each other, putting $\theta = 0$, or 2π into Eq. (4.3.10), we know that the group velocity is equal to the phase velocity. Then from Eq. (4.3.8) we have the phase velocity

$$u_{\pm} = \frac{c}{n(\nu)} \pm f_1 v, \tag{10.3.3a}$$

with f_1 being defined by Eq. (4.3.6b)

$$f_1 = 1 - \frac{1}{n^2} + \frac{\nu}{n} \frac{dn}{d\nu}, \tag{10.3.3b}$$

where the subscripts "$\pm$" represent the cases of the parallel and opposite parallel directions of the light ray relative to the direction of motion of the medium, respectively, ν is the frequency of light in the medium as measured in the laboratory frame F. In this kind of experiments while a main part of the liquid is running in the tube, a certain part of the liquid is at rest in the ends of the tube. Thus the surfaces of the liquid are at rest as seen from the laboratory. This implies that $v_z = 0$ in Fig. 5, and thus equation (4.4.6b) gives $\nu = \nu_0$, i.e., the frequencies in the moving medium and rest medium are equal to each other and at the same time, the frequencies in the rest medium and empty space are also equal to each other. Therefore the frequency ν in Eqs. (10.3.3) is just the frequency in the empty space ν_0. The coefficient f_1 defined by Eq. (10.3.3b) is called *Lorentz's drag coefficient*, which differs from Fresnel's drag coefficient by a dispersion term. This coefficient was firstly obtained in 1895 by Lorentz from his electron theory.

The difference between the traveling times of the two light rays propagating in opposite directions is given by

$$\Delta t = \frac{2l}{u_-} - \frac{2l}{u_+}, \tag{10.3.4a}$$

where l is the length of the path in the medium. The corresponding difference between the phases of the light waves when they meet again is then

$$\delta = \frac{c}{\lambda_0} \Delta t = \frac{2lc}{\lambda_0} \left(\frac{1}{u_-} - \frac{1}{u_+} \right) \simeq 4n^2 l f_1 \frac{v}{\lambda_0 c}, \tag{10.3.4b}$$

where Eqs. (10.3.3a) is used. The difference δ is just a relative shift between the position of the interference fringes while the liquid is at rest in the tube and that when a liquid current is sent through the tube. When the direction of motion of the liquid is inverted, i.e., $v \to -v$, the corresponding shift is

$$\Delta = 2\delta = 8ln^2 f_1 \frac{v}{\lambda_0 c}. \tag{10.3.5}$$

In order to test Fresnel's ether theory, in 1851, H.L. Fizeau performed his experiment [157] in which he used the arrangement as shown in Fig. 37 with the running water. The total length of the path of light was $l \simeq 1.5$ m. The velocity of the water current v was about 7 m/sec. The source emitted white light rays. The observed shift of the central position of interference fringes, when the direction of $\mathbf{v}$ was inverted, was equal to 4.6 parts of the distance between two neighboring fringes. A possible dispersion effect is within the experimental error, and therefore, could not be observed. When the dispersion term can be neglected, Lorentz's drag coefficient f_1 defined by Eq. (10.3.3b) reduces to Fresnel's drag coefficient f defined by Eq. (10.3.1b). Under the conditions of this experiment, the formula (10.3.5) gives the prediction $\Delta = 0.404$ which is in agreement with the observed value 0.46 within the experimental error.

Table 10.1. *Results of Zeeman's flowing water experiment*

λ_0 (Å)	Δ_f	Δ_{f_1}	Δ_{ob}
4500	0.786	0.852	0.826± 0.007
4580	0.771	0.808	0.808± 0.005
5461	0.637	0.660	0.656± 0.005
6870	0.500	0.513	0.511± 0.007

Thirty-five years later, Michelson and Morley [159] in 1886 performed a similar experiment where the length of the water tuber $l = 10$ m, the velocity of water $v = 1$ m/sec, the wave length $\lambda_0 = 5700$ Å, for the sodium spectral D1 the index of refraction of water $n = 1.33$, the observed shift $\Delta_{ob} = 0.1840$. Putting these data into Eq. (10.3.5) one can obtain Fresnel's coefficient $f_{ob} = 0.434 \pm 0.02$, while the theoretical value is $f_{(theor)} = 1 - (1/n^2) = 0.437$ and Lorentz's coefficient is $f_{1(theor)} = 1 - (1/n^2) - (\nu_0/n)(dn/d\nu_0) = 0.451$. We see that the observed value is in agreement with both the theoretical values $f_{(theor)}$ and $f_{1(theor)}$ within the experimental error. This is to say that the experiment cannot distinguish $f_{(theor)}$ from $f_{1(theor)}$.

The fringe shift is sometimes called *Fizeau's effect*. In the above experiments the source emits white light, so that it is difficult to determine the value of the wave length λ_0. Therefore the experimental errors are large. To distinguish f_1 from f a more accurate experiment is needed.

From 1914 to 1922, Zeeman [160–163] performed a set of experiments in which a water current and a moving transparent solid rod were used.

In the experiment [160] in which a water current was used, $2l = 604.0$ cm, $v = 553.6$ cm/sec, the monochromatic light rays are yellow, green and violet. In table 10.1, the observed values are compared with the corresponding calculation results, where a value of Δ_{f_1} represents a theoretical value given by Eqs. (10.3.5) and (10.3.3b), a value of Δ_f represents a theoretical value without dispersion given by Eqs. (10.3.5) and (10.3.1b), as well as a value of Δ_{ob} represents an observed fringe shift. By using the values in table 10.1, we can obtain the corresponding theoretical and observed values of Fresnel's and Lorentz's coefficients. The results are given in table 10.2 where a value of f or f_1 represents a theoretical value obtained from Eq. (10.3.1b) or Eq. (10.3.3b), as well as a value of $f_{1(obs)}$ corresponds to an observed value Δ_{ob}, i.e., $f_{1(obs)} = (c\lambda_0/8ln^2v)\Delta_{ob}$. It is shown from the tables that the experimental accuracy (1–2%) is enough to distinguish f_1 from f. These observed values are in agreement with the theoretical values of Lorentz's drag coefficient f_1. Thus this experiment is a test of Eqs. (10.3.5) and (10.3.3b).

Table 10.2. *Results of Zeeman's flowing water experiment*

λ_0 (Å)	f	f_1	$f_{1(obs)}$
4500	0.443	0.464	0.465
4580	0.442	0.463	0.463
5461	0.439	0.454	0.451
6870	0.435	0.447	0.445

Other Zeeman's experiments [161–163] were performed by use of moving solid bodies, instead of water. In order to compare these experiments with Einstein's theory of special relativity, we now first discuss theoretical prediction for the experiments. These experiments differ from the above kind of experiments by the moving surfaces of the solid bodies. In the present case we know that $v_x = 0, v_z = v$ and $\cos\gamma = 1$ (i.e., $\gamma = 0°$) in Eqs. (4.4.10), so that for the velocity of light in a solid body with the index of refraction n we have

$$u_\pm = \frac{c}{n} \pm f_2v, \qquad (10.3.6a)$$

with

$$f_2 = 1 - \frac{1}{n^2} + \frac{v_0}{n^2}\frac{dn}{dv_0}, \qquad (10.3.6b)$$

where $n = n(v_0)$ is a function of the wave length of light in empty space, "$\pm$" have the same meaning as those in Eq. (10.3.3a) and f_2 is called Laub's coefficient which was obtained firstly by Laub in 1908 from the relativistic Doppler effect. Let the length of the arms PM_1 and M_3M_2 of the interferometer (see Fig. 37 where the tube is replaced by moving solid bodies) be L, and the length of a moving solid rod be l. The path of light in an arm consists of three parts: The first and third parts are the paths in the empty space, the lengths of which being L_1 and L_3 ; The second is a path in the solid rod, being denoted by L_2. Thus we have $L = L_1 + L_2 + L_3$. For the light ray parallel to the direction of motion of the rod, the phase velocity is given by u_+ of Eq. (10.3.6a). Let t_+ represents the time passing through the path L. Then, we have $t_+ = t_1 + t_2 + t_3$, where $t_1 + t_3 = (L_1 + L_3)/c$ and $t_2 = L_2/u_+$. Note that L_2 is the path of light in the rod, which is longer than the length of the rod l due to the motion. Thus we have $L_2 - l = vt_2$ or $L_2 = l + vt_2$, and then to first-order

$$t_2 = \frac{L_2}{u_+} = \frac{l + vt_2}{u_+} \simeq \frac{l}{u_+} + n\frac{v}{c}t_2,$$

i.e.,

$$t_2 = \frac{l}{u_+ (1 - nv/c)}. \tag{10.3.7}$$

By substituting Eq. (10.3.6a) for u_+ in equation (10.3.7), we have

$$t_2 = \frac{nl}{c}\left[1 + \frac{v}{c}(1 - f_2)\right]. \tag{10.3.8}$$

From the result we write

$$t_3 = \frac{L_3}{c} = \frac{L - L_1 - L_2}{c} = \frac{L}{c} - \frac{L_1}{c} - \frac{vt_2}{c} - \frac{l}{c} = \frac{L - l}{c} - \frac{L_1}{c} - \frac{nl}{c}\frac{v}{c}. \tag{10.3.9}$$

The total time t_+ is then given by

$$t_+ = t_1 + t_2 + t_3 = \frac{L - l}{c} + \frac{nl}{c}\left\{1 - \frac{v}{c} + \frac{nv}{c}(1 - f_2)\right\}. \tag{10.3.10}$$

Furthermore the time t_- spent by the light ray propagation in opposite direction can be obtained from Eq. (10.3.10) by the replacement $v \to -v$:

$$t_- = \frac{L - l}{c} + \frac{nl}{c}\left\{1 + \frac{v}{c} - n\frac{v}{c}(1 - f_2)\right\}. \tag{10.3.11}$$

The phase difference of the two rays propagating in opposite directions is given by

$$\delta = \frac{c}{\lambda_0}(t_- - t_+) = \frac{2lv}{c\lambda_0}[n + n^2(f_2 - 1)]. \tag{10.3.12}$$

Inverse of the direction of motion, i.e., $v \to -v$, leads to a shift of the interference fringes

$$\Delta = 2\delta = \frac{4lv}{c\lambda_0}[n + n^2(f_2 - 1)]. \tag{10.3.13}$$

By using Eq. (10.3.6b) in Eq. (10.3.13), we get

$$\Delta = \frac{4vl}{c\lambda_0}\left[n - 1 + \nu_0\frac{dn}{d\nu_0}\right]. \tag{10.3.14}$$

This is the theoretical formula to be compared with those experiments involving moving solid bodies.

In the experiment involving moving quartz rods by Zeeman and Snethlage (1920) [162], end surfaces of the rods was perpendicular to the light axis of the quartz, while the light ray was propagating in the direction of the light axis. The length of the rod was 100 cm or 140 cm. The velocity of the rod $v \sim 1000$ cm/sec. The wave lengths of incident light rays were, respectively, 4750 Å, 5380 Å and 6510 Å. The comparisons between the observed results and theoretical prediction Eq. (10.3.14) are given in table 10.3.

Another experiment involving a moving flint glass rod was performed by Zeeman *et al.* in 1922 [163], in which $l = 120$ cm and $v = 1000$ cm/sec. They made eighty-three measurements and obtained the average value of the fringe shift 0.242 ± 0.004. However a white light beam was still used in the experiment. A effective wave length was determined as 4750 Å. From these data, equation (10.3.14) gives a theoretical value 0.242, which is in agreement with the above observed value.

Table 10.3. Results of Zeeman's moving quartz experiment

λ_0 (Å)	Δ_{ob}	Δ_{theor}
4750	0.156 ± 0.007	0.166
	0.156 ± 0.008	
5380	0.148 ± 0.006	0.143
	0.148 ± 0.012	
6510	0.125 ± 0.007	0.115
	0.123 ± 0.014	

10.3.2. The Transverse "Drag" Experiment

R.V. Jones (1970, 1972) [164] observed the transverse Fresnel drag experienced when light passes through a refracting medium moving at right angles to the original direction of the light. Later, his other experiment (1975) [165] had enough accuracy for the detection of dispersion. The arrangement of the type is shown in Fig. 38.

Light from the source S is focused by the condenser lens L_1 through the disk D and the lens L_2, forming an image of S mid-way between the two mirrors M_1 and M_2 that together constitute a corner reflector. The light then comes back on a parallel path through lens L_3 (similar to L_2) and again through D to L_4 (similar to L_1). A grid G_1 is interposed between L_1 and L_2 in the principal focal plane (allowing for the presence of D) of L_2; a unit magnification image of G_1 is therefore formed in the plane G_2, where an identical grid is placed. When the disk D moves transversely, the light passing through the disk will be dragged sideways relative to its incident direction. In this case the position of the images of G_1 in the plane of G_2 will therefore then be

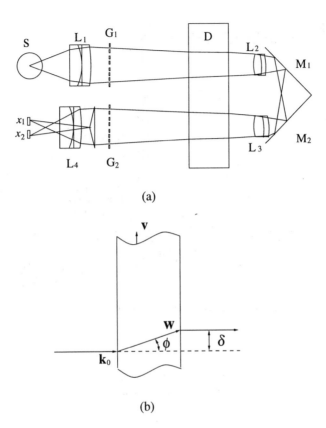

(a)

(b)

Fig. 38. The arrangement of Jones' transversal experiments.

displaced sideways. Because of the double passages of the light through the disk, the displacement of the images of G_1 will be twice the amount due to a single passage.

Let us now calculate the transverse displacement (δ) of the light beam through the disk D as shown in Fig. 38 (b) by means of the electrodynamics for moving bodies. The direction of the light ray inside the rotating disk is just the direction of group velocity $\mathbf{W}$. For the present experiment the incident angle $\theta_i = 0$. Equation (4.4.12) then gives the refraction angle $\theta = 0$, i.e., the wave vector $\mathbf{k}$ inside the disk is still perpendicular to the tangential velocity $\mathbf{v} = (v, 0, 0)$. Equation (4.3.10) then gives the direction of the light ray inside the disk,

$$\tan \psi = n f_1 \frac{v}{c}, \tag{10.3.15a}$$

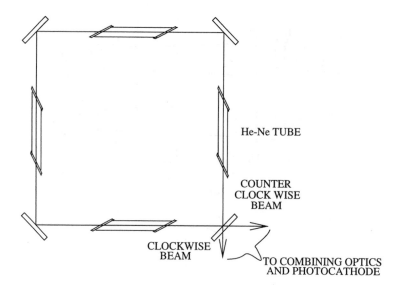

Fig. 39. Traveling wave ring laser schematic diagram.

where

$$f_1 = 1 - \frac{1}{n^2} + \frac{\nu_0}{n} \frac{dn}{d\nu_0}. \tag{10.3.15b}$$

and

$$v = v_x = |\mathbf{\Omega} \times \mathbf{r}|, \tag{10.3.15c}$$

with $\mathbf{r}$ being the radial vector of incident point on the surface of the disk, and $\mathbf{\Omega}$ being the angular velocity. By definition of the displacement δ as shown in Fig. 38 (b), we have

$$\delta = l \tan \psi = nl \left(1 - \frac{1}{n^2} + \frac{\nu_0}{n} \frac{dn}{d\nu_0} \right), \tag{10.3.16}$$

where l is the thickness of the moving disk. For the arrangement shown in Fig. 38 (a) the total displacement is 2δ due to the double passage through the disk.

In Jones' primary experiment [164(a)], a change in rotation speed from 600 to 1800 rpm of a glass disk, of 19.1mm thickness and refractive index 1.51, produced a transverse displacement of about 1.5×10^{-6} mm in a light beam passing twice through the disk at a radius of 110 mm from the axis of rotation. This result agrees with the prediction 2δ, δ being given by (10.3.16) without the dispersion term, to within the 10% accuracy of the experiment.

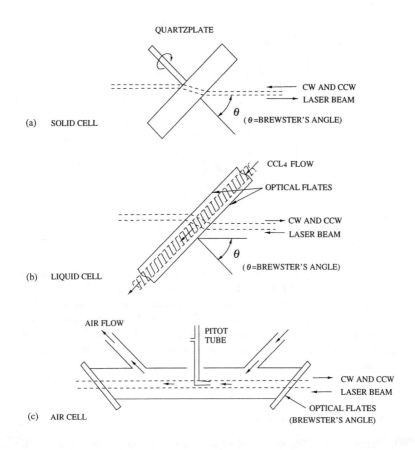

Fig. 40. Moving solid, liquid, and gas cell holders.

Later, Jones performed another measurement with high accuracy [164(b)]. For light traversing a disk of $l = 2.456$ cm thickness and refractive index $n = 1.524$ at an operating radius of $r = 13.75$ cm, when the disk was reversed from $+1501.9$ to -1501.9 rev/min the observed displacement was 6.175 nm $\pm$ 0.016 nm standard deviation (1nm$= 10^{-9}$m). This result is in good agreement with the expected value of 6.174 nm given by the theoretical formula (10.3.16) without the dispersion term.

In 1975, Jones [165] performed a more accurate experiment in which a glass disk with large dispersion was used. The observed result is in agreement with a prediction by Eq. (10.3.16), a dispersion effect being observed.

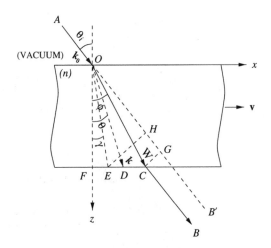

Fig. 41. Path length of the light ray in the ring laser. The total cavity path length of the ring laser is $L = AB'A$. $\mathbf{k_0}$ and $\mathbf{k}$ are the wave vectors, respectively, in the empty space and medium flat . θ_i and θ are incident and refractive angles, respectively. The direction of velocity $\mathbf{v}$ of the medium flat is along x-axis. $\mathbf{W}$ is a group velocity. The path length of the light ray in the medium flat is $l'' = \overrightarrow{OC}$. $l = \overrightarrow{OE}$ is the path of light when the medium is at rest, γ is the angle of refraction. $l' = OF$ is the thickness of the medium flat.

10.3.3. Other "Drag" Experiments

In 1964, Macek *et al.* [166] measured phase changes in optical length as a difference between resonant frequencies for oppositely directed waves by use of a ring laser shown in Fig. 39.

In figure 39 the ring laser used the 1.153 μ line of a He-Ne gas system in a meter square optical resonator, with gas tubes in three legs and with the fourth leg containing the moving medium. Figure 40 illustrates the various means of inserting the moving media into the optical path of the laser cavity. For the solid, a quartz optical flat was tilted to Brewster's angle [see Eq. (4.4.11b) for the definition] to minimize reflection losses, and then rotated about an axis normal to the flat surface. A component of quartz motion along the ring laser optical path exists whenever the path of the refracted light ray through the flat and the axis of rotation of the flat are noncoplanar. For the gas, dry air flowed through an axially oriented tube. For the liquid, CCl_4 flowing between two optical flats inserted in the cavity at Brewster's angle produced a velocity component of the moving medium along the refracted light path.

Since the cw oscillation is obtained in the ring laser configuration by directing the traveling waves around the complete optical circuit instead of retroreflecting them, the clockwise and counter-clockwise modes are independent. These modes were extracted through one of the corner mirrors, rendered collinear by external combining optics, and mixed on a photocathode. The frequency splitting arises from the removal of the mode degeneracy existing for the two oppositely traveling waves, due to the differential cavity pathlength change produced by the moving media within the laser cavity. The resulting mode split is, therefore, proportional in frequency to the axial velocity component of the moving media and the ratio of the path length through the media to the total cavity path length.

We now calculate the frequency splitting by making use of the electrodynamics equations in moving media as given in Secs. 4.3 and 4.4. The geometry for this calculating is shown in Fig. 41.

It is shown from the geometry in Fig. 41 that the total geometric path of the light ray in the ring laser is $L'' = L_0 + l''$, where $L_0 = AO + CBA$ is the path length in empty space, which can be expressed as $L_0 = L - l'' \cos(\theta_i - \psi)$ with L being the total cavity path of the laser, and $l'' = OC = |l''|$ is the path length in the medium flat. Thus the total path length of light is written as

$$\Delta = n_0 L_0 + \Delta'' = [L - l'' \cos(\theta_i - \psi)] + \Delta'', \tag{10.3.17}$$

where $n_0 \Delta_0$ is the path length of light in the empty space, $n_0 = 1$ being assumed, and Δ'' is the path length of light in the medium flat. By definition of a path length of light, the wave number $(k_0 \Delta''/2\pi)$ involving at the length Δ'' in vacuum is equal to the wave number $(\mathbf{k} \cdot l''/2\pi)$ involving at the length l'' in the medium. Thus we have

$$\Delta'' = \frac{1}{k_0} \mathbf{k} \cdot \mathbf{l}'' = \frac{k l''}{k_0} \cos(\psi - \theta). \tag{10.3.18}$$

By inserting $v_x = v$ and $v_z = 0$ in Eqs. (4.4.10a), (4.3.10), and (4.4.12), we have

$$\frac{k}{k_0} = \frac{c}{u_\pm} = n(1 \mp f_1 vn \sin\gamma), \tag{10.3.19}$$

$$\tan\psi = \frac{\sin\theta \pm nf_1 v}{\cos\theta}, \tag{10.3.20}$$

$$\sin\theta = \sin\gamma(1 \pm nf_1 v \sin\gamma). \tag{10.3.21}$$

Using Eqs. (10.3.19)–(10.3.21) in (10.3.18), we obtain the path length for two oppositely directed waves in the medium

$$\Delta_\pm = \Delta_0' \mp l n^2 f_1 \frac{v_m}{c}, \tag{10.3.22}$$

where Δ_0' is the path length of light in the rest medium (i.e., $v = 0$), which is approximately equal to the total cavity path length (L) of the laser due to L being

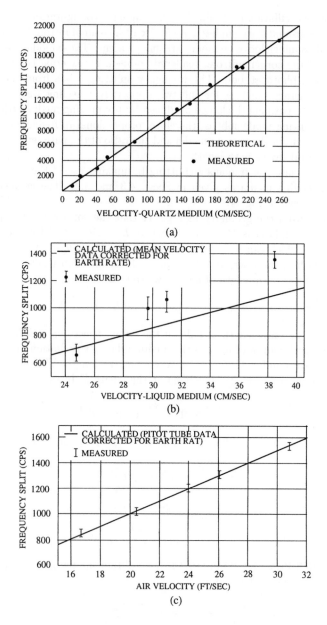

Fig. 42. Fresnel drag frequency mode split curves; (a) solid medium (quartz), (b) liquid medium (CCl$_4$), (c) gaseous medium (dry air).

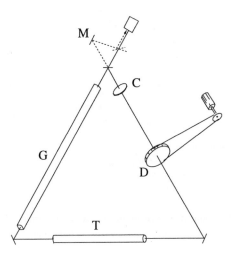

Fig. 43. Schematic of triangular ring laser with He-Ne plasma tube T (bottom arm), drag disk D, compensator flat C (right arm), and an open glass tube G in the left arm to reduce unwanted Fresnel drag by moving air. The beam mixer is behind the mirror M. The drive for the drag disk is mounted off the granite support and coupled to the disk by a cotton thread "belt."

much larger than $l = OE$, the path length in the rest medium, and $v_m = v \sin \gamma$ is the component of $\mathbf{v}$ in the direction of $\mathbf{l} = \overrightarrow{OE}$.

By use of the oscillation condition for the ring laser, $\Delta = N\lambda$ with N being an integral number, we obtain the ring laser mode split or beat frequency $\delta \nu$ from (10.3.22)

$$\frac{\delta \nu}{\nu} = -\frac{\delta \lambda}{\lambda} = -\frac{\delta \Delta}{\Delta}$$

or explicitly

$$\delta \nu \cong \frac{2l v_m n^2}{L \lambda_0} \left(1 - \frac{1}{n^2} + \frac{\nu_0}{n} \frac{dn}{d\nu_0} \right). \qquad (10.3.23)$$

The experiments by Macek *et al.* (1964) [166] were not enough accurate to detect a possible dispersion effect, so that, equation (10.3.23) without the dispersion term was used to compare with the experiments. The experimental results are plotted in Fig. 42. The observed values for the rotating quartz flat and the dry air flow are quite consistent with the predicted values given by Eq. (10.3.23) without dispersion [see Fig. 42(a,c)]. Note that for the gas, v_m represents the velocity of the air flow and l is its length [see Fig. 40(c)]. The flowing liquid experiments did not yield good

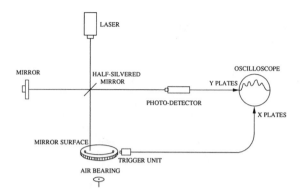

Fig. 44. The arrangement for synchronizing the display from a Michelson interferometer in order to investigate any frequency shift in light reflected from a transversely moving surface.

correlation between theory and results [see Fig. 42(b)], due primarily to the overly simplified procedure employed to monitor the liquid flow velocity.

Later, Bilger and Zavadny (1972) [167] performed more accurate measurements of the Fresnel drag in a triangular ring laser (see Fig. 43) at a wavelength of 0.2368 μ. The experiment is in agreement with Eq. (10.3.23). The observed result was given by a value of Lorentz coefficient f_1: In fused silica, the measured value is

$$f_{1(obs)} = 0.541 \pm 0.003,$$

while the theory gives

$$f_{1(theor)} = 1 - \frac{1}{n^2} + \beta \frac{\nu_0}{n} \frac{dn}{d\nu_0} = 0.5423, \qquad \text{for} \quad \beta = 1.$$

The coefficient β is thus determined as $\beta = 0.87 \pm 0.22$, which includes the theoretical value $\beta = 1$.

10.4. Reflection at Moving Mirrors

The reflection of light by moving mirrors and the refraction of light between moving media can be treated in a similar way on the basis of the electrodynamics in moving media, as done in Sec. 4.4. Equation (4.4.8b) shows that $\omega_r = \omega_0$ in case of $v_z = 0$. This result is accurate. To do this, let us re-derive the reflection law of light on a surface of a moving medium in Fig. 5 where $v_z = 0$ is now assumed. By using the transformations for the wave vector $\mathbf{k}$ and frequency ω, Eqs. (2.10.6), in the reflection law (4.4.2) on a rest surface, we have

$$(k_r)_x - (k_0)_x = (\omega_r - \omega_0)\frac{v_x}{c^2}, \tag{10.4.1}$$

$$\omega_r - \omega_0 = v_x[(k_r)_x - (k_0)_x]. \tag{10.4.2}$$

Putting (10.4.1) into (10.4.2), we obtain

$$\omega_r = \omega_0. \tag{10.4.3}$$

This is to say that when the surface of a medium is parallel with the direction of motion itself, the reflection frequency ω_r will be equal to its incident frequency ω_0.

Experiments for the reflection of light by moving mirrors have been performed by Sagnac and others (1913, 1914) [170]. The results of these experiments are in accordance with the relativistic Doppler formula.

An experiment on the reflection from a transversely moving mirror was done by Jennison and Davies (1974) [171] by use of an arrangement shown in Fig. 44. A rotating mirror, supported by an air bearing, is in one arm of a Michelson interferometer. The other arm has a mirror which can be moved in the direction of the reference light beam by a crystal transducer. The transverse speeds at the incident point of light on the rotating mirror were of 5–20 m s^{-1}. Observation showed no evidence of any shift in the position of the fringes in the display, the accuracy being better than one part in 10^{16}.

CHAPTER 11

THE TESTS OF RELATIVISTIC MECHANICS

11.1. The Test of Variation of Mass with Velocity

The first experimental investigation concerning the dependence of mass on velocity was carried out by Kaufmann (1901) [172] with the aid of the parabola method. In determining the mass-to-charge ratio e/m of the β rays, he first found that the value of m varied with the velocity of the electron if the independence of the charge e on the velocity is assumed. In order to give an explanation of this phenomenon, many expressions for mass–velocity relation were derived from different models in the ether theory and Lorentz's theory of electron. Abraham (1903) [173] assumed that an electron is a rigid sphere moving with a velocity u, and then obtained the mass formula,

$$m = \frac{3}{4} \frac{m_0}{\beta^2} \left[\frac{1+\beta^2}{2\beta} \ln \frac{1+\beta}{1-\beta} - 1 \right], \tag{11.1.1}$$

where $\beta = u/c$. Lorentz (1904) [174] assumed that the mass of an electron is produced by electromagnetic interaction and the volume of the electron will contract along direction of its motion, and then obtained

$$m = \frac{m_0}{\sqrt{1 - u^2/c^2}}, \tag{11.1.2}$$

which has the same form as that of the relativistic mass equation (3.1.6).

Kaufmann's experiment yielded no decision between the competing theories of Abraham and Lorentz. Later, some measurements were performed, which agree with the predictions of Lorentz–Einstein's formula (11.1.2).

11.1.1. Deflection of Charged Particles in Electric and Magnetic Fields

(I) *Magnetic Deflection.*

Let us consider a particle with a rest mass m_0 and a charge e moves in a magnetostatic field $\mathbf{B}$. The force acting on the particle by the magnetic field then is

$$\mathbf{F} = \frac{e}{c} \mathbf{u} \times \mathbf{B} = \frac{e}{c} \mathbf{u}_\perp \times \mathbf{B}, \tag{11.1.3}$$

where $\mathbf{u}_\perp$ is the component of the particle velocity $\mathbf{u}$ in the vertical direction of $\mathbf{B}$. This equation shows that the force $\mathbf{F}$ is perpendicular to the velocity $\mathbf{u}$ so that

$\mathbf{F} \cdot \mathbf{u} = 0$. Equation of motion (3.1.9) then reduces to

$$m\frac{d\mathbf{u}}{dt} = \mathbf{F} = \frac{e}{c}\mathbf{u}_\perp \times \mathbf{B}, \tag{11.1.4}$$

where m is the relativistic mass given by Eq. (11.1.2). In the present case, equations (3.1.10) and (3.1.16b) give $dT/dt = dE/dt = 0$, i.e., that the energy $E = mc^2$ is constant in time. The mass m and velocity u are, therefore, constant in time.

Equation (11.1.4) shows that $\mathbf{F} \perp \mathbf{B}$ so that the component $u_\parallel$ of the velocity in the direction of $\mathbf{B}$ is constant in time. Thus the vertical component $u_\perp$ is also constant in time. These show that the orbit of motion for a charged particle in a uniform magnetostatic field is a helix about the direction of $\mathbf{B}$, the projection of which on a plane vertical to the direction of $\mathbf{B}$ is a circle. Let ρ denote the radius of the circle. Equation (11.1.4) reduces to

$$\frac{mu_\perp^2}{\rho} = \frac{e}{c}u_\perp B, \tag{11.1.5}$$

or

$$p_\perp = mu_\perp = \frac{e}{c}B\rho, \tag{11.1.6}$$

with $B = |\mathbf{B}|$. In case of the direction of $\mathbf{u}$ being perpendicular to the direction of $\mathbf{B}$, the equation above is simply written as

$$p = mu = \frac{e}{c}B\rho. \tag{11.1.7}$$

This shows that by measuring the radius ρ of the orbit for a charged particle in a known magnetic field $\mathbf{B}$, one could get the momentum p of the particle. Furthermore, if the velocity u (or the kinetic energy T) is known, then the mass m could be found from the equation $p = mu$ [or $T = (m - m_0)c^2$].

(II) *Electric Deflection*

The force ($\mathbf{F} = e\mathbf{E}$) acting on a charged particle in a uniform electrostatic field $\mathbf{E}$ is a constant quantity. A particle, if its initial velocity $\mathbf{u}$ is parallel to the direction of the force, will move along a straight line which is treated as the x-axis. In this case, equation (3.1.8) becomes

$$\frac{d}{dt}\left\{\frac{u}{\sqrt{1 - u^2/c^2}}\right\} = g, \tag{11.1.8}$$

where

$$g \equiv \frac{F}{m_0} = e\frac{E}{m_0}. \tag{11.1.9}$$

At the initial moment ($t = 0$), the velocity of the particle is assumed to be zero. By integrating Eq. (11.1.8) we then have

$$\frac{u}{\sqrt{1 - u^2/c^2}} = gt, \tag{11.1.10}$$

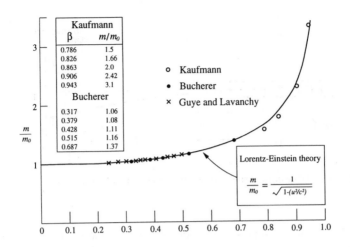

Fig. 45. Variation of electron mass with velocity.

or

$$u = \frac{dx}{dt} = \frac{gt}{\sqrt{1 + (gt/c)^2}}.$$ (11.1.11)

Let $x = 0$ at $t = 0$. Integration of the above equation gives

$$x = \frac{c^2}{g}\sqrt{1 + \left(\frac{gt}{c}\right)^2} - \frac{c^2}{g},$$ (11.1.12)

or

$$\left(x + \frac{c^2}{g}\right)^2 - c^2 t^2 = \frac{c^4}{g^2}.$$ (11.1.13)

For $(gt)^2 \ll c^2$, neglecting higher terms more than $(gt/c)^2$, equation (11.1.12) reduces to the law of motion,

$$x = \frac{1}{2}gt^2.$$ (11.1.14)

If the initial velocity $\mathbf{u}$ of the particle is perpendicular to the uniform electrostatic field $\mathbf{E}$, in the lowest order approximation, equation (11.1.14) is valid for the displacement perpendicular to the initial velocity. Thus we have

$$\Delta X = \frac{1}{2}gt^2 = \frac{1}{2}\frac{eE}{m}\frac{l^2}{u^2},$$ (11.1.15)

where ΔX represents the deviation of the motion of the particle from the direction of initial velocity, $t = l/u$ and $g = eE/m$ are used, and m is the relativistic mass of the

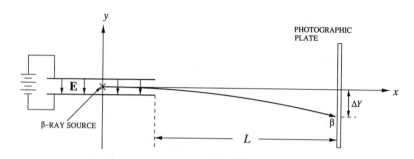

Fig. 46. Scheme of the experimental arrangement used by Butchere.

particle. This equation represents the deflection of a charged particle in a uniform electrostatic field. The quantities $\Delta X, E$ and l can be all directly measured in the laboratory.

(III) *Measurements for Electric and Magnetic Deflections of Electrons.*

Equation (11.1.7) shows that by measuring the deflection of a charged particle in a known magnetic field, we can determine the momentum mu. On the other hand, equation (11.1.15) indicates that by measuring the deflection of a charged particle in a known electric field we can find a value of the quantity mu^2. Therefore, from measurements for the deflections of a charged particle in electric and magnetic fields we can decide the dependence of the mass on the velocity for the particle. We call this way the electromagnetic deflection method. Kaufmann (1901) [172] used this method to determine the relation between the mass-to-charge ratio and velocity for β-particles. His results are given in Fig. 45.

By use of a similar method, Butchere (1909) [175] measured again the mass-to-charge ratio of β-particles. The arrangement is shown in Fig. 46. In the center between the circular plates of a condenser, there is a source of β-rays. The system is placed in a magnetic field $\mathbf{B}$ (the direction of $\mathbf{B}$ is assumed to be parallel to the positive direction of z-axis) which is perpendicular to the electric field $\mathbf{E}$ (along the negative direction of y-axis) between the condenser plates. An electron, with the charge $-e$, moving in the direction of x-axis will be acted on by two kinds of forces: an electric force $(e\mathbf{E})$ in the positive direction of y-axis, and a magnetic force $(eu\mathbf{B}/c)$ in the negative direction of y-axis. If the magnitudes of the two forces are not the same, then motion of the electron, due to the action of a non-zero resultant of forces, would deviate from the direction of x-axis and thus reaches the plate of the condenser. Therefore only such electrons can leave the condenser, whose velocity has a certain value defined by the electric and magnetic field intensities. In other words, they are those electrons on which the effects of the two kinds of fields are just compensated each other. This implies that the equilibrium condition must be satisfied,

$$\frac{e}{c}uB = eE,$$

or

$$u = \frac{cE}{B}. \tag{11.1.16}$$

It is the electron with a velocity equal to cE/B that can leave the condenser. We see that the function of the condenser is like that of a selector of velocity. Outside the condenser the motion of the electrons is influenced only by the magnetic field and thus its curvature radius ρ is given by Eq. (11.1.7). By inserting Eq. (11.1.16) in Eq. (11.1.7), we obtain

$$\rho = \left(\frac{c^2}{e}\right)\frac{mE}{B^2}. \tag{11.1.17}$$

In order to observe the displacement ΔY for those electrons from the condenser, a photographic plate parallel with the x-y plane is placed outside the condenser. Let L denote the distance between the photographic plate and condenser. From the geometry as shown in Fig. 46, we have the relationship,

$$\Delta Y(2\rho - \Delta Y) = L^2,$$

or

$$\rho = \frac{L^2 + \Delta Y^2}{2\Delta Y} \simeq \frac{1}{2}\frac{L^2}{\Delta Y}. \tag{11.1.18}$$

By putting this equation into Eq. (11.1.17) we obtain

$$\frac{e}{m} = \left(\frac{2\Delta Y}{L^2 + \Delta Y^2}\right)\frac{c^2 E}{B^2} \simeq \frac{2c^2 E \Delta Y}{L^2 B^2}. \tag{11.1.19}$$

This result shows that one can determine the mass-to-charge ratio of an electron by measuring the quantities E, B, L, and ΔY. Results obtained by Butchere (1909) in this way are shown in Fig. 45. Later, Neumann (1914) [176] repeated the experiment of the Butchere type.

The first experiment on variation of mass of artificially accelerated electrons was carried out in 1908 by Hupka [177]. The electron beam was deflected by the magnetic field of a pair of coils. The deflecting field was varied in such a way that for each value of the accelerating voltage, the deflection was kept constant. Since the deflecting coils did not contain iron, the deflecting field intensity was proportional to the coil current. By measuring the accelerating voltage and coil current, the relation between mass and velocity could be determined with the aid of the expression of the momentum and kinetic energy. Due to the lower accuracy, Hupka's results could not distinguish Eq. (11.1.2) from Eq. (11.1.1).

Guye and Lavanchy (1915) [178] measured the electric and magnetic deflections of the electron beam from a cathode ray tube. The experimental principle is the following: The electron beam is deflected once by an electric field E and once by a magnetic field B. The two kinds of deflections were measured separately. The electric field E was produced by a voltage V on a deflection plate, E being proportional to the voltage V. The magnetic field B was caused by a hollow coil, and thus B was proportional to the current I. Then, from Eqs. (11.1.15), (11.1.16), and (11.1.7) we get the electric deflection:

$$\Delta X = A_1 \frac{V}{mu^2}, \qquad (11.1.20)$$

and the magnetic deflection:

$$\Delta Y = A_2 \frac{I}{mu}, \qquad (11.1.21)$$

where A_1 and A_2 are two proportional constants depending on the geometry of the experimental arrangement. The V and I are adjusted to such strength that the path of the electrons under investigation coincides with the path followed by a reference beam of low-speed electrons. This is to say that the high-speed and low-speed beams have the same deflections ΔX and ΔY. Thus from the above equations we have

$$\frac{u}{u'} = \frac{I'V}{IV'}, \qquad (11.1.22)$$

and

$$\frac{m}{m'} = \frac{V'I^2}{V(I')^2}, \qquad (11.1.22)$$

where the unprimed quantities represent those quantities for the high-speed electrons, and the primed quantities denote the quantities of the reference electrons. In this way, Guye and Lavanchy (1915) made all 2000 separate determinations of m/m' for electrons with speeds ranging from 0.26c to 0.48c. A few of their determinations are plotted in Fig. 45. As a group their experiments may be interpreted as confirming the Lorentz–Einstein law, Eq. (11.1.2), to one part in two thousand.

Rogers, McReynolds and Rogers (1940) [179] made three kinds of β-particles with three different speeds, which were emitted from the RaB source, enter perpendicularly to a uniform magnetic field. The orbits of the electrons were circles, the radii of which are given by Eq. (11.1.7). In this way they measured the quantity:

$$B\rho = \frac{mu}{c}. \qquad (11.1.23)$$

At the same time they made the same electrons move in a radial electrostatic field E, and measured the quantity:

$$ER = \frac{mu^2}{e}, \qquad (11.1.24)$$

where R was the radius of circular orbit for the electrons. From the observed values of $B\rho$ and ER in the above equations, they obtained the values of the mass and velocity of the electrons:

$$\frac{m}{c} = \frac{B^2\rho^2}{c^2 ER}, \tag{11.1.25}$$

$$u = \frac{cER}{B\rho}. \tag{11.1.26}$$

The velocities of the three electrons were $0.63c, 0.69c$ and $0.75c$. The measurements have an accuracy of 1.0% that is less than one tenth of the difference between Eqs. (11.1.1) and (11.1.2). Their measurements agree with the predictions of Lorentz–Einstein's equation (11.1.2), as shown in Tab. 11.1 and Fig. 47.

Table 11.1 *Comparisons between the measurements of Rogers et al. and predictions*

Speeds of electrons	m/m_0 measured	m/m_0 from (11.1.2)	m/m_0 from (11.1.1)
0.6337c	1.298	1.293	1.220
0.6961c	1.404	1.393	1.290
0.7496c	1.507	1.511	1.369

11.1.2. High Energy Synchro-Cyclotron

Equation (11.1.7) shows that the orbit of a charged particle moving perpendicularly to the magnetic field **B** is a circle of radius ρ. The angular velocity of the particle is

$$\omega = \frac{u}{\rho}. \tag{11.1.27}$$

By use of Eq. (11.1.7), the above equation becomes

$$\omega = \left(\frac{e}{m}\right)\frac{B}{c}. \tag{11.1.28}$$

Here m is the relativistic mass for the particle. The frequency of an accelerating voltage should be equal to the angular frequency ω (synchronization). Thus the great success of constructing high energy synchro-cyclotrons has provided a powerful test for equation (11.1.2).

Equation (11.1.28) shows that by directly measuring the angular frequency ω and magnetic field B we can obtain the mass for a charged particle moving in a synchrocyclotron. The measurements of this kind for 385-MeV protons (corresponding to

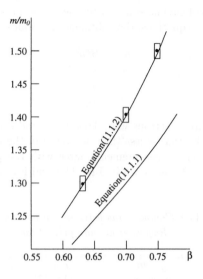

Fig. 47. Comparisons between the measurements of Rogers *et al.* and predictions.

a velocity of $u \sim 0.7c$) were performed by Grove and Fox (1953) [180] . From the measured values of ω and B they calculated the quantity,

$$\left(\frac{e}{m}\right)_1 = \frac{\omega c}{B}. \tag{11.1.29}$$

At the same time, by putting the value of the orbital radius ρ into Eq. (11.1.27) and then using (11.1.2), they calculated the relativistic prediction:

$$\left(\frac{e}{m}\right)_2 = \frac{e}{m_0}\sqrt{1 - \frac{u^2}{c^2}} = \frac{e}{m_0}\sqrt{1 - \frac{\omega^2\rho^2}{c^2}}. \tag{11.1.30}$$

From the 64 measurements they obtained

$$\frac{(e/m)_1 - (e/m)_2}{(e/m)_2} = -0.0006 \pm 0.001.$$

This is in agreement with the prediction to within an accuracy of 0.1%.

11.1.3. Other Measurements

Meyer *et al.* (1963) [181] reported an experiment for testing the mass–momentum relation of the electrons at values of the velocity from $0.987c$ to $0.990c$, who used such a method by which the electrons and non relativistic protons moved on identical orbits in a cylindrical electric field. The momentum of the electrons was still determined by the magnetic deflection. The experimental result was given by an average value $(\bar{Y})$ of the quantity

$$Y = \frac{m/m_0}{\sqrt{1 + p^2/m_0^2 c^2}}.$$

The result was $\bar{Y} = 0.00037 \pm 0.00036$ thereby agreeing with the value of unity given by the Lorentz–Einstein law, Eq. (11.1.2), to this accuracy.

Geller and Koliarits (1972) [182] measured the mass increase with velocity of relativistic electrons. A beta-ray spectrometer was used to momentum-analyze electrons emitted from a Tl^{204} source, and a semiconductor detector was used for measuring the energy of the electrons. From Eqs. (3.1.18) and (3.1.16a) one arrives at

$$\frac{p^2}{2T} = m_0 + \frac{T}{2c^2}.$$

Here the momentum is given by Eq. (11.1.7):

$$p = \frac{e}{c} B\rho.$$

Thus, by use of the experimental values of B, ρ, and T for plotting the quantity $(p^2/2T)$ as a function of T, they obtained the values of the rest mass m_0 of the electron and the speed of light c: $m_0 = (8.99 \pm 0.30) \times 10^{-31}$ gram; $c = (2.99 \pm 0.11) \times 10^8$ m/sec. These agree with the usual values.

In interpreting all the above experiments, the motion law for charged particles in electric and magnetic fields are used. Thus these experiments indeed test the relativistic mechanics as well as the law of motion of a charged particle in electric and magnetic fields. After the independence of charge on velocity is assumed [1], one could come to the conclusion of the validity of the Lorentz–Einstein relation (11.1.2).

11.1.4. The Measurements of Flight-Time for Moving Particles

In order to separate the law of electromagnetic motion of charged particles from the mass–velocity law, Bertozzi (1964) [184] performed an experiment in which the speeds of electrons with kinetic energies in the range 0.5–15 MeV were determined by measuring the time required for the electrons to traverse a given distance and the

---

[1] Many measurements have shown that atoms and molecules have electrical neutrality. For instance, an experiment by Dylla and King (1973) [183] yielded the upper limit for the charge per molecule (ϵ of SF_6): $|\epsilon| \leq 2 \times 10^{-19} e$ and $|\epsilon|/M \leq 1 \times 10^{-21} e$, where e is the magnitude of the electronic charge and M is the total number of nucleons. This is consistent with the fundamental belief of zero charge in balance. Due to the motion of electrons inside atoms, this kind of measurements is usually regarded as the experimental proofs of the independence of charge on velocity.

kinetic energy was determined by calorimetry. His result shows that the dependence of the kinetic energies on their velocities of the electrons agrees with the following relation obtained from Eq. (3.1.14a):

$$\frac{u^2}{c^2} = 1 - \left(\frac{m_0 c^2}{m_0 c^2 + T}\right)^2. \tag{11.1.31}$$

If the kinetic energy T is much greater than the proper energy $m_0 c^2$, to the lowest order in $(m_0 c^2/T)^2$, the above equation reduces to

$$\frac{c - u}{c} \simeq \frac{m_0^2 c^4}{2T^2}. \tag{11.1.32}$$

Brown *et al.* (1973) [185] compared the velocity of 11-GeV electrons to the velocity of visible light by use of time-of-flight techniques. Their result was given by

$$\frac{c - u}{c} = \frac{\Delta t}{t} = (-1.3 \pm 2.7) \times 10^{-6},$$

where t was the time required for the visible light to traverse a given distance, Δt represented the time difference of the electron and light to traverse the same distance. This implies that the velocity of the 11-GeV electrons was found to be equal to the velocity of visible light to an accuracy of 10^{-4}. The prediction given by the relativistic equation (11.1.32) for the difference should be approximately equal to $m_0 c^2/aT^2 \simeq 10^{-9}$, which agrees with the observed value above. Later, Guiragossián *et al.* (1975) [186] performed a similar measurement at the Stanford Linear Accelerator Center. In this experiment the relative velocities of $\sim$ 15-GeV γ-rays and electrons with energies ranging from 15 to 20.5 GeV were measured by using a time-of-flight technique with 1-psec sensitivity and a flight path of $\sim$ 1 km. No significant difference in the velocities of light and electrons was observed to within $\sim$ 2 parts in 10^7. According to the relativistic equation (11.1.32), the expected difference should be $(c - u)/c \sim 5 \times 10^{-10}$. Thus the observed result agrees with the prediction.

11.1.5. Elastic Collisions

Lewis and Tolman (1909) [16] proposed a thought experiment concerning the elastic collision between two spheres with the same mass. This experiment shows that the relativistic mass–velocity relation (3.1.6) would be obtained provided that the conservation law of momentum is assumed to be valid in any inertial frames and that Einstein's law of the addition of velocities is used. We have known in Chap. 10 that Einstein's law of the addition of velocities can be tested by the electromagnetism experiments for moving bodies. Investigating the elastic collision among moving bodies with high speeds would then give a test for the conservation law and thus the relativistic mass–velocity relation. The first experiment of this kind was carried out by

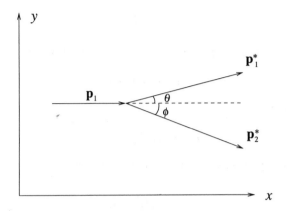

Fig. 48. Elastic collision between two particles.

Champion (1932) [187], who observed the scattering of β-particle beam from electrons at rest in a Wilson chamber. Since the velocities of the incident electron beam were very high, the binding energy of electrons in atoms at rest in the chamber can be neglected, so that the electrons could be regarded as at rest and thus the collisions between the electrons and incident electron beam can be regarded as those between two free particles. Under this approximation we can use the conservation laws and transformation equations for momentum and energy to analyze the scattering angle [13]. Let us assume that the cloud chamber is at rest in the laboratory frame F, in which the velocity and momentum of the electron 1 (an incident electron) be denoted by $\mathbf{u}_1$ and $\mathbf{p}_1 = m_0\mathbf{u}_1/\sqrt{1 - u_1^2/c^2}$, respectively, and the velocity and momentum of the electron 2 (an electron at rest in the chamber) be $\mathbf{u}_2 = \mathbf{p}_2 = 0$. Let $\mathbf{p}_1^*$ and $\mathbf{p}_2^*$ be the momenta of the two electrons after the collision, and θ and ϕ the angles between the direction of the incident electron and the vectors $\mathbf{p}_1^*$ and $\mathbf{p}_2^*$, respectively (cf. Fig. 48). For convenience we choose the Cartesian axes in the laboratory frame F in such a way that $\mathbf{p}_1$ is parallel to the x-axis and $\mathbf{p}_1^*$ lies in the xy-plane.

Then, if the momentum is conserved (z-component of the total momentum is zero before the collision, and then is also zero after the collision), $\mathbf{p}_2^*$ must also lie in the xy-plane. We now introduce the center of gravity system F' in which the total momentum $\mathbf{p}' = \mathbf{p}_1' + \mathbf{p}_2' = 0$. F' is moving relative to F with a certain velocity v in the direction of the x-axis. From the transformation equation (3.2.9a) between the x-components of the total momenta before the collision as seen in F and F', i.e., p_x and p_x',

$$\frac{p_x - vE/c^2}{\sqrt{1 - v^2/c^2}} = p_x' = 0,$$

we obtain

$$v = \frac{c^2 p_x}{E}.$$ (11.1.33)

Here the total momentum is $\mathbf{p} = \mathbf{p}_1 + \mathbf{p}_2 = \mathbf{p}_1 = (p_x, 0, 0)$, where

$$p_x = m_0 u_1 / \sqrt{1 - u_1^2/c^2};$$

and the total energy is

$$E = E_1 + E_2 = m_0 c^2 / \sqrt{1 - u_1^2/c^2} + m_0 c^2.$$

By putting these expressions into the above equation, we have

$$v = \frac{u_1}{1 + \sqrt{1 - u_1^2/c^2}}.$$ (11.1.34)

Since $\mathbf{p}_2' = -\mathbf{p}_1'$, the two electrons have the same initial velocity u' relative to F' and move in opposite directions parallel to the x-axis. Since, furthermore, the electron 2 was initially at rest in F ($\mathbf{u}_2 = 0$), from Einstein's law of addition of velocities, we know that the velocity u' relative to F' must be equal to the relative velocity v of F' and F, i.e.,

$$u' = v, \qquad E_1' = E_2' = \frac{m_0 c^2}{\sqrt{1 - v^2/c^2}} = \frac{1}{2} E'.$$ (11.1.35)

By use of the theorems of conservation of momentum and energy we have the momenta and energies for the two electrons after the collision:

$$\mathbf{p}_2'^* = -\mathbf{p}_1'^*, \qquad E_1'^* = E_2'^* = \frac{1}{2} E' = \frac{m_0 c^2}{\sqrt{1 - v^2/c^2}},$$ (11.1.36)

$$p_2'^* = p_1'^* = \sqrt{(E_1'^*)^2/c^2 - m_0 c^2} = m_0 v / \sqrt{1 - \frac{v^2}{c^2}}.$$ (11.1.37)

Consequently the two electrons will have the same velocity v relative to F' after the collision also and, according to (11.1.36) and (11.1.37), we have

$$\frac{v^2 (E_1'^*)^2}{c^4} = (p_1'^*)^2.$$ (11.1.38)

From Fig. 48 we see at once that

$$\tan \theta = \frac{p_{1y}^*}{p_{1x}^*}, \qquad \tan \phi = \frac{-p_{2y}^*}{p_{2y}^*}.$$ (11.1.39)

Using the transformations (3.2.9) and (3.2.8) for the momenta of the two electrons after the collision, we obtain by means of Eqs. (11.1.36)–(11.1.39)

$$\tan\theta\tan\phi = \frac{p_{1y}^*(-p_{2y}'^*)(1-v^2/c^2)}{(p_{1x}'^* + vE_1'^*/c^2)(p_{2x}'^* + vE_2'^*/c^2)}$$

$$= \frac{(p_{1y}'^*)^2(1-v^2/c^2)}{v^2(E_1'^*)^2/c^4 - (p_{1x}'^*)^2}$$

$$= 1 - \frac{v^2}{c^2}.$$

$$(11.1.40)$$

Using Eq. (11.1.34) in the equation above, we get

$$\tan\theta\tan\phi = \frac{2}{1+\gamma_1}. \qquad (11.1.41a)$$

where

$$\gamma_1 = \frac{1}{\sqrt{1 - u_1^2/c^2}}. \qquad (11.1.41b)$$

Equation (11.1.41a) shows that the product of $\tan\theta$ and $\tan\phi$, for the elastic collision between two particles with the same mass, is independent of the rest masses of the electrons and is a function of the velocity of the incident electron alone.

In the limit $c \to \infty$, equations (11.1.41) reduce to the corresponding formula of Newtonian mechanics:

$$\gamma_1 = 1, \qquad \tan\theta\tan\phi = 1,$$

i.e.,

$$\tan(\theta+\phi) = \frac{\tan\theta + \tan\phi}{1 - \tan\theta\tan\phi} = \infty,$$

or

$$\theta + \phi = \frac{\pi}{2}, \qquad (11.1.42)$$

Equation (11.1.42) shows that according to Newtonian mechanics the directions of motion of two particles after the collision should be perpendicular to each other.

When the velocity of the incident electron approaches the velocity of light, however, $\gamma_1 > 1$ and thus

$$\tan\theta\tan\phi < 1,$$

which shows that the angle between the directions of motion of the electrons after the collision in this case will be smaller than 90°. The first experiment for testing the relativistic equation (11.1.41) was carried out by Champion (1932) [187], who directly determined θ and ϕ by use of photograph concerning the collision between the β-particles and the electrons at rest in chamber, and found agreement with Eq.

(11.1.41). This may be regarded as an experimental proof of the theorems of conservation of momentum and energy in collisions between electrons, and thus also of Lorentz–Einstein's formula, Eq. (11.1.2), for the variation of mass with velocity[2].

11.1.6. The Fine Structure of Atomic Spectra

In order to explain the fine-structure splitting of hydrogen spectral lines, Sommerfeld (1916) applied the relativistic correction for the mechanical model of Bohr theory and claimed that the relativistic correction is the origin of the fine-splitting of spectral lines. He proved that the relativistic expressions for momentum and energy could give an explanation of the fine-structure splitting of hydrogen spectral lines (see, e.g., Ref. [190] for the detail).

Glitscher (1917) [191] investigated in detail the problem concerning the fine-structure splitting of atomic spectral lines, who, by use of Bohr theory, obtained different expressions of the fine-structure splitting according to different mass–velocity relations. In the limit of the mass of atom going to infinity, the splitting is given by

$$\Delta \nu = 2\gamma_1 \left(\frac{z}{2}\right)^2 R_\infty \alpha^2, \qquad (11.1.43)$$

where $\alpha = e^2/\hbar c$ is the fine structure constant, R_∞ is Rydberg constant, and γ_1 is defined in the expansion of the mass as a function of velocity,

$$m = m_0 \left(1 + \gamma_1 \frac{u^2}{c^2} + \gamma_2 \frac{u^4}{c^4} + \cdots\right). \qquad (11.1.44)$$

$\gamma_1 = 1/2$ for Lorentz–Einstein's formula (11.1.2), while $\gamma_1 = 2/5$ for Abraham's formula (11.1.1).

Glitscher calculated a value of $\alpha = e^2/\hbar c$ by use of the known values of e, c, and $\hbar$, as well as values of α by putting an observed value of the fine-structure splitting $\Delta \nu$ for He^+ spectra and the value of R_∞ into equation (11.1.43) for $\gamma_1 = 1/2$ or $2/5$. Comparison between the calculated values of the two kinds showed that the two kinds of calculations agreed with each other only if $\gamma_1 = 1/2$. Later, Faragó and Jánossy (1957) [188] analyzed again the fine splitting by use of observed values with an higher accuracy, which indicated the validity of Lorentz–Einstein's formula (11.1.2) to within an accuracy of 0.05%.

However, the application of Sommerfeld's theory to other spectra failed. After quantum mechanics was developed, introducing the spin of electrons could give a

---

[2]Faragó and Jánossy (1957) [188] analyzed Champion's experiment and claimed that Champion has greatly overestimated the accuracy of his measurements and thus these measurements do not add much to our knowledge of the law of the change in mass with velocity. Raboy and Trail (1958) [189] claimed that Champion's measurements could not distinguish Lorentz–Einstein's formula from Abraham's formula.

satisfactory explanation to the fine structure splitting. In the relativistic quantum mechanics, the Dirac equation for electrons contains a term of electron spin effect. A calculation of energy level for a hydrogen atom made by use of the Dirac equation would yield the same result as that of Sommerfield. Thus validity of Sommerfield's result comes from the cancelation between the error in Bohr's theory and the omission of the electron spin.

11.2. Relation of Mass and Energy

Relation (equivalence) of mass and energy is one of the most important results of the special theory of relativity. It should be noted that the mass in this relation is the inertial mass. A free particle with the energy E certainly has the inertial mass m; The connection of the magnitudes of E and m is given by Einstein's famous formula

$$E = mc^2. \tag{11.2.1}$$

In experiments, the directly observable quantities are the changes ΔE and Δm, but not the absolute magnitudes, of the energy E and mass m, which are shown by experiments to have the similar relation:

$$\Delta E = c^2 \Delta m. \tag{11.2.2}$$

It is well known that this relation for any physical system could be derived from relativistic mechanics by use of the relation $E = mc^2$ and conservation law of energy. Thus a test of the relation between mass and energy is also a test of the conservation law.

This relation shows that a certain amount of change (transfer) in energy must be accompanied by a corresponding change (transfer) in mass, and *vice versa*. This is the physical meaning of the relation (or equivalence) of mass and energy.

As an example to clarify the meaning on the relation between mass and energy, we now consider a nuclear reaction. Let A and B represent two nuclei as initial states of a reaction, and a and b be two nuclei produced through the reaction:

$$A + B \to a + b. \tag{11.2.3}$$

We shall use below a symbol with a subscript $i = A, B, a, b$ to denote the physical quantity corresponding to the nuclei A, B, a, b. For instance, E_A and E_B represent the total energies of the nuclei A and B respectively, T_a denotes the kinetic energy of the nucleus a, and so on.

According to the conversation law of energy, the total energies before and after the reaction should be equal to each other:

$$E_A + E_B = E_a + E_b. \tag{11.2.4}$$

By use of the definition of energy, Eq. (3.1.16a), the above equation gives the law of energy transfer:

$$\Delta T = \Delta E_0, \tag{11.2.5a}$$

where

$$\Delta E_0 = E_{0A} + E_{0B} - E_{0a} - E_{0b}, \tag{11.2.5b}$$

$$\Delta K = K_a + K_b - K_A - K_B. \tag{11.2.5c}$$

This indicates that after the reaction the increase ΔT in the total kinetic energy equals to the decrease ΔE_0 in the total proper energy (i.e., the energy transfers from one type into other type).

Putting the relation (11.2.1) of mass and energy into the conservation law (11.2.4), we have

$$m_A + m_B = m_a + m_b. \tag{11.2.6}$$

This is the law of conservation of relativistic mass. This shows that in the special theory of relativity, the conservation law of energy and the conservation law of mass are no longer two parallel laws, but are two laws equivalent to each other.

The relation of mass and velocity shows that the inertia of a missive particle is no longer measured by its proper mass, but should be measured by its total (i.e., relativistic) mass. Let us introduce the concept of the moving mass m_T which is the difference between the relativistic mass m and proper mass m_0:

$$m_T = m - m_0. \tag{11.2.7}$$

Comparison of this equation with the definition (3.1.14a) shows that the moving mass m_T is related to the kinetic energy T as follows:

$$T = m_T c^2. \tag{11.2.8}$$

By inserting Eq. (11.2.7) in Eq. (11.2.6), we obtain the law of mass transfer:

$$\Delta m_T = \Delta m_0, \tag{11.2.9a}$$

where

$$\Delta m_0 = m_{0A} + m_{0B} - m_{0a} - m_{0b}, \tag{11.2.9b}$$

$$\Delta m_T = m_{TA} + m_{TB} - m_{Ta} - m_{Tb}. \tag{11.2.9c}$$

This result indicates that after the reaction, the decrease Δm_0 (i.e., the so-called "mass defect") in the total rest mass is equal to the increase Δm_T in the total kinetic mass, i.e., that the mass transfers from one type into other type. Due to the relations (3.1.17) and (11.2.8), equations (11.2.9) and (11.2.5) are not independent each other. In fact, by putting $E_0 = m_0 c^2$ into Eq. (11.2.5b), equation (11.2.5a) then becomes

$$\Delta T = c^2 \Delta m_0. \tag{11.2.10}$$

Similarly, putting $T = m_T c^2$ into Eq. (11.2.5c), then Eq. (11.2.5a) becomes

$$c^2 \Delta m_T = \Delta E_0. \tag{11.2.11}$$

In experiments the observed quantities are the rest masses m_{0i} and the kinetic energies T_i for $i = A, B, a, b$. From these observed values one can calculate the changes Δm_0 and ΔT in the total mass and kinetic energy, and then compare the results with the mass–energy relation (11.2.10). Thus it is the law of energy transfer, Eq. (11.2.10), that is used for direct comparison with experimental results.

A nucleus consists of protons and neutrons. The rest mass of a nucleus is less than the sum of the rest masses of the involved protons and neutrons, the difference between them is the "mass defect". The mass defect for a nucleus (A) consisting of N nucleons (or Z protons and $A - Z$ neutrons) is given by

$$\Delta m_0 = Z m_{0(H)} + (A - Z) m_{0(n)} - M_{0(Z,A)},$$

where $m_{0(H)} = 1.008142$ u is the rest mass of a hydrogen atom, $m_{0(n)} = 1.008982$ u the rest mass of a free neutron, and $M_{0(Z,A)}$ the rest mass of a nucleus consisting of Z protons and $A - Z$ neutrons. The binding energy of nucleus ΔE_0 is just the energy corresponding to the mass defect Δm_0:

$$\Delta E_0 \text{ (MeV)} = 931 \left[1.008142 \, Z + 1.008982 \, (A - Z) - M_{0(Z,A)} \right].$$

The average binding energy per nucleon $\Delta E_0/A$ is the required work to separate a nucleon from the nucleus.

The first test of the mass–energy relation was carried out by Cockcroft and Walton (1932) [192] by use of nuclear reaction: $_1H^1 + _3Li^7 \rightarrow 2\,\alpha$. The observed values by means of a mass-spectrometer are as follows:

$$m_0(_3Li^7) = 7.01818 \text{ u},$$

$$m_0(_1H^1) = 7.01818 \text{ u},$$

$$m_0(_2He^4) = 7.01818 \text{ u}.$$

By putting these values into Eq. (11.2.9b), one arrives at the change in the proper energy, $\Delta E_0 = c^2 \Delta m_0 = 17.25$ MeV, which should be equal to the difference of the kinetic energies of the two α particles and incident proton ΔT. In Cockcroft and Walton's experiment the kinetic energy for the incident proton was 0.25 MeV and after reaction the kinetic energy of α particle was measured to be 8.6 MeV. The $_3Li^7$ target was at rest and hence its kinetic energy $T(_3Li^7) = 0$. From these values one obtains the increase in the kinetic energy after the reaction compared to that before the reaction should be $\Delta T = (2 \times 8.6 - 0.25)$ MeV $= 16.95$ MeV.

Smith (1939) [193] repeated the experiment of Cockcroft and Walton with higher accuracy and measured an energy liberation of $\Delta T = 17.28 \pm 0.03$ MeV which agrees well with the expected value $\Delta E_0 = 17.25$ MeV.

A lot of experiments using various nuclear reactions have been performed, and have verified the mass–energy relation with very high accuracy [190, 194, 195].

The process of a typical energy (or mass) transfer is that of annihilation of a pair of e^- and e^+, in which the total rest mass transfers the moving mass and at the same time, the total proper energy transfers the radiation energy. Experiments have shown that a positive electron (i.e., positron), when slowed, will be attracted by a negative electron to form a positronium similar to a hydrogen atom; A singlet positronium is such a state in which the spin directions of the electron and positron are oppositely parallel to each other, while a triplet positronium is such a state in which the spin directions of the electron and positron are parallel to each other. A singlet positronium, the lifetime of which is about 10^{-10} sec, will annihilate into two photons. Due to the conservation law of momentum and energy, the energy of each photon should be $m_0c^2 = 0.511$ MeV. That has been proved by experiments. A triplet positronium, the lifetime of which is about 1.5×10^{-7} sec, will annihilate into three photons. In this process the spin and angular momentum keep constant. Experiments showed that the directions of motion for three photons lay in the same plane and that the conservation law of momentum was valid. An inverse process, in which radiation energy (or kinetic mass) will transfer into proper energy (or rest mass), is that a photon in a electric field of nucleus will transfer into a pair of positive and negative electrons. Experiments proved that such process might occur only when the energy ($h\nu$) of the photon is larger than the proper energy ($2m_0c^2$) of the positive and negative electrons.

The gravitational red shift could be regarded as a test of the conservation law of energy and the mass–energy relation: The red shift implies a change in energy of photon; The prediction of red shift given by the conservation law of energy and the mass–energy relation agrees with the experimental result [196].

Table 11.2. *Evidences for variation of mass with velocity*

Investigator	Method	Particle	Speed	Result
Kaufmann (1901)[172]	EM deflection	β rays	0.79–0.94c	cf. Fig.45
Butchere (1909)[175]	EM deflection	β rays	0.32–0.69c	cf. Fig.45
Hupka (1910)[177]	Magnetic deflection	Electrons		An lower accuracy
Guye & Lavanchy (1915)[178]	EM deflection	cathode rays	0.26–0.48c	cf. Fig.45
Rogers et al.(1940)[179]	EM deflection	β rays	0.63, 0.69, 0.75c	agrees with mass formula of SR to 1%
Champion (1932)[187]	Elastic collision	Electrons		Scattering angle $< 90°$
Meyer et al. (1963)[181]	Magnetic deflection	Electrons	0.987–0.990c	agrees with mass formula of SR to 0.04%
Faragó & Jánossy (1957)[188]	Fine structure splitting	Hydrogen		An accuracy of 0.05%
Grove & Fox (1953)[180]	Cyclotron frequency	Proton	$\sim 0.7c$	An accuracy of 0.1%
Geller and Koliarits (1972)[182]	β-ray spectrometer & semiconductor detector	β rays	$\sim 0.7c$	agrees with predictions of SR
Bertozzi (1964)[184]	Time-of-flight technique & calorimetry	Electrons	0.88c, 0.91c, 0.96c	agrees with predictions of SR
Brown et al. (1973)[185]	Time-of-flight techniques	Electrons		agrees with predictions of SR
Guiragossián et al. (1975)[186]	Time-of-flight techniques	Electrons		agrees with predictions of SR

CHAPTER 12

THE UPPER BOUNDS ON PHOTON MASS

In special theory of relativity, any particles moving with the vacuum speed c of light certainly possess a vanishing proper mass. This conclusion can be obtained from the following different ways. The mass-velocity relation (3.1.6) or the mass–energy relation (3.1.16a) shows that the rest mass m_0 of a particle with a velocity of $u = c$ should be zero. On the other hand, the U(1) gauge invariant of Maxwell's electromagnetic field equations refuses the presence of a mass term in Lagrangian of electromagnetic field, and thus the electromagnetic field is a massless vector field. Furthermore, the quantum electrodynamics indicates that a relativistic, quantized electromagnetic field of frequency ν is recognized as an assembly of photon particles with energy $h\nu$. These light quanta travel at the velocity c and hence, have zero rest mass.

The success of quantum electrodynamics in predicting experiments to six or more decimal places is regarded as a powerful proof of the zero rest mass of photons. In addition, some other kinds of experiments for testing the zero rest mass of photons were carried out [197], which we shall introduce in this chapter.

12.1. Dispersion Effect of Velocity of Light in Vacuum

In Sec. 5.3 we have discussed the dispersion effect of Proca (massive) electromagnetic waves: Equation (5.3.4) shows that the group velocity of a massive, electromagnetic wave packet in vacuum depends on its frequency. For two packets we then have

$$-\frac{\Delta v}{c} \equiv \frac{v_{g_1} - v_{g_2}}{c} \simeq \frac{1}{2}\mu^2 c^2 \left(\frac{1}{\omega_2^2} - \frac{1}{\omega_1^2}\right), \tag{12.1.1}$$

where the terms of higher order more than $(\mu^2 c^2/\omega^2)^2$ are neglected. In the same approximation, equation (5.3.2) yields

$$\frac{\mu^2 c^2}{\omega^2} \simeq \frac{\mu^2}{k^2} = \frac{\mu^2 \lambda^2}{4\pi^2}. \tag{12.1.2}$$

By means of Eq. (12.1.2) the difference of light speed Δv can be expressed in terms of wavelengths as

$$-\frac{\Delta v}{c} \simeq \frac{\mu^2}{8\pi^2}(\lambda_2^2 - \lambda_1^2). \tag{12.1.3}$$

If two waves with the wavelengths λ_1 and λ_2 pass through the same distance L, the difference between their traveling times then is given by

$$\Delta t = \frac{L}{v_{g_1}} - \frac{L}{v_{g_2}} \simeq \frac{L}{8\pi^2 c}(\lambda_2^2 - \lambda_1^2)\mu^2. \tag{12.1.4}$$

Equations (12.1.1)–(12.1.4) will be used to determine the photon rest mass.

12.1.1. Measurements of the Velocity of Light

Equation (12.1.1) indicates that the photon rest mass could be determined by measuring the speeds of electromagnetic waves with different frequencies.

The speeds in vacuum for various electromagnetic waves with different frequencies were measured by different methods, such as cavity resonator, radar, radio interferometer, spectrum, crystal modulator, etc. For the electromagnetic waves with frequencies from 10^8 Hz to 10^{15} Hz, the observed velocity c is a constant to within an accuracy of about $10^{-5} \sim 10^{-6}$ [198]. The lowest frequency measurement with an accuracy of 10^{-5} is that at $\nu = 1.73 \times 10^8$ Hz [199]; The highest frequency is 10^{15} Hz. This is to say that the measurements did not show the dispersion effect for the electromagnetic waves with the frequencies of $\nu = 1.73 \times 10^8$ Hz $\sim 10^{15}$ Hz to within the accuracy of 10^{-5}. This means $(-\Delta v/c) \leq 10^{-5}$. From equation (12.1.1) where $\omega_2 = 2\pi\nu_2 = 3.46\pi \times 10^8$ Hz and $\omega_1 = 2\pi\nu_1 = 2\pi \times 10^{15}$ Hz, noting that $1/\omega_1 \ll 1/\omega_2$ and neglecting the term of $1/\omega_1^2$, one arrives at [197]

$$\mu \simeq \frac{2\pi\nu}{c}\sqrt{2\left(\frac{-\Delta v}{c}\right)} \leq 1.6 \times 10^{-4}\text{cm}^{-1} \equiv 5.6 \times 10^{-42}\text{g}. \tag{12.1.5}$$

12.1.2. Arrival Time of Light Ray from Stars

Equation (12.1.4) shows that a limit on the photon mass can be obtained by measuring the difference in arrival time of radiation of different frequencies with the same origin. De Broglie (1940) suggested a measurement for the arrival time of starlight.

It should be emphasized that the dispersion effect of starlight can be explained by not only the photon mass but also the nonlinear effect of electromagnetic field or the dispersion of light traveling through the interstellar plasma in a magnetic field. This is the main difficulty for determining the photon mass by means of the starlight [200].

In Maxwell's electromagnetic field theory, the dispersion equation for a electromagnetic wave of frequency ω traveling through the plasma is [197]

$$k^2 = \left(\frac{\omega^2}{c^2}\right)\left(1 - \frac{\omega_p^2}{\omega^2 \pm \omega\omega_B}\right), \tag{12.1.6a}$$

$$\omega_p^2 = \frac{4\pi ne^2}{m}, \qquad \omega_B = \frac{eB}{mc}\cos\alpha. \tag{12.1.6b}$$

Here n and m are the electron density and mass respectively, and α is the angle between the magnetic field (B) and the direction of propagation of the wave. In the interstellar space the magnetic field B is small, and hence the quantity ω_B can be neglected. Then Eqs. (12.1.6) lead to

$$v_g = \frac{d\omega}{d|\mathbf{k}|} = c\sqrt{1 - \frac{\omega_p^2}{\omega^2}}. \tag{12.1.7}$$

Comparison between Eqs. (12.1.7) and (5.3.4) indicates that the characteristic frequency ω_p of the plasma would produce the same effect as the photon mass μ, i.e., that the effect of ω_p is the same as that of μc on the dispersion in the speed of starlight. Thus if the density of the interstellar plasma cannot be independently determined, the photon mass effect then could not be separated from the plasma effect. This means that we are restricted in determining the photon mass by use of the dispersion of starlight.

(I) *Observations of Binary Star*

De Broglie (1940) [201] suggested that a limit on the photon rest mass could be determined by using light from a star emerging from behind its dark binary companion. De Broglie considered the case $\lambda_2^2 - \lambda_1^2 = 0.5 \times 10^{-8}$ cm^2 (for example, a red light ray of wavelength $\lambda_2 \sim 8000$ Å, and the blue light ray of wavelength $\lambda_1 \sim 4000$ Å), $L = 10^3$ light years, and $\Delta t \leq 10^{-3}$ sec. Then from Eq. (12.1.4) one can get [1].

$$\mu \simeq \left[\frac{8\pi^2 c \Delta t}{L(\lambda_2^2 - \lambda_1^2)} \right]^{\frac{1}{2}} \leq 2.3 \times 10^{-2} \text{cm}^{-1} \equiv 0.8 \times 10^{-39} \text{g}. \tag{12.1.8}$$

(II) *Observations of Pulsars*

The discovery of pulsars provides a way for determining the photon rest mass by means of a measurement of dispersion in arrival time of radio signals from the pulsars.

Staelin and Reifenstein III (1968) [203] observed the dispersion of radio signals from the pulsating radio source NP 0527, and gave (in assumption of zero photon mass) an effective average plasma density of $\bar{n}_e \leq 2.8 \times 10^{-2}$ electrons cm^3. Feinberg (1969) [204] assumed that the dispersion is partly the photon mass effect. From comparison between Eqs. (12.1.7) and (5.3.4) one has $\omega_p/c = 4\pi e^2 \bar{n}_e/mc$, and then gets the photon rest mass,

$$\mu = \frac{4\pi e^2 \bar{n}_e}{mc} \leq 3 \times 10^{-7} \text{cm}^{-1} = 10^{-44} \text{g} = 6 \times 10^{-12} \text{eV}. \tag{12.1.9}$$

Feinberg (1969) [204] pointed out that this determination is supplementary to Schrödinger's external field method (cf. Sec. 12.3.1).

---

[1]In the last step of his calculation, de Broglie made a mistake of order 10^5, quoting a limit of 10^{-44} g. After 28 years, Kobzarev and Okun' (1968) [202] made the correction.

12.2. The Tests of Coulomb's Law

In Maxwell's electromagnetic field theory, the force between two point charges (e_1 and e_2) is inversely proportional to the square of the distance (r) between them:

$$F \sim \frac{e_1 e_2}{r^2}. \qquad (12.2.1)$$

This is the famous Coulomb's law, the inverse square force law for electricity.

Deviations from the Coulomb's law can be expressed as [205]

$$F \sim \frac{e_1 e_2}{r^{2+q}}, \qquad (12.2.2)$$

where q is an arbitrary constant.

The inverse square force law was first announced in 1785 by Coulomb. Actually, the discovery of the law precedes Coulomb by quite a number of years. In 1769, S.J. Robison made the first experimental determination of the law. By balancing electrical and gravitational forces acting on a sphere, Robison obtained a result of $q \sim 0.06$. In order to test the inverse square force law, Cavendish (1773) carried out an experiment by use of two concentric conducting spheres connected by a wire. He charged the outer sphere, and then disconnected the wire. If there were a deviation from the inverse square force law, then upon removing the outer sphere, one would find a calculable charge on the inner sphere. By use of an electrometer for measuring the potential on the inner sphere, however, Cavendish did not find any charge on the sphere, and hence concluded that the charge on the inner sphere must be less than 1/60 of the total charge on the whole arrangement. This yielded $|q| < 1/50$. Maxwell (1873) improved the result in a new experiment. The only modification was that the outer shell was grounded instead of removed, and the inner globe was tested for charge through a small hole. The null result of the experiment gives a limit of $q < 1/21600$.

In the above experiments the measured quantity is the electric potential on the conducting sphere. Derivation of a value of q was made from a relation between q and potential, which was given by Maxwell [205].

Let the electrostatic force between two unit charges be an arbitrary function of the distance s among them. The electrostatic potential then is given by

$$U(r) = \int_r^\infty F(s)ds. \qquad (12.2.3)$$

In case of a uniform distribution of charge Q_2 on a shell of radius R_2, the equation above can be used to the potential at a given position separated by a distance r from the center of the shell. The surface density of charge on the sphere is $\sigma = Q_2/4\pi R_2^2$, and then the total charge on the element of area $dA = R_2^2 \sin\theta d\theta d\phi$ is $\sigma dA = (Q_2/4\pi)\sin\theta d\theta d\phi$ which produces the potential at the point P,

$$dV = \frac{Q_2}{4\pi}\sin\theta d\theta d\phi U(s), \qquad (12.2.4)$$

where s is the distance between the point P and the element of area dA. In case of the coordinate axis passing through the point P, we have

$$s^2 = r^2 + R_2^2 - 2R_2 r \cos \theta. \tag{12.2.5}$$

By taking derivative of Eq. (12.2.5), we get

$$s ds = R_2 r \sin \theta d\theta. \tag{12, 2, 6}$$

The potential produced by Q_2 is an integral of Eq. (12.2.4) over a range of $\phi = 0 \to 2\pi$ and $\theta = 0 \to \pi$. Substituting Eq. (12.2.6) into Eq. (12.2.4) and then integrating the result, we obtain

$$\begin{aligned}
V(r) &= \frac{Q_2}{4\pi R_2 r} \int_0^{2\pi} d\phi \int_{|r-R_2|}^{r+R_2} s U(s) ds \\
&= \frac{Q_2}{2 R_2 r} [f(r+R_2) - f(|r-R_2|)],
\end{aligned} \tag{12.2.7}$$

where

$$f(r) \equiv \int_0^r s U(s) ds. \tag{12.2.8}$$

Let us now consider experiments of Cavendish's type. The radii of two concentral spheres are R_1 and R_2 ($R_1 < R_2$). The charges on the inner and outer shells are denoted by Q_1 and Q_2. Using Eq. (12.2.7), one can get the potential on the inner shell,

$$V(R_1) = \frac{Q_1}{2R_1^2} f(2R_1) + \frac{Q_2}{2R_1 R_2} [f(R_1 + R_2) - f(R_2 - R_1)], \tag{12.2.9}$$

and the potential on the outer shell,

$$V(R_2) = \frac{Q}{2R_2^2} f(2R_2) + \frac{Q}{2R_1 R_2} [f(R_1 + R2) - f(R_2 - R_1)]. \tag{12.2.10}$$

After the outer shell is charged, a part of charge will pass through the connected wire into the inner shell until equilibrium occurs: $V(R_1) = V(R_2) \equiv V_0$. In this case we can get a solution to the equilibrium equation of (12.2.9) and (12.2.10),

$$Q_1 = 2R_1 V_0 \frac{R_1 f(2R_2) - R_2 [f(R_2 + R_1) - f(R_2 - R_1)]}{f(2R_1) f(2R_2) - [f(R_2 + R_1) - f(R_2 - R_1)]^2}. \tag{12.2.11}$$

In Cavendish's experiment, the outer shell after being charged was removed to infinity. From (12.2.9) the potential on the inner shell becomes

$$V_1(R_1) = \frac{Q_1}{2R_1^2} f(2R_1). \tag{12.2.12}$$

In the repeated experiment by Maxwell, the wire was disconnected and the outer shell was grounded instead of removed. For this case, by putting $V(R_2) = 0$ into Eq. (12.2.10) to find Q_2 and then inserting the result in Eq. (12.2.9), and then by use of Eq. (12.2.11), the charge Q_2 can be expressed in terms of V_0. In this way the potential on the inner shell can be expressed as

$$V_2(R_1) = V_0 \left\{ 1 - \left(\frac{R_2}{R_1} \right) \frac{f(R_2 + R_1) - f(R_2 - R_1)}{f(2R_2)} \right\}. \tag{12.2.13}$$

By putting the expression of the electrostatic force between two unit charges, the Eq. (12.2.2) with $e_1 = e_2 = 1$, and $|q| < 1$ into Eq. (12.2.3), we get

$$U(r) = \frac{1}{1 + q^2} r^{-1-q}.$$

Then inserting the result in Eq. (12.2.8) we obtain

$$f(r) = \frac{1}{1 - q^2} r^{1-q}.$$

Since $|q| \ll 1$, so that

$$f(r) \cong r\{1 - q \ln r\}. \tag{12.2.14}$$

Using this result in Eqs. (12.2.12) and (12.2.13), we obtain the expected formula:

$$V_1(R_1) \cong - \left(\frac{R_2}{R_2 - R_1} \right) qM(R_1, R_2), \tag{12.2.15}$$

$$V_2(R_1) \cong -qM(R_1, R_2), \tag{12.2.16}$$

where

$$M(R_1, R_2) \equiv \frac{1}{2} V_0 \left[\ln \frac{4R_2^2}{R_2^2 - R_1^2} - \frac{R_2}{R_1} \ln \frac{R_2 + R_1}{R_2 - R_1} \right]. \tag{12.2.17}$$

In the above we have obtained the expected expressions to be compared with the static experiments of Cavendish and Maxwell. A nonstatic experiment was carried out by Plimpton and Lawton (1936). They took two concentric conducting spheres of radii $R_1 = 2.0$ ft and $R_2 = 2.5$ ft, grounded them, and then charged the outer sphere to $V_0 = 3000$ V. Actually, for technical reasons, the voltage was quasistatic, having a frequency of 130 cycles/min. Later, the performed experiments all used a high frequency alternating voltage. By use of Eqs. (12.2.7) and (12.2.14), we get the relative voltage between the two spheres for these cases,

$$\frac{V(R_2) - V(R_1)}{V(R_1)} = -qM(R_1, R_2), \tag{12.2.18}$$

where

$$M(R_1, R_2) \equiv \frac{1}{2} \left[\frac{R_1}{R_2} \ln \left(\frac{R_2 + R_1}{R_2 - R_1} \right) - \ln \left(\frac{4R_1^2}{R_2^2 - R_1^2} \right) \right]. \tag{12.2.19}$$

Here we have obtained the expected relation between q and the voltage V. In order to determine the photon rest mass μ from experimental results, it is also needed to have a relation of μ and V.

Let the alternating voltage on the outer sphere be denoted by $V_0 \exp(i\omega t)$. Inside the sphere there is no charge ($\rho = 0$). In this case, a solution to the massive electromagnetic field equation (5.2.8b) can be written as $\phi(r, t) = \phi(r)e^{i\omega t}$. Then the wave equation becomes

$$(\nabla^2 + k^2)\phi(r) = 0, \qquad (12.2.20a)$$

$$k^2 = \frac{\omega^2}{c^2} - \mu^2. \qquad (12.2.20b)$$

A solution to equation (12.2.20a) is

$$\phi(r) = \frac{1}{r}e^{\pm ikr}. \qquad (12.2.21)$$

From this solution we can obtain the expected solution satisfying the experimental conditions above:

$$\phi(r) = V_0 \frac{R_2}{r} \frac{e^{ikr} - e^{-ikr}}{e^{ikR_2} - e^{-ikR_2}} \qquad (r \le R_2). \qquad (12.2.22)$$

We now integral another Proca field equation (5.2.4b) over the range inside the sphere:

$$\int dV \nabla \cdot \mathbf{E} = -\mu^2 \int dV \cdot \phi. \qquad (12.2.23)$$

Putting the solution (12.2.22) into the right-hand side of (12.2.23), using Gauss theorem, and noting $\mathbf{E} = E(r)\hat{r}$ where $\hat{r} = \mathbf{r}/r$, the equation above becomes

$$\int dV \nabla \cdot \mathbf{E} = \oint \mathbf{E} \cdot d\sigma = 4\pi r^2 E,$$

$$\int dV \phi = -\frac{4\pi}{R^2} \frac{V_0 R_2}{e^{ikR_2} - e^{-ikR_2}} \left\{ ikr(e^{ikr} + e^{-ikr}) - (e^{ikr} - e^{-ikr}) \right\}.$$

From these results we get the electric field inside the sphere:

$$\begin{aligned}
\mathbf{E}(r, t) = \ &\hat{r} \left(\frac{\mu^2}{R^2 r^2} \right) \frac{V_0 R_2}{e^{ikR_2} - e^{-ikR_2}} \cdot \\
&\cdot \{ ikr(e^{ikr} + e^{-ikr}) - (e^{ikr} - e^{-ikr}) \} e^{i\omega t}.
\end{aligned}$$

$$(12.2.24)$$

The right-hand side of this equation can be approximated as

$$\mathbf{E}(r, t) \cong -\frac{\mu^2}{3}(V_0 e^{i\omega t})\mathbf{r}, \qquad (12.2.25)$$

where such terms of orders k^2r^2 and $k^2R_2^2$ are neglected because of $kr < 1$ and $\omega/c > \mu$ for the performed experiments. This leads to $\nabla \cdot \mathbf{E} = 0$, and hence $\oint \mathbf{E} \cdot d\mathbf{l} = 0$. Then the potential difference between the inner and outer spheres is obtained:

$$
\begin{aligned}
V(R_1) - V(R_2) &= \int_{R_1}^{R_2} d\mathbf{l} \cdot \mathbf{E} = -\frac{1}{6}(V_0 e^{i\omega t})\mu^2(R_2^2 - R_1^2) \\
&= -\frac{1}{6}\mu^2(R_2^2 - R_1^2)V(R_2).
\end{aligned}
$$

Then we obtain the expected expression for the relative voltage:

$$
\frac{V(R_1) - V(R_2)}{V(R_2)} = -\frac{1}{6}\mu^2(R_2^2 - R_1^2). \tag{12.2.26}
$$

This result shows that the relative voltage $\Delta V/V$ is independent of the frequency of the applied alternating voltage, i.e., that the boundary condition problem in the nonstatic experiments is the same as that in the static experiments [207,208].

In the experiment performed by Plimpton and Lawton (1936) [206], $R_2 \sim 75$ cm, $R_1 \sim 60$ cm, and $V = 3000$ V. The accuracy for the measurement of voltage was 10^{-6} V, and thus the null result implies $\Delta V \leq 10^{-6}$ V. By putting these values into Eqs. (12.2.18), (12.2.19), and (12.2.26), we have

$$
q \leq 2.0 \times 10^{-9}, \tag{12.2.27}
$$

$$
\mu \leq 10^{-6}\text{cm}^{-1} \equiv 3.4 \times 10^{-44}\text{g}. \tag{12.2.28}
$$

Later, some experiments for testing Coulomb's law were carried out, results of which are shown in Table 12.1.

Table 12.1. *Experimental results for testing Coulomb's law*

Experiment	ω (Hz)	q	μ
Robison (1769)	0	$\leq 6 \times 10^{-2}$	
Cavendish (1773)	0	$\leq 2 \times 10^{-2}$	
Coulomb (1785)	0	$\leq 4 \times 10^{-2}$	
Maxwell (1873)	0	$\leq 4.9 \times 10^{-5}$	
Plimpton and Lawton (1936) [206]	2	$\leq 2 \times 10^{-9}$	$\leq 3.4 \times 10^{-44}$
Cochran and Franken (1968) [209]	10^2–10^3	$\leq 9.2 \times 10^{-12}$	$\leq 3 \times 10^{-45}$
Barlett, Goldhagen and Phillips (1970) [207]	2.5×10^3	$\leq 1.3 \times 10^{-13}$	$\leq 3 \times 10^{-46}$
Williams, Faller, and Hill (1970) [208]	4×10^6	$\leq 6 \times 10^{-16}$	$\leq 1.6 \times 10^{-47}$

12.3. The Magnetostatic Effect of Photon Mass

In the present section we will introduce the magnetostatic effect of the photon rest mass which was investigated by Schrödinger (1943) [210] as a method for determining the photon rest mass.

By his discussion on the geomagnetic surveys, Schmidt pointed out that the geomagnetic field contains three types of magnetic fields: *"dipole field"*, *"external field"*, and *"nonpotential field"*. The magnetic dipole field points towards the south geomagnetic pole (not coinciding with the south geographic pole [211]), which is produced by a magnetic dipole moment of **m**; The "external field" antiparallel to the dipole moment **m** is a uniform field over the surface of the earth, the origin of which could not be inside the earth; The nonpotential field is produced by a constant current (i.e., the earth-air current). In 1923, Bauer analyzed the 1922 magnetic surveys, and showed that within the geomagnetic fields a great part is caused by inner sources (i.e., the magnetic dipole), the external field amounts to about 3%, and the nonpotential field makes up about 3%. In 1924, Schmidt also discussed the 1922 magnetic surveys, and obtained a dipole field of 31089γ where $1\gamma = 10^{-5}$ G and an external field of 539γ. However, the geomagnetic field is more likely such field produced by an eccentric dipole. He obtained that the position of the eccentric dipole is at 6.5 degrees north latitude and 161.8 degrees east longitude, and found out a "hemispheric current effect". Schrödinger pointed out that a photon rest mass effect could give an explanation of the "external field" and "nonpotential field", and, in this way, gave an upper limit of 2×10^{-47} g on the photon rest mass [210].

12.3.1. Schrödinger's (External Field) Method

In 1943, Schrödinger [212] developed a unification theory describing the interaction among graviton, meson, and electromagnetic field. In case of neglecting gravitational field, the equations of motion in vacuum for nonstrong electromagnetic field are the same as the Proca field equations. In order to explain the "external field effect" by using the photon mass, let us consider such a field produced by a magnetic dipole.

Let a stationary current **J** distribute in a small area V, and generate a constant magnetic field. Thus Proca's wave equation (5.2.8a) reduces to

$$(\nabla^2 - \mu^2)\mathbf{A} = -\frac{4\pi}{c}\mathbf{J}. \tag{12.3.1}$$

A solution to the equation is

$$\mathbf{A} = \frac{1}{c}\int_V d\mathbf{r}' G(\mathbf{r} - \mathbf{r}')\mathbf{J}(\mathbf{r}'), \tag{12.3.2}$$

where Green's function G is a function of Yukawa potential type:

$$G(\mathbf{r} - \mathbf{r}') = \frac{e^{-\mu|\mathbf{r}-\mathbf{r}'|}}{|\mathbf{r} - \mathbf{r}'|}. \tag{12.3.3}$$

If the dimension of the area V is small, The multipole expansion of the equation above is

$$
\begin{aligned}
G(\mathbf{r} - \mathbf{r}') &= G(r) + x'_K \left(\frac{\partial G}{\partial x'_K} \right)_{x'_K = 0} + \cdots \\
&= G(r) + \mathbf{r}' \cdot (\nabla' G(\mathbf{r} - \mathbf{r}'))_{\mathbf{r}'=0} + \cdots,
\end{aligned}
$$

(12.3.4)

By substituting this formula into Eq. (12.3.2), one arrives at

$$
\mathbf{A} = \mathbf{A}^{(0)} + \mathbf{A}^{(1)} + \cdots,
\tag{12.3.5a}
$$

$$
\mathbf{A}^{(0)} = \frac{1}{c} G(r) \int_V d\mathbf{r}' \mathbf{J}(\mathbf{r}') = 0,
\tag{12.3.5b}
$$

$$
\mathbf{A}^{(1)} = \frac{1}{c} \int_V d\mathbf{r}' \mathbf{J}(\mathbf{r}')[\mathbf{r}' \cdot (\nabla' G(\mathbf{r} - \mathbf{r}'))_{\mathbf{r}'=0}],
\tag{12.3.5c}
$$

i.e.,

$$
\mathbf{A} \simeq \mathbf{A}^{(1)} = \frac{1}{c} \int_V d\mathbf{r}' \mathbf{J}(\mathbf{r}')[\mathbf{r}' \cdot (\nabla' G(\mathbf{r} - \mathbf{r}'))_{\mathbf{r}'=0}].
\tag{12.3.6}
$$

This is a magnetic dipole potential which can be written as

$$
\mathbf{A}^{(1)} = \nabla \times (\mathbf{m} G) = \nabla \times \left(\mathbf{m} \frac{e^{-\mu r}}{r} \right).
\tag{12.3.7}
$$

Here the magnetic dipole moment is defined by

$$
\mathbf{m} = \frac{1}{2c} \int_V (\mathbf{r}' \times \mathbf{J}(\mathbf{r}')) d\mathbf{r}'.
\tag{12.3.8}
$$

The corresponding magnetic dipole field is given by

$$
\begin{aligned}
\mathbf{H} \simeq \mathbf{H}^{(1)} &= \nabla \times \mathbf{A}^{(1)} = \nabla \times \nabla \times \left[\mathbf{m} \frac{e^{-\mu r}}{r} \right] \\
&= \frac{e^{-\mu r}}{r^3} \left\{ \left(1 + \mu r + \frac{1}{3} \mu^2 r^2 \right) (3\mathbf{m} \cdot \hat{r}\hat{r} - \mathbf{m}) - \frac{2}{3} \mu^2 r^2 \mathbf{m} \right\}.
\end{aligned}
$$

(12.3.9)

Let $\mathbf{m} = m\hat{z}$ with $\hat{z} \equiv \mathbf{z}/z$, then equation (12.3.9) becomes

$$
\mathbf{H}^{(1)} = \frac{m'}{r^3} (3\hat{z} \cdot \hat{r}\hat{r} - \hat{z}) - \frac{2}{3} \frac{e^{-\mu r}}{r^3} \mu^2 r^2 \mathbf{m},
\tag{12.3.10}
$$

where m' is defined by Eq. (12.3.11b) below. This indicates that the dipole field $\mathbf{H}^{(1)}$ contains two parts: One is the same as the magnetic dipole field in Maxwell's theory, i.e.,

$$
\mathbf{H}_D = \frac{m'}{r^3} (3\hat{z} \cdot \hat{r}\hat{r} - \hat{z})
\tag{12.3.11a}
$$

with

$$m' \equiv e^{-\mu r} \left(1 + \mu r + \frac{1}{3}\mu^2 r^2 \right) m; \tag{12.3.11b}$$

The other is a new field:

$$\mathbf{H}_{ext} \equiv -\frac{2}{3}\frac{e^{-\mu r}}{r^3}\mu^2 r^2 \mathbf{m}, \tag{12.3.12}$$

which is antiparallel to the magnetic dipole moment $\mathbf{m}$, and is a uniform field on the surface of the earth (here $r \simeq R = $ constant). Schrödinger explained this new field as the "external field" in the geomagnetic field analyses. The ratio of the "external field" (H_{ext}) to the dipole field (H_D) at the equator (i.e., $\hat{z} \cdot \hat{r} = 0$) of the earth is

$$\frac{H_{ext}}{H_D} = \frac{(2/3)\mu^2 R^2}{1 + \mu R + \mu^2 R^2/3}. \tag{12.3.13}$$

As mentioned above, by use of the 1922 magnetic surveys, Schmidt (1924) obtained

$$\frac{H_{ext}}{H_D} = \frac{539}{31089}. \tag{12.3.14}$$

By using the 1922 magnetic surveys and Eq. (12.3.13), Schrödinger (1943) obtained

$$\mu R = 0.176,$$

where the radius of the earth is $R \simeq 6378$ km/sec. This yielded a photon rest mass

$$\mu = 2.76 \times 10^{-10}\text{cm}^{-1} = 1.0 \times 10^{-47}\text{g}. \tag{12.3.15}$$

Later, Bass and Schrödinger (1955) [210] argued that multiplying this result by a factor of 2 would give a reliable upper limit:

$$\mu \leq 2.0 \times 10^{-47}\text{g}. \tag{12.3.16}$$

Goldhaber and Nieto (1968) [213] improved on Schrödinger's results by using Cain's fit to geomagnetic data from earthbound and satellite measurements. For epoch 1960.0, Cain obtained the values of

$$m = 31044\gamma R^3,$$

$$\mathbf{H}_{ext} \cdot \hat{m} = (21 \pm 5)\gamma,$$

$$\mathbf{H}_{ext} \cdot \hat{y} \equiv \frac{\mathbf{H}_{ext} \cdot (\hat{s} \times \hat{m})}{|\hat{s} \times \hat{m}|} = (14 \pm 5)\gamma,$$

$$\mathbf{H}_{ext} \cdot (\hat{y} \times \hat{m}) = (8 \pm 5)\gamma. \tag{12.3.17}$$

Here $\hat{s}$ is a unit vector pointing towards the south geographic pole, $\hat{y}$ is a unit vector perpendicular to $\hat{m} \equiv \mathbf{m}/m$, and R is the radius of the earth.

To obtain a mass limit from these numbers, the contributions of the "true external" sources to $\mathbf{H}_{ext}$ must be subtracted from the values of Eq. (12.3.17). These include $\sim 9\gamma$ from the quit-time proton belt, perhaps 15–30 γ due to currents in the geomagnetic tail, and $\sim 15\gamma$ from the hot component of the plasma in the magnetosphere, which are all parallel to the dipole moment. In addition, there is a true external field antiparallel to the dipole of $\sim 20\gamma$ at the equator, which is due to the compression of the geomagnetic field by the solar wind. Finally, the interplanetary field of $\sim 5\gamma$ points in an unknown direction at the earth's surface. Thus, the total external field parallel to $\mathbf{m}$ due to known sources is $\leq 40\gamma$. By subtracting the total external field from $\mathbf{H}_{ext} \cdot \hat{m} = (21 \pm 5)\gamma$, Goldhaber and Nieto obtained $H_{ext} \leq 20\gamma$. In addition, they considered the reliability of the numbers and argued the possibility of some errors in the estimates of true external fields. Thus, they added 100 γ to their estimate ($H_{ext} \leq 20\gamma$) and got $H_{ext} \leq 120\gamma$. On the other hand, the first formula of Eqs. (12.3.17) yielded $H_D = m/R^3 = 31044\gamma$. Thus they obtained

$$H_{ext}/H_D = 120/31044 \simeq 3.9 \times 10^3. \qquad (12.3.18)$$

Substituting this number into Eq. (12.3.13), they gave a limit on the photon rest mass of

$$\mu \leq 1.15 \times 10^{-10} \text{cm}^{-1} \equiv 4.0 \times 10^{-48} \text{g}. \qquad (12.3.19)$$

The corresponding Compton wavelength is $\lambda = 2\pi/\mu \sim 81R$ where R is the radius of the earth.

Schrödinger's external field method can also be applied for the magnetic dipole fields of other planets. For instance, Davis, Jr., Goldhaber, and Nieto (1975) [214] gave an upper limit on the photon rest mass of $\mu \leq 2 \times 10^{-11}$ cm$^{-1} \equiv 6 \times 10^{-16}$ eV $\equiv 8 \times 10^{-49}$ g by using the external field method to the observations of Jupiter's magnetic field.

12.3.2. Altitude-Dependence of Geomagnetic Field

The magnetic dipole field $\mathbf{H}$ given by Eq. (12.3.9) will exponentially decay with increase of the radial coordinate (i.e., the altitude) r. Thus, one might use satellite measurements at varying altitudes to detect the exponential decay and to determine the photon mass.

Expanding Eq. (12.3.9) in a power series of μr and then neglecting cubic terms in μr, one arrives at

$$
\begin{aligned}
H^{(1)} &\cong H_D \left\{ 1 + \frac{(\mu r)^2 (1 - 5\cos^2 \theta)}{2(1 + 3\cos^2 \theta)} \right\} \\
&\equiv F(\mu, r, \theta) H_D,
\end{aligned}
$$

$$(12.3.20a)$$

where

$$F = 1 + \frac{(\mu r)^2 (1 - 5 \cos^2 \theta)}{2(1 + 3 \cos^2 \theta)}, \qquad (12.3.20b)$$

$$H_D = \frac{m}{r^3} |(3\hat{m} \cdot \hat{r}\hat{r} - \hat{m})| = \frac{m}{r^3} (1 + 3 \cos^2 \theta)^{\frac{1}{2}}. \qquad (12.3.20c)$$

An advantage of the altitude-dependent method is that the photon mass effect could be distinguished from an effect of true external sources which is nearly independent of the altitude. However, the main limitation on this method is that external perturbations become quite significant beyond $\sim 3R$.

Gintsburg (1963) [200], the first to apply the altitude-dependent method, used the assumption $F = 1 - (\mu r)^2$ to obtain a mass limit ($\mu < 3 \times 10^{-48}$ g) from magnetic measurements at varying altitudes by Vanguard, Explorer, and Pioneer satellites. Goldhaber and Nieto (1968) [213] argued that the same conservative error estimation that they had used for the external field method would make Gintsburg's limit (6–10)$\times 10^{-48}$ g.

12.3.3. Eccentric Dipole or "Vertical Current" Effect

As mentioned above, the geomagnetic field is more likely such a field produced by an eccentric dipole. The position of the eccentric dipole is at 6.5 degrees north latitude and 161.8 degrees east longitude. The distance between the eccentric dipole origin and the geocenter is $|\mathbf{b}| = 342$ km. Schrödinger (1943) pointed out an effect due to the displacement $\mathbf{b}$ of the eccentric dipole origin. From Eq. (12.3.7), the lines of vector potential are circles around the magnetic dipole axis, and hence would intersect the surface of the earth. From Eq. (5.2.4a), this means that there is a "feigned" vertical current from the component of $\mathbf{A}$ perpendicular to Earth's surface. That is, the integral of $\mathbf{H}$ along a closed path on the surface would fail to vanish. From $\nabla \times \mathbf{H} = -\mu^2 \mathbf{A}$, this current can be obtained to be $i_\perp = (\mu^2/4\pi) A_\perp$.

From Eq. (12.3.7), the central dipole potential is

$$\mathbf{A} = \left[\nabla \left(\frac{e^{-\mu r}}{r} \right) \right] \times \mathbf{m} = -\frac{(1 + \mu r)}{r^2} e^{-\mu r} \hat{r} \times \mathbf{m},$$

or

$$A_{max} = \frac{(1 + \mu r)m}{r^2} e^{-\mu r}. \qquad (12.3.21)$$

For an eccentric dipole with an eccentric distance of $b \ll R$, the maximum of vertical component of vector potential $\mathbf{A}$ on the earth's surface (a sphere of radius R) is

$$A_{\perp max} = A \sin \theta \simeq \frac{1 + \mu R}{R^2} e^{-\mu R} m \frac{b}{R}.$$

The corresponding current then is given by

$$i_{\perp max} = \frac{(\mu R)^2}{4\pi} \left[\frac{(1 + \mu R)e^{-\mu R} m}{R^3} \right] \frac{b}{R^2} \simeq H_D \left(\frac{\mu^2 R^2}{4\pi} \right) \frac{b}{R^2}. \qquad (12.3.22)$$

Here $\sin\theta \simeq \theta \sim b/R = 342/6378 \simeq 1/19$, and θ is the angle between $\mathbf{A}$ and the earth's surface. H_D, the magnetic field at the earth's equator, is given by Eq. (12.3.11) (neglecting terms in $\mu^2 R^2$), which was obtained by Schmidt to be $\simeq 31089\gamma$. Schrödinger (1943) substituted a value ($\mu R = 0.176$) from using the external field method into Eq. (12.3.22) to obtain the maximum vertical current:

$$i_{max\ vert} \simeq 6.4 \times 10^{-3} \text{A/km}^2, \tag{12.3.23}$$

and predicted ascending currents in the Eurasian–African hemisphere and descending currents in the American–Pacific one, which agreed with the earlier "hemisphere current effect" (Schmidt, 1924). However, Schröginger's value (12.3.23) was much smaller than Schmidt's one (50.8×10^{-3} A/km^2). According to the correction by Bass and Schrödinger (1955) [210], i.e., $\mu R \simeq 2 \times 0.176$, the expected maximum current will be greater than the one in Eq. (12.3.23) by a factor of four:

$$i_{max\ vert} \simeq 26 \times 10^{-3} \text{A/km}^2. \tag{12.3.24}$$

Later, observations showed that this kind of current was smaller than the above.

However, there is a more elementary reason for giving up this approach. The effect produced by a massive photon via its "eccentric dipole" compared to that produced via its "external field" is intrinsically smaller by a number of order (b/R), which for the earth is $\simeq 1/19$. Thus the "eccentric dipole effect" will not yield as good a mass limit as that to be obtained by a similar experimental effort using the "external field effect", even if the real currents could be separated out.

12.4. The Magnetohydrodynamic Effects

The elementary equations in magnetohydrodynamics come from the ordinary hydro-dynamics and Maxwell's electromagnetic field equations including the interaction be-tween flowing fluid and magnetic field. A cold plasma model might describe a nondis-sipative fluid, which has been successfully applied for plasma in laboratory and our universe. Many effects, such as a hydromagnetic wave, predicted by the cold plasma model exist within a hot plasma. However, non-zero photon mass would produce an effect on the predictions due to its effects on the ordinary Maxwell's electromagnetic field equations. Thus, magnetohydrodynamic effects within interplanetary plasma would yield a limit on the photon rest mass.

12.4.1. Hydromagnetic Waves

Consider a cold, nondissipative plasma moving in a uniform magnetic field $\mathbf{H}$. When small perturbations occur in such a plasma, two types of magnetic waves would be formed: the acoustic wave and the Alfvén wave (the purely magnetic wave). The magnetosonic wave, a longitudinal wave, propagates perpendicular to the direction of the static magnetic field **H**, with a velocity of

$$V_A = \sqrt{\frac{H^2}{4\pi\rho}}.$$

Here ρ is the mass density, and V_A is called the Alfvén velocity. If the plasma is not cold, the acoustic velocity S compared to the Alfvén velocity could not be neglected. Thus, the velocity of magnetosonic waves will be $\sqrt{S^2 + V_A^2}$, and the dispersion relation for these waves is given by

$$k^2 = \frac{\omega^2}{V_A^2}. \tag{12.4.1}$$

The Alfvén wave, a transverse wave (i.e., its oscillating field points in the vertical direction of **H**), propagates with the Alfvén velocity V_A. The dispersion equation for the Alfvén waves is

$$k^2 \cos^2\theta = \frac{\omega^2}{V_A^2}. \tag{12.4.2}$$

Here θ is the angle between the wave vector **k** and the magnetic field **H**.

The above results are based on the ordinary hydrodynamics in which the photon rest mass is zero. However, if the photon has a finite mass of μ, then Maxwell's equations should be replaced by the Proca equations. The corresponding dispersion relations then are given by

$$k^2 = \frac{\omega^2}{V_A^2} - \mu^2 \tag{12.4.3}$$

for the magnetosonic waves, and

$$k^2 \cos^2\theta = \frac{\omega^2}{V_A^2} - \mu^2 \tag{12.4.4}$$

for the Alfvén waves. These results show that in case of $\omega = \omega_c = \mu V_A$, the wave vector k vanishes, i.e., that a wave with the frequency ω_c is a non-propagating field. The frequency ω_c is called the critical frequency: The hydromagnetic waves with frequencies $\omega > \omega_c$ would propagate without decay ($k^2 > 0$), while the waves with the frequencies $\omega < \omega_c$ would exponentially decay ($k^2 < 0$). Thus, the photon mass ($\mu = \omega_c/V_A$) could be obtained by measuring the critical frequency ω_c, while a limit on the photon mass ($\mu < \omega/V_A$) could be yielded by measuring the hydromagnetic waves with the frequencies of $\omega \geq \omega_c$. The method for determining the photon rest mass by measuring the hydromagnetic waves was first suggested by Gintsburg (1963) [200].

Patel and Cahill (1964) [215] analyzed the data of measurement of the magnetic field in the earth's magnetosphere by use of Explorer XII and found that hydromagnetic waves having a period of 200 sec were produced at about 50 000 km from the center of the earth and then traveled along the magnetic field lines to the earth's surface in about 1.5 min, during which time their amplitude was attenuated by one-third. Later, Patel (1965) [216] regarded these waves as being at or above the critical frequency. In the earth's magnetosphere, the average magnetic field is $H = 10^{-3}$ G and the density of the plasma is $\rho = 50$ ions (or electrons) cm^{-3}. By using these values in Eqs. (12.4.3) and (12.4.4) for $k = 0$, Patel obtained

$$\mu = \frac{\omega}{V_A} = \frac{2\pi}{T}\sqrt{\frac{H_0^2}{4\pi nM}} \simeq 10^{-9} \text{cm}^{-1},$$

where M is the mass of a proton. He then took this value as the upper limit to the photon mass:

$$\mu < 10^{-9} \text{cm}^{-1} \equiv 4 \times 10^{-47} \text{g}. \qquad (12.4.5)$$

However, Patel pointed out that it could easily be off by an order or two of magnitude because of uncertainties in the values assumed for the plasma density and the magnetic field strength. The obvious uncertainty follows from their assumption that the damping of these waves comes from a massive photon but not from other dissipative mechanisms. Thus, a convincing mass limit would require the solution of many difficulties.

Hollweg (1974) [217] used the new observations of the Alfvén waves in the interplanetary medium provided by spacecraft to set an upper limit $\mu \le 3.6 \times 10^{-11}$ cm^{-1} $\equiv 1.3 \times 10^{-48}$ g , and a stronger, but less certain upper limit $\mu < 3.1 \times 10^{-12}$ cm^{-1} $\equiv 1.1 \times 10^{-49}$ g.

Observations of the center region of the Crab Nebula (see Ref. [218] and references therein) indicate that the plasma consists of an ultrarelativistic electron component and a tenuous lower-energy background component, embedded in a magnetic field that is relatively uniform on the scale $\sim 10^{18}$ cm. Quasiperiodic disturbances ($\omega \sim 10^{-6}$ sec^{-1}) generated in the vicinity of the pulsar propagate across the magnetic field out into the nebula. This series of "wisps" has been identified as a sequence of magnetoacoustic waves in which the wave compressions produce local enhancements of the synchrotron radiation. Barnes and Scargle (1975) [219] gave the dispersion relation for these magnetoacoustic waves:

$$\left(\frac{\omega}{kc}\right)^2 \left[1 + \frac{4\pi}{B^2}(\varepsilon + P_\perp)\right] = 1 + \frac{2\pi P_\perp}{B^2}(4 - \zeta) + \frac{\mu^2}{k^2}, \qquad (12.4.6)$$

where B is the background magnetic field strength, $P_\perp$ is the total plasma pressure transverse to **B**, ε is the total energy density (including proper energy) of matter in the plasma, and ζ is a numerical factor between 0 and 1. Equation (12.4.6) implies wave evanescence ($k^2 < 0$) unless

$$\omega > \omega_c = \frac{\mu c}{\sqrt{1 + (4\pi/B^2)(\varepsilon + P_\perp)}}. \qquad (12.4.7)$$

They used the observations for hydromagnetic waves in the Crab Nebula to obtain an upper limit to the photon rest mass of $\mu \simeq 10^{-16} - 10^{-15}$ cm$^{-1} \equiv 3 \times 10^{-54} - 3 \times 10^{-53}$ g (uncertainty < factor 100). However, they pointed out that there remains a theoretical difficulty. An essential assumption in this method is that the background plasma is infinite and uniform. The Proca equations, however, show that, in general, either a static magnetic field varies appreciably over distances of order μ^{-1}, or there is a large background current $\mathbf{J}_0 \simeq (c/4\pi)\mu^2 \mathbf{A}(J_0 \gg (c/4\pi)|curl \ \mathbf{B}|)$. In either case, the dispersion relation breaks down for $|\mathbf{k}| \leq \mu$. Then the upper limit of μ given by them and in Refs. [217,220] are open to doubt.

12.4.2. Dissipation of the Interplanetary Magnetic Field

The usual theory of hydromagntics shows that conducting medium moving in a magnetic field would generate an induced e.m.f. (electromotive force). If the medium is conducting perfectly and there are no outside influences, e.m.f. then vanishes. In this case, there is no relative motion of the medium and magnetic field lines. In other words, the magnetic field lines move together with the medium, i.e., the magnetic field lines are permanently frozen in. On the other hand, if the medium has a finite conductivity, the field decreases at a rate determined by the dimensions of the supporting plasma and by its conductivity. However, if the photon were to have a finite mass the situation would be different. Williams and Park (1971) [221] proposed a method for setting an upper limit on the photon mass by use of this dissipation effect.

Williams and Park imagined a galactic arm straightened out so as to form a long filament of plasma with a magnetic field running down it. They assumed that the plasma is electrically neutral and that there is no electric field along the filament, so that the currents flow perpendicular to its length. They also assumed that the spatial distribution of the galactic plasma does not change significantly for $\sim 10^6$ years. Then they needed to deal only with the damping force and the electromagnetic force acting on electrons and ions, but not with the hydrodynamic equations. Ignoring inertial forces, the electrodynamic forces on the electrons are equal to the damping forces exerted by their interaction with the ions and the atoms. From this they obtained the relation between the electric field $\mathbf{E}$ and current $\mathbf{J} = n_e e(\mathbf{v}_i - \mathbf{v}_e)$:

$$E_x = \left(\sigma_e^{-1} + \sigma_i^{-1}\right) j_x + \frac{H j_y}{n_e e c},$$

$$E_y = \left(\sigma_e^{-1} + \sigma_i^{-1}\right) j_y - \frac{H j_x}{n_e e c}, \qquad (12.4.8a)$$

with

$$\sigma_e^{-1} = \frac{m}{n_e e^2}\left(\frac{1}{\tau_{ie}} + \frac{1}{\tau_{ae}}\right), \qquad \sigma_i^{-1} = \frac{\tau_{ia} H^2}{n_e M_i c^2}, \qquad (12.4.8b)$$

where m and M_i are the masses of the electron and ion respectively, n_e and n_i represent the number densities of the electrons and ions respectively, τ_{ie} is the relaxation time for the collisions between the electrons and ions, and τ_{ae} is the relaxation time for the collisions between the electrons and atoms, and so on. The known properties of "cool" plasma (H I zones of the Galaxy) give $\sigma_e \approx 10^{10}$ sec^{-1}, $\sigma_i \approx 10^{-3}$ sec^{-1}, so that the dissipation of energy proceeds mostly from ions to atoms and the electronic contribution can be ignored. By use of Eq. (12.4.8) and the Proca equations, they obtained the dissipation (W) of the magnetic energy in the arm of the Galaxy:

$$W \sim exp\left[-\frac{2\nu}{l^2}\left(1 + \mu^2 l^2\right)t\right],$$

where $\nu = c^2/4\pi\sigma_i \approx 10^{23}$ cm^2/sec, l is a length of the order of the radius of the spiral arm. Thus the dissipation time of the magnetic energy is

$$\tau \sim \frac{l^2}{2\nu}\left(1 + \mu^2 l^2\right)^{-1},$$

which reduces to the usual result in case of $\mu = 0$. The flux of primary cosmic rays has remained roughly constant, on the average, over the last 10^6 years. This yields $\tau \sim 10^6$ years. By assuming the field constant over 10^6 yr, Williams and Park gave an upper limit to the photon mass of $\mu < 10^{-8}$ cm$^{-1} \equiv 3.5 \times 10^{-56}$ g.

Byrne and Burman (1972) [222] re-examined dissipation of large-scale magnetic fields in the Galaxy and claimed that Williams and Park (1971) [221] used a common mis-interpretation of the tensor conductivity. Byrne and Burman considered three-fluid plasma which consist of electrons, protons and identical neutral particles. Ignoring inertial effects and pressure gradients, they gave the equilibrium equations for forces acting on the electrons and protons:

$$-\frac{e}{m_e}\left(\mathbf{E} + \frac{1}{c}\mathbf{u}_e \times \mathbf{H}\right) + \nu_{ei}(\mathbf{u}_i - \mathbf{u}_e) + \nu_{en}(\mathbf{u}_n - \mathbf{u}_e) + \mathbf{g} = 0, \qquad (12.4.9)$$

$$\frac{e}{m_i}\left(\mathbf{E} + \frac{1}{c}\mathbf{u}_i \times \mathbf{H}\right) + \nu_{ie}(\mathbf{u}_e - \mathbf{u}_i) + \nu_{in}(\mathbf{u}_n - \mathbf{u}_i) + \mathbf{g} = 0, \qquad (12.4.10)$$

where the subscripts e, i and n denote quantities pertaining to the electron, proton and neutral fluids, respectively, e is the charge on a proton, E and H are the electric and magnetic fields, g is the gravitational acceleration, m and u denote particle masses and fluid velocities, and ν_{ab} represents the momentum relaxation frequency for collisions of particles of component fluid a with particles of component fluid b. For a quasi-neutral plasma, the current density $\mathbf{j}$ is

$$\mathbf{j} = ne(\mathbf{u}_i - \mathbf{u}_e), \qquad (12.4.11)$$

where n is the electron or proton number density. By using the conservation law of momentum $m_i\nu_{ie} = m_e\nu_{ei}$ and Eqs. (12.4.9) and (12.4.10), one arrives at

$$\mathbf{E} + \frac{1}{c}\mathbf{u}_H \times \mathbf{H} = \sigma^{-1}\mathbf{j} + \frac{m_i m_e}{e(m_i + m_e)}(\nu_{in} - \nu_{en})(\mathbf{u}_e - \mathbf{u}_n), \qquad (12.4.12)$$

where the conductivity σ is given by

$$\sigma^{-1} = \frac{m_e}{ne^2}\left\{\nu_{ei} + \left(\frac{m_i}{m_i+m_e}\right)^2\nu_{en} + \frac{m_i m_e}{(m_i+m_e)^2}\nu_{in}\right\}, \tag{12.4.13}$$

and

$$\mathbf{u}_H = \frac{m_i\mathbf{u}_e + m_e\mathbf{u}_i}{m_i+m_e}, \qquad \mathbf{u}_e = \frac{m_i\mathbf{u}_i + m_e\mathbf{u}_e}{m_i+m_e}. \tag{12.4.14}$$

The last term in Eq. (12.4.12) represents friction between the combined electron–proton fluid and the neutral fluid, and can be neglected because, in interstellar space, the electron–proton collision frequency is much larger than the electron–neutral and proton–neutral collision frequencies. Then Eqs. (12.4.12) and (12.4.13) become

$$\mathbf{E} + \frac{1}{c}\mathbf{u}_H \times \mathbf{H} = \sigma^{-1}\mathbf{j}, \tag{12.4.15}$$

$$\sigma = \left(\frac{m_e}{ne^2}\nu_{ei}\right)^{-1} = \frac{e^2}{m_e}\frac{T^{3/2}/5.5}{\ln(220T/n^{1/3})}. \tag{12.4.16}$$

Putting Eq. (12.4.15) in Faraday's law, one has

$$\frac{\partial \mathbf{H}}{\partial t} = -c\nabla \times \mathbf{E} = \nabla \times (\mathbf{u}_H \times \mathbf{H}) - c\sigma^{-1}\nabla \times \mathbf{j}. \tag{12.4.17}$$

Then by using the Proca equation (5.2.4a), equation (12.4.17) becomes

$$\begin{aligned}
\frac{\partial \mathbf{H}}{\partial t} &= \nabla \times (\mathbf{u}_H \times \mathbf{H}) - c\sigma^{-1}\left(\frac{c}{4\pi}\nabla \times (\nabla \times \mathbf{H}) + \frac{c}{4\pi}\mu^2\nabla \times \mathbf{H}\right) \\
&= \nabla \times (\mathbf{u}_H \times \mathbf{H}) + \frac{c^2}{4\pi}\sigma^{-1}(\nabla^2\mathbf{H} - \mu^2\mathbf{H}).
\end{aligned}$$

$$\tag{12.4.18}$$

Here the displacement current $(\partial\mathbf{E}/\partial t)$ and the conductivity gradients are neglected. It is shown from Eq. (12.4.18) that when the plasma is conducting perfectly, i.e. $(\sigma \to \infty)$, we then have

$$\frac{\partial \mathbf{H}}{\partial t} = \nabla \times (\mathbf{u}_H \times \mathbf{H}),$$

which is the same as that in the usual hydromagnetics, i.e., that the magnetic field is "frozen" in the medium. This shows that the last two terms in Eq. (12.4.18) represent dissipation effects. If L denotes a distance over which H varies significantly, then $\nabla^2\mathbf{H} \sim L^2 H$. A spatial integral of the scalar product between $\mathbf{H}$ and both sides of Eq. (12.4.18) gives the dissipation effect for total magnetic energy:

$$\frac{dW}{dt} = \frac{d}{dt}\int H^2 d^3x,$$

or

$$W \sim exp\left[-\frac{c^2}{4\pi\sigma}(L^{-2}+\mu^2)t\right]. \tag{12.4.19}$$

This implies that the decay time is

$$\tau \sim \frac{4\pi\sigma}{c^2}(L^{-2}+\mu^2)^{-1}. \tag{12.4.20}$$

For $L^2 \gg \mu^{-2}$, this equation reduces to

$$\mu \simeq (4\pi\sigma/c^2\tau)^{\frac{1}{2}}. \tag{12.4.21}$$

Equation (12.4.16) shows that σ depends only slightly on n but strongly on T. Then Eq. (12.4.21) indicates that the variation of μ with T is more important than its variation with τ. Thus the best upper limit for μ to be obtained from considering the dissipation of large-scale magnetic fields will result from studying cool regions with long-lived fields.

Byrne and Burman (1972) [222] gave their arguments for determining an upper limit on the photon mass μ: For galactic HI regions, $T \leq 10^2$K and $n \sim 10^{-3}$ to 10^{-2} cm^{-3} so that $\sigma \leq 5 \times 10^9$ sec^{-1}. Hence if the existence of long-scale magnetic fields with $\tau \geq 10^6$ yr is established, then Eq. (12.4.21) gives $\mu \leq 10^{-12}$ cm$^{-1} \equiv 4 \times 10^{-50}$ g; and so on. Furthermore, they gave a correction of an earlier treatment, and then concluded that the most that can be said appears to be that $\mu \leq 3 \times 10^{-11}$ cm^{-1}.

12.4.3. Instability of Interstellar Plasma

It has been shown (cf. the references in Ref. [223]) that a plasma becomes unstable and locally evacuated, with a corresponding large fall in conductivity, when the electron drift speed V exceeds the electron thermal speed U_e. In addition, instabilities would occur when V exceeds the phase velocities of some wave types.

If V_m denotes the maximum electron drift speed that the plasma can support stably, then the maximum current is $j_m = neV_m$ where n is the electron number density and e is the magnitude of the electronic charge.

Goldhaber and Nieto (1971) [197] suggested a method for determining the photon mass: since there is a maximum current density that a plasma can support, knowledge of magnetic fields in the Galaxy can be used to obtain a limit on μ. By use of the data, $H \sim 10^{-6}$ G, $V_m \leq 10^5$ cm/sec, $n \leq 1/$cm^3, and $R \simeq 10^{21}$cm, they obtained a limit of $\mu \leq 10^{-15}$ cm$^{-1} \equiv 3.5 \times 10^{53}$ g. Following the method suggested by Goldhaber and Nieto (1971) [197], Byrne and Burman (1973) [223], by discussing various instabilities in the interstellar medium of the galactic HI regions, established a limit on the photon mass: $\mu \leq 3 \times 10^{-15}cm^{-1} \equiv 10^{-52}$ g.

12.5. Other Methods

Yamaguchi (1959) [224] argued that there are hydro-magnetic turbulences of huge scale motions in interstellar gases, etc., and notably in Crab nebula. Such turbulent motions suggest that the Compton wavelength $1/\mu$ for the photon should not be smaller than the linear dimension D of magnetic turbulence: $\mu^{-1} \geq D$. Using the dimension of the Crab nebula $D \sim 10^{17}$cm, he gave an upper limit to the photon mass: $\mu \leq 10^{-17}$cm$^{-1} \equiv 4 \times 10^{-55}$ g. Later, he pointed out that the same technique applied to the field in one of the spiral arms of our galaxy could yield a limit $\mu \leq 10^{-21}$cm$^{-1} \equiv 4 \times 10^{-59}$ g (cf. Ref. [197]).

Franken and Ampulshi (1971) [225] pointed out that measurements with vary low-frequency parallel resonance circuits could set an upper limit to μ. They assumed that the relation between the resonance frequency of the circuit ω' and the photon mass μ:

$$\omega'^2 = \omega_0{}^2 + \omega_c{}^2, \tag{12.5.1}$$

where ω_0 is the resonance frequency in the massless electromagnetic case, ω' is the observed resonance frequency in their experiments, and $\omega_c \equiv \mu c^2/\hbar$. Using this method they gave $\mu < 3 \times 10^{-12}$ cm$^{-1} \equiv 10^{-49}$ g.

However, as pointed out in Refs. [226–228], if the concept of a photon mass μ is introduced by Proca equations, then equation (12.5.1) is not correct. It is shown that the difference between Proca's and Maxwell's electromagnetic fields is of second-order in μD where D is the dimension of an experimental arrangement. Thus, for a resonance circuit the relative change in the resonance frequency caused by μ is

$$\frac{\delta\omega'}{\omega'} = O\left((\mu D)^2\right), \tag{12.5.2}$$

where $\delta\omega' = \omega' - \omega_0$. It is shown that the relative difference depends only on μD but not on ω'. Thus, for the experiments by Franken and Ampulshi, equation (12.5.2) will yield an upper limit of $\mu \leq 10^{-41}$ g.

In addition, a measurement of longitudinal photons would yield a limit to the photon mass. However, their observation will be difficult if not impossible, and their effects are negligible, even on an astronomical scale [197].

12.6. Summary

Table 12.2. *Upper limit on photon rest mass*

Year	Investigator	Method	$\mu(\text{cm}^{-1})$	$m_0(\text{g})$
1971	cf. Ref. [197]	Speed of light	1.6×10^{-4}	5.6×10^{-42}
1940	de Broglie [201]	Dispersion of starlight	2.3×10^{-2}	0.8×10^{-39}
1969	Feinberg [204]	Dispersion of starlight	3×10^{-7}	10^{-44}
1936	Plimpton and Lawton [206]	Coulomb's law	10^{-6}	3.4×10^{-44}
1968	Cochran and Franken (1968) [209]	Coulomb's law	9×10^{-8}	3×10^{-45}
1970	Barlett, Goldhagen and Phillips (1970) [207]	Coulomb's law	1.3×10^{-8}	3×10^{-46}
1970	Williams, Faller, and Hill (1970) [208]	Coulomb's law	6×10^{-10}	1.6×10^{-47}
1943	Schrödinger [212]	External field (geomagnetic field)	5.6×10^{-10}	2.0×10^{-47}
1968	Goldhaber and Nieto [213]	External field (geomagnetic field)	1.15×10^{-10}	4.0×10^{-48}
1975	Davis, Jr., Goldhaber, and Nieto [214]	External field (Jupiter's magnetic field)	2×10^{-11}	8×10^{-49}
1963	Gintsburg [200]	Altitude-dependence of geomagnetic field	10^{-11}	3×10^{-48}
1965	Patel [216]	Alfvén waves in Earth's magnetosphere (dispersion effect)	10^{-9}	4×10^{-47}

Table 12.2. *Upper limit on photon rest mass (continued)*

Year	Investigator	Method	$\mu(\text{cm}^{-1})$	$m_0(\text{g})$
1974	Hollweg [217]	Alfvén waves in interplanetary medium (dispersion effect)	3.6×10^{-11}	1.3×10^{-48}
1975	Barnes and Scargle [219]	Magnetoacoustic waves in Crab Nebula (dispersion effect)	$10^{-16}-$ 10^{-15}	$3 \times 10^{-54}-$ 3×10^{-53}
1972	Byrne and Burman [222]	Dissipation of large-scale magnetic fields in Galaxy	10^{-12}	4×10^{-50}
1971	Goldhaber and Nieto [197]	Maximum current density in plasma of Galaxy	10^{-15}	4×10^{-53}
1973	Byrne and Burman [223]	Maximum current density in plasma of Galaxy	3×10^{-15}	10^{-52}
1959	Yamaguchi [224]	Dimension of hydro-magnetic turbulence in Crab nebula	10^{-17}	4×10^{-55}
1971	c.f. [197]	Dimension of magnetic field in arms of Galaxy	10^{-21}	4×10^{-59}

CHAPTER 13

THE TESTS OF THOMAS PRECESSION

13.1. The Fine Structure of Atomic Spectra

An electron in an electric field of nucleus moves along a closed orbit around the nucleus. Due to the magnetic moment of the electron, the Thomas precession given by Eq. (2.12.17) would effect the interaction energy of the electron in the atom. According to the assumption by Uhlenbeck and Goudsmit, the relation between the electron moment μ and spin angular momentum $\mathbf{s}$ is

$$\mu = -\frac{ge}{2m_0 c}\mathbf{s}, \tag{13.1.1}$$

where g is called g factor (e.g., $g = 2$ for the electron), $-e$ is the charge of the electron, m_0 is the rest mass of the electron, and c is the speed of light in vacuum.

Let an electron move with a velocity of $\mathbf{v}$ in electric and magnetic fields $\mathbf{E}$ and $\mathbf{B}$. Equation of motion for the spin angular momentum of the electron in the fields is given by

$$\frac{d\mathbf{s}}{dt} = \mu \times \mathbf{B}' + \omega_T \times \mathbf{s}, \tag{13.1.2}$$

where $\mathbf{B}'$ is the magnetic induction in the rest frame of the electron, and ω_T is the angular velocity of the Thomas precession defined by Eq. (2.12.17). From Eqs. (4.2.2) and (4.2.5) we have

$$\mathbf{B}' \cong \mathbf{B} - \frac{\mathbf{v}}{c} \times \mathbf{E}. \tag{13.1.3}$$

Substituting Eqs. (13.1.1) and (13.1.3) into Eq. (13.1.2), one arrives at

$$\frac{d\mathbf{s}}{dt} = \mathbf{s} \times \left[\frac{-ge}{2m_0 c}\left(\mathbf{B} - \frac{\mathbf{v}}{c} \times \mathbf{E}\right) - \omega_T\right]. \tag{13.1.4)}$$

This result implies that the electron has the interaction energy

$$U = -\mathbf{s} \cdot \left[\frac{-ge}{2m_0 c}\left(\mathbf{B} - \frac{\mathbf{v}}{c} \times \mathbf{E}\right) - \omega_T\right]. \tag{13.1.5}$$

For an atom with an electron, the electric force acting on the electron is

$$-e\mathbf{E} = -\nabla V(r),$$

and, in general, reduces approximately to

$$-e\mathbf{E} = \frac{\mathbf{r}}{r}\frac{dV}{dr}, \tag{13.1.6}$$

where $V(r)$ is the mean potential energy. By using Eqs. (13.1.6) and (2.12.17) in (13.1.5), we obtain

$$U = \frac{ge}{2m_0c}\mathbf{s}\cdot\mathbf{B} + \frac{g}{2m_0^2c^2}(\mathbf{s}\cdot\mathbf{L})\frac{1}{r}\frac{dV}{dr} - \frac{1}{2c^2}\mathbf{s}\cdot(\mathbf{v}\times\mathbf{a}), \qquad (13.1.7)$$

where

$$\mathbf{L} = m_0(\mathbf{r}\times\mathbf{v})$$

is the orbital angular momentum of the electron. Equation (13.1.6) shows that the acceleration of the electron is

$$\mathbf{a} \simeq \frac{-e\mathbf{E}}{m_0} = -\frac{\mathbf{r}}{m_0 r}\frac{dV}{dr}. \qquad (13.1.8)$$

By use of the result in Eq. (13.1.7) we have

$$U = \frac{ge}{2m_0c}\mathbf{s}\cdot\mathbf{B} + \frac{(g-1)}{2m_0^2c^2}\mathbf{s}\cdot\mathbf{L}\frac{1}{r}\frac{dV}{dr}. \qquad (13.1.9)$$

Due to $g = 2$ for an electron, the second term on the right-hand side of the above equation is the spin–orbit interaction energy for an electron in an atom. It is well known that equation (13.1.9) gives an explanation of Zeeman effect and that of the fine-structure splitting of hydrogen spectral lines.

13.2. Measurements for $(g-2)$ Factor of Leptons

Consider the motion of an electron in a uniform magnetic field $\mathbf{B}$. Let the velocity $\mathbf{v}$ of the electron be perpendicular to the magnetic field $\mathbf{B}$. In this case the electron will move along a circular orbit with the angular frequency given by Eq. (11.1.28),

$$\omega_e = \frac{eB}{mc} = \frac{eB}{\gamma m_0 c}, \qquad (13.2.1a)$$

with

$$\gamma = \frac{1}{\sqrt{1 - v^2/c^2}}. \qquad (13.2.1b)$$

From Eq. (13.1.4) we know that the precession of the electron spin is

$$\frac{d\mathbf{s}}{dt} = \omega_s \times \mathbf{s},$$

where

$$\omega_s = \frac{ge}{2m_0c}\left(\mathbf{B} - \frac{\mathbf{v}}{c}\times\mathbf{E}\right) + \omega_T = \frac{ge}{2m_0c}\mathbf{B} + \omega_T \qquad (13.2.2)$$

is the precession frequency of the spin. Here $\mathbf{E} = 0$ is used.

The angular frequency of Thomas precession is given by Eq. (2.11.16):

$$\omega_T = \left(1 - \frac{1}{\sqrt{1 - v^2/c^2}}\right) \frac{\mathbf{v} \times \mathbf{a}}{v^2}, \qquad (13.2.3)$$

where **a** is the acceleration of the electron moving along a circular orbit in the magnetic field, i.e.,

$$\mathbf{a} = \frac{-e}{\gamma m_0 c} \mathbf{v} \times \mathbf{B}. \qquad (13.2.4)$$

By putting the equation into Eq. (13.2.3) and noting that **v** is perpendicular to **B**, we get

$$\omega_T = (1 - \gamma) \frac{-e}{\gamma m_0 c} \frac{\mathbf{v} \times (\mathbf{v} \times \mathbf{B})}{v^2} = (1 - \gamma) \frac{e}{\gamma m_0 c} \mathbf{B}. \qquad (13.2.5)$$

By using the result in Eq. (13.2.2) we obtain the precession frequency of the electron spin in the uniform magnetic B,

$$\omega_s = \left(\frac{g - 2}{2}\right) \frac{eB}{m_0 c} + \omega_c, \qquad (13.2.6a)$$

or

$$\omega_a = \omega_s - \omega_c = \left(\frac{g}{2} - 1\right) \frac{eB}{m_0 c}. \qquad (13.2.6b)$$

Here ω_c is the electronic (orbit) angular frequency defined by Eq. (13.2.1). The assumption by Uhlenbeck and Goudsmit, equation (13.1.1), means that $g = 2$ for an electron. However, experiments showed that the g factor is slightly different from 2. The difference, the $(g - 2)$ factor, corresponds to the anomalous magnetic moment of an electron.

Equation (13.2.6b) gives the difference between the electronic spin precession frequency ω_s and its orbital frequency ω_c predicted by Einstein's theory of special relativity, which can be directly measured. Therefore, by use of an observed value of the quantity $eB/m_0 c$ in Eq. (13.2.6b), one can find out a value of the factor $(g - 2)/2$. Experimental results for measuring the $(g - 2)$ factor of leptons are shown in Table 13.1.

Table 13.1. *Experimental measurements for $g - 2$ of leptons*

Year	Investigator	Lepton	γ	$(g - 2)/2$
1977	Van Dyck, Schwinberg and Dehmelt [229]	Electron (e^-)	$1 + 10^{-9}$	0.00115965241(20)
1971	Wesley and Rich [230]	Electron (e^-)	1.2	0.00115965770(350)
1979	Cooper *et al.* [231]	Electron (e^-)	2.5×10^4	0.0011622(200)
1972	Bailey *et al.* [232]	Muon $(\mu^\pm)$	12	0.00116616(31)
1977	Bailey *et al.* [233]	Muon $(\mu^\pm)$	29.2	0.001165922(9)

The $g - 2$ factor can be calculated by use of quantum electrodynamics which is based on Einstein's theory of special relativity. This calculation to the magnitude of order α^3 gives [234]

$$\frac{1}{2}(g - 2) = 1 + \frac{\alpha}{2\pi} - 0.328479 \left(\frac{\alpha}{\pi}\right)^2 + (1.49 \pm 0.25) \left(\frac{\alpha}{\pi}\right)^3,$$

where

$$\alpha = \frac{1}{137.03604(11)}$$

is the fine structure constant. This theoretical result agrees with experiments to whitin a very high accuracy.

We now consider comparison between Eq. (13.2.6) and the $g - 2$ measurements, which is not relevant to the quantum electrodynamics. Table 13.1 shows a fact that the values of $g - 2$ are independent of the velocities of the measured particles, which is in agreement with the prediction of Eq. (13.2.6). Then this can be regarded as a direct test of Thomas precession. In order to estimate relative accuracy of these measurements, a theory or a model different from Einstein's special relativity is needed. For this purpose, Newman *et al.* (1978) [235] suggested a model in which the momentum p and energy E of a free electron has a relation $E = E(p)$ and then the rest mass of the electron is given by

$$\frac{1}{m_0} = \lim_{p \to 0} \frac{1}{p} \frac{dE}{dp}. \tag{13.2.7}$$

The cyclotron frequency of the electron in a uniform magnetic field B is

$$\omega_c = \frac{eB}{\tilde{\gamma} m_0 c}, \tag{13.2.8}$$

with

$$\tilde{\gamma} = \frac{p}{m_0} \frac{dp}{dE}. \tag{13.2.9}$$

Then the precession frequency of the electron spin, (13.2.6a), should be changed to

$$\omega_s = \frac{geB}{2m_0 c} + (1 - \gamma)\omega_c = \omega_c \left(\frac{1}{2}g\tilde{\gamma} - \gamma + 1\right),$$

with

$$\gamma = \frac{1}{\sqrt{1 - v^2/c^2}}, \qquad \frac{v}{c} = \frac{1}{c} \frac{dE}{dp}, \tag{13.2.10}$$

and hence the difference between the cyclotron and precession frequencies is

$$\omega_D = \omega_s - \omega_c = \left(\frac{g}{2} - \frac{\gamma}{\tilde{\gamma}}\right) \frac{eB}{m_0 c}. \tag{13.2.11}$$

Newman *et al.* regarded the quantity $(1 - \gamma)\omega_c$ as a relativistic effect, i.e., Thomas effect, where γ has the usual form. On the other hand, the $\tilde{\gamma}$ in the expression of ω_c comes from dynamics and thus may be different from the usual γ. The result (13.2.11) from this model gives a relation between the frequency ω_D and the electron velocity. Newman *et al.* used two experimental results for determining the factor $(g/2) - \gamma/\tilde{\gamma}$ on the right-hand side of Eq. (13.2.11). One is the experiment carried out by Wesley and Rich (1971) [230] (see Tab. 13.1), in which the magnetic field was $B = 1.2 \times 10^3$G, the kinetic energy of electron was 110 keV (the corresponding velocity of electron was $v/c = 0.75$, i.e., $\gamma = 1.2$), and the difference in frequency ω_D was directly measured. By putting these values in Eq. (13.2 11), they obtained

$$\left(\frac{g}{2} - \frac{\gamma}{\tilde{\gamma}}\right) = 0.00115965770(350). \tag{13.2.12}$$

The other is the experiment performed by Van Dyck, Schwinberg, and Dehmelt (1977) [229] (see Tab. 13.1), in which electrons were nonrelativistic ($\gamma - 1 = 10^{-9}$) and the observed difference in frequency was

$$\frac{\omega_s - \omega_c}{\omega_c} = 0.00115965241(20). \tag{13.2.13}$$

By making comparison of Eqs. (13.2.12) and (13.2.13), they yielded

$$1 - \frac{\gamma}{\tilde{\gamma}} = (5.3 \pm 3.5) \times 10^{-9}. \tag{13.2.14}$$

This shows that these measurements agree with the prediction ($\gamma = \tilde{\gamma}$) of Einstein's special relativity to within an accuracy of 5×10^{-9}. As pointed out by Combley *et al.* (1979) [236], of course, a relative accuracy from this method is very much dependent upon the model adopted for the breakdown of special relativity.

Cooper *et al.* (1979) [231] used the above model and assumed that the quantity $\gamma/\tilde{\gamma}$ can be expanded in the power series of $(\gamma - 1)$:

$$\frac{\gamma}{\tilde{\gamma}} = 1 + C_1(\gamma - 1) + \cdots.$$

Here a value of the efficient C_1 was obtained from two experimental results through

$$C_1 = \frac{a^{(1)} - a^{(2)}}{\gamma^{(1)} - \gamma^{(2)}},$$

where

$$a \equiv \frac{1}{2}\left(g - \frac{\gamma}{\tilde{\gamma}}\right) = \omega_D\left(\frac{eB}{m_0c}\right)^{-1}.$$

The upperscripts (1) and (2) correspond to two experiments with different leptonic velocities. Their results are shown in Table 13.2.

Table 13.2. *Summary of lepton $g - 2$ relativity tests*

Method	References	$\gamma^{(1)}$	$\gamma^{(2)}$	C_1
μ^-, μ^+ g factor	[232] and [233]	12	29.2	$(1.4 \pm 1.8) \times 10^{-8}$
e^- g factor	[229] and [230]	1	1.2	$(-2.6 \pm 1.8) \times 10^{-8}$
e^- g factor	[229] and [231]	1	2.5×10^4	$(-1.0 \pm 8.0) \times 10^{-10}$

REFERENCES

[1] A. Einstein, *Ann. der Phys.* (Leipzig **17** (1905) 891.

[2] W.F. Edwards, *Am. J. Phys.* **31** (1963) 482–489.

[3] H.P. Robertson, *Rev. Mod. Phys.* **21** (1949) 378.

[4] R. Mansouri and R.U. Sexl, *Gen. Relativ. Grav.* **8** (1977) 497.

[5] J.A. Winnie, *Phil. Sci.* **37** (1970) 81.

[6] A.A. Michelson, *Am. J. Sci.* **22** (1881) 120.

[7] A.A. Michelson, and E.W. Morley, *Am. J. Sci.* **34**n (1887) 333; *hil. Mag.* **24** (1887) 449.

[8] H. Poincaré, *Rc. Circ. mat Palermo* **21** (1906) 129.

[9] H. Minkowski, Raum und Zeit. *Phys. Z.* **10** (1909) 104.

[10] O.M. Bilaniuk, and F.C.G. Sudarshan, *Phys. Today* **22(5)** (1969) 43.

[11] O.M. Bilaniuk, V.K. Deshpande, and F.C.G. Sudarshan, *Am. J. Phys.* **30** (1962) 718.

[12] O.M. Bilaniuk, *et al.*, *Phys. Today* **22**(12) (1969) 47.

[13] C. Møller, *The Theory of Relativity* Clarendon Press, Oxford, 1972.

[14] L.T. Thomas, *Phil. Mag.* **3** (1927) 1.

[15] G.P. Fisher, *Am. J. Phys.* **40** (1972) 1772.

[16] G.N. Lewis and R.C. Tolman, *Phil. Mag.* **18** (1909) 510.

[17] M. Wilson and H.A. Wilson, *Proc. R. Soc.* **A89** (1913) 99.

[18] H. Minkowski, *Nachr. Ges. Wiss. Göttingen 53* (1908)

[19] Y.Z. Zhang, *Acta Physica Sinica* **4** (1975) 180.

[20] Y.Z. Zhang, *Acta Physica Sinica* **26** (1977) 455.

[21] W.F. Parks and J.T. Dowell, *Phys. Rev. A* **9** (1974) 565.

[22] Reichenbach, *The Philosophy of Space and Time* (Dover Publications, Inc., New York, 1958), p.142.

[23] A. Grunbaum, *Logical and Philosophical Foundations of the Special Theory of Relativity* in *Philosophy of Science*, A. Danto and S. Morgenbesser, eds. (Meridian Books, New York, 1960).

[24] M. Ruderfer, *Proc. IRE* **48** (1960) 1661.

[25] Y.Z. Zhang, *Gen. Relativ. Grav.* **27** (1995) 475.

[26] B. Bertotti, *Radio Sci.* **14** (1979) 621.

[27] D.W. MacArthur, *Phys. Rev.* A **33** (1986) 1.

[28] M.P. Haugan and C.M. Will, *Phys. Today* **40**(5) (1987) 69.

[29] G. Abolghasem, M.R.H. Khajehpour and R. Mansouri, *Phys. Lett.* A **132** (1988) 310.

[30] E. Riis *et al.*, *Phys. Rev. Lett.* **60** (1988) 81.

[31] E. Riis *et al.*, *Phys. Rev. Lett.* **62** (1989) 842.

[32] Z. Bay and J.A. White, *Phys. Rev. Lett.* **62** (1989) 841.

[33] M.D. Gabriel and M.P. Haugan, *Phys. Rev.* D **41** (1990) 2943.

[34] T.P. Krisher *et al.*, *Phys. Rev.* D **42** (1990) 731.

[35] C.M. Will, *Phys. Rev.* D **45** (1992) 403.

[36] B.W. Petley, *Nature* **303** (1983) 373.

[37] O. Römer, *Mem. Acad.* **10** (1676) 575;

[38] L. Karlov, *Austral. J. Phys.* **23** (1970) 243.

[39] R. Mansouri and R.U. Sexl, *Gen. Relativ. Grav.* **8** (1977) 515.

[40] A. Brillet and J.L. Hall, *Phys. Rev. Lett.* **42** (1979) 549.

[41] J.J. Larmor, *Aether and Matter* (Cambridge, 1900), p. 167.

[42] M. Ruderfer, *Phys. Rev. Lett.* **7** (1961) 361.

[43] A. A. Tyapkin, *Lett. Nuovo. Cimento* **7** (1973) 760.

[44] H.A. Lorentz, *Verh. K. Akad. Wet.* **1** (1892) 74.

[45] O. Lodge, *Phil. Trans. R. Soc.* A **184** (1893) 727.

[46] R. J. Kennedy, *Proc. Natl. Acad. Sci.* **12** (1926) 621.

[47] K. K. Illingworth, *Phys. Rev.* **30** (1927) 692.

[48] G. Joos, *Ann. Physik.* **7** (1930) 385.

[49] D. C. Miller, *Revs. Mod. Phys.* **5** (1933) 203.

[50] R. S. Shankland *et al.*, *Revs. Mod. Phys.* **27** (1955) 167.

[51] J. Shamir and R. Fox, *Nuovo. Cimento.* **62B** (1969) 258.

[52] L. Essen, *Nature* **175** (1955) 793.

[53] T. S. Jaseja, A. Javan, J. Murray, and C.H. Townes, *et al.*, *Phys. Rev.* A **133** (1964) 1221.

[54] E. W. Silvertooth, *J. Opt. Soc. Amer.* **62** (1972) 1330.

[55] W. S. N. Trimmer, R.F. Baierlein, J.E. Faller, and H.A. Hill *Phys. Rev.* **D8** (1973) 3321.

[56] R. J. Kennedy and E. M. Thorndike, *Phys. Rev.* **42** (1932) 400.

[57] M. Ruderfer, *Phys. Rev. Lett.* **5** (1960) 191.

[58] C. Møller, *Proc. Roy. Soc.* **A270** (1962) 306.

[59] J. P. Cedarholm, G.F. Bland, B.L. Havens, and C.H. Townes, *Phys. Rev. Lett.* **1** (1958) 342.

[60] D. C. Champeney, G.R. Isaak, and A.M. Khan, *Phys. Lett.* **7** (1963) 241.

[61] D. C. Champeney and P. B. Moon, *Proc. Phys. Soc.* **77** (1961) 350.

[62] R. Cialdea, *Lett. Nuovo. Cimento* **4** (1972) 821.

[63] D.F. Comstock, *Phys. Rev.* **30** (1910) 267.

[64] W. de Sitter, *Proc. Acad. Sci. Amst.* **15** (1913) 1297.

[65] W. Zurhellen, *Astr. Nachr* **198** (1914) 1.

[66] H. Dingle, *Month. Not. Roy. Astro. Soc.* **119** (1959) 67.

[67] G. Van Biesbroeck, *Astrophys. J.* **75** (1932) 64.

[68] O. Heckmann, *Ann. D'. Astrophys.* **23** (1960) 410.

[69] Q. Majorana, *Phil. Mag.* **37** (1919) 285.

[70] R. Tomaschek, *Ann. d. Phisik* **73** (1924) 105.

[71] R. C. Tolman, *Phys. Rev.* **31** (1910) 33.

[72] W. Kantor *Nuovo Cimento, Ser. II*, **9B** (1972) 69.

[73] A.M. Bonch–Bruewich, *Optika Spektrosk,* **9** (1960) 134.

[74] A.A. Michelson *Astrophys. J.* **37** (1913) 37.

[75] Q. Majorana *Comp. Rend.* **165** (1917) 424; *Comp. Rend.* **167** (1918) 71.

[76] P. Beckmann and P. Mandies *Radio. Sci.* **69D** (1965) 623.

[77] W. Kantor *J. Opt. Soc. Amer.* **52** (1962) 978.

[78] G. C. Babcock and T. G. Bergman *J. Opt. Soc. Amer.* **54** (1964) 147.

[79] P. Beckmann and P. Mandies *Radio. Sci.* **68D** (1964) 1265.

[80] J. F. James and R. S. Sternberg *Nature* **197** (1963) 1192.

[81] F. B. Rotz *Phys. Lett.* **7** (1963) 252.

[82] R. O. Waddoups, W. F. Edwards, and J.J Merrill, *J. Opt. Soc. Amer.* **55** (1965) 142.

[83] J. Zahejsky and V. Kolesnikov, *Nature,* **212** (1966) 1227.

[84] W. Kantor, *Nuovo Cimento, Ser II,* **B11** (1972) 93.

[85] J. G. Fox, *Am. J. Phys.* **30** (1962) 297.

[86] D. Sadeh, *Phys. Rev. Lett.* **10** (1963) 271.

[87] T. Alväger, A. Nilsson, and J. Kjellman *Nature* **197** (1963) 1191; *Arkiv Fysik,* **26** (1964) 209.

[88] T. A. Fillippas and J. G. Fox, *Phys. Rev.* **B135** (1964) 1071.

[89] T. Alväger, F.J.M. Farley, J. Kjellman, and L. Wallin, *Phys. Lett.* **12** (1964) 260; *Arkiv Fysik,* **31** (1966) 145.

[90] W. Pauli, *Theory of Relativity.* (Pergamon Press, New York 1958.)

[91] D. Hils and J.L. Hall, *Phys. Rev. Lett.* **64** (1990) 1697.

[92] Y.Z. Zhang, *Phys. Rev. Lett.* **A 95** (1983) 225.

[93] F.T. Trouton, *Trans. Roy. Dub. Soc.* **7** (1902) 379;   F.T. Trouton and M.R. Noble, *Plil. Trans.* **202** (1903) 165.

[94] R. Tomaschek, *Ann. de Phys.* **78** (1926) 743;   **80** (1926) 509.

[95] C.T. Chase, *Phys. Rev.* **28** (1926) 378.

[96] J.W. Butler, *Am. J. Phys.* **36** (1968) 936.

[97] G. Cocconi and E. Salpeter, *Nuovo Cimento* **10** (1958) 646;   *Phys. Rev. Lett.* **4** (1960)

[98] V.W. Hughs *et al.,* *Phys. Rev. Lett.* **4** (1960) 342.

[99] C.W. Sherwin it et al., *Phys. Rev. Lett.* **4** (1960) 399.

[100] R.W.P. Drever *et al., Phil. Mag.* **6** (1961) 683.

[101] T.E. Chupp, *et al., Phys. Rev. Lett.* **63** (1989) 1541; and references therein.

[102] L.B. Rédei, *Phys. Rev.* **162** (1962) 1299.

[103] M. Dardo, G. Navarra and P. Penengo, *Nuovo Cimemto* **61A** (1969) 219.

[104] L.-E. Lundberg and L.B. Rédei, *Phys. Rev.* **169** (1968) 1012.

[105] A.J. Greenberg *et al., Phys. Rev. Lett.* **23** (1969) 1267.

[106] A. Einstein, *Naturwiss.* **6** (1918) 697.

[107] R.C. Tolman, *Relativity, Thermodynamics and Cosmology* (Oxford U. P., New York, 1966), pp. 194–197.

[108] E.G. Cullwick, *Electromagnetism and Relativity* (Longman and Green, London, 1957).

[109] P. cornille, *Phys. Lett.* A **131** (1988) 156.

[110] S.J. Prokhovnik, *Found. Phys.* **19** (1989) 541.

[111] W.A. Rodrigues, Jr. and M.A.F. Rosa, *Found. Phys.* **19** (1989) 705.

[112] A. Schild, *Amer. Math. Monthly* **66** (1959) 1.

[113] R. A. Muller, *Am. J. Phys.* **40** (1972) 966.

[114] E. Lowry, *Amer. J. Phys.* **31** (1963) 59.

[115] E. Taylor and J. Wheeler, *Spacetime Physics.* (Freeman, San Francisco, 1966), pp. 94–95.

[116] Y.S. Huang, *Helv. Phys. Acta.* **66** (1993) 346.

[117] J.C. Hafele and R. E. Keating, *Science* **177** (1972) 166; *idem* **177** (1972) 168.

[118] J.C. Hafele, *Nature,* **227** (1970) 270; *Nature. Phys. Sci.* **229** (1971) 238; *Am. J. Phys.* **40** (1972) 81.

[119] P. Cornille, *Phys. Lett.* A **131** (1988) 156.

[120] H. E. Ives and G. R. Stilwell, *J. Opt. Soc. Am.* **28** (1938) 215.

[121] H. E. Ives, *J. Opt. Soc. Am.* **27** (1937) 389.

[122] H. E. Ives and G. R. Stilwell, *J. Opt. Soc. Am.* **31** (1941) 369.

[123] G. Otting, *Physik. Z.* **40** (1939) 681.

[124] W. Kantor, *Spectr. Letters* **4** (1971) 61.

[125] H. I. Mandelberg and L. Witten, *J. Opt. Soc. Am.* **52** (1962) 529.

[126] D. Hasselkamp, E. Mondry, and A. Scharmann, *Z. Physik* A **289** (1979) 151.

[127] A. Olin, T.K. Alexander, O. Häusser, A.B. McDonald, and G.T. Ewan, *Phys. Rev.* D8 (1973) 1633.

[128] B. D. Josephson, *Phys. Rev. lett.* **4** (1960).

[129]] R. V. Pound and G. A. Rebka, *Phys. Lett.* **4** (1960) 274.

[130] R. V. Pound, G.B. Benedek, and R. Drever, *Phys. Rev. Lett.* **7** (1961) 405.

[131] C. W. Sherwin, *Phys. Rev.* **120** (1960) 17.

[132] H.J. Hay, J.P. Schiffer, T.E. Cranshaw, and P.A. Egelstaff, *Phys. Rev. Lett.* **4** (1960) 165.

[133] A. D. Dos Santos, *Nuovo Cimento* **B32** (1976) 519.

[134] D. C. Champeney, G.R. Isaak, and A.M. Khan, *Nature,* **198** (1963) 1186; *Proc. Phys. Soc.* **85** (1965) 583.

[135] W. Kündig, *Phys. Rev.* **129** (1963) 2371.

[136] J. J. Snyder and J. L. Hall, *Laser Spectroscopy* (Proc. and Intern. Conf. Megeve, June 23-27, 1975), P. 6.

[137] M. Kaivola, O. Poulsen, E. Riis, and S.A. Lee, *Phys. Rev. Lett.* **54** (1985) 255.

[138] (a) R.F.C. Vessot and M.W. Levine, *Gen. Relativ. Grav.* **10** (1979) 181.
     (b) R.F.C. Vessot *et al., Phys. Rev. Lett.* **45** (1980) 2081.

[139] B. Rossi and D. B. Hall, *Phys. Rev.* **59** (1941) 223.

[140] D. H. Frisch and J. H. Smith, *Am. J. Phys.* **5** (1963) 342.

[141] D. M. Lederman *et al., Phys. Rev.* **83** (1951) 685.

[142] R. P. Durbin *et al., Phys. Rev.* **88** (1952) 179.

[143] J. D. Jackson, *Classical Electrodynamics* (New York, Wiley, 1962.)

[144] A. J. Greenberg *et al., Phys. Rev. Lett.* **23** (1969) 1267.

[145] D. S. Ayres *et al., Phys. Rev.* **D3** (1971) 1051.

[146] F. J. M. Farley *et al.*, *Nuovo Cimento*, **45** (1966) 281; J. Bailey *et al.*, *Phys. Lett.* B **28** (1968) 287; Nuovo Cim. A**9** (1972) 369.

[147] J. Bailey *et al.*, *Nature* **268** (1977) 301.

[148] J. Bailey *et al.*, *Nucl. Phys.* B **150** (1979) 1.

[149] H. C. Burrowes *et al.*, *Phys. Rev. Lett.* **2** (1959) 117.

[150] M. Faraday, *Experimental Researches in Electricity,* Vol. I. London, 1855, Paragraphs 225–230.

[151] R. Becker, *Electromagnetic fields and interactions,* Vol. I: Electromagnetic theory and Relativity, Blaisdell, New York, 1964, §§71, 86, 87, and 88.

[152] W. K. H. Panofsky and M. Phillips, Classical Electricity and Magnegtism, Addison-Wesley, Reading, Mass., 1955.

[153] A. Sommerfield, Electrodynamics, Academic Press, New York, 1952.

[154] J. Djurić, *J. Appl. Phys.* **46** (1975) 679.

[155] E. Whittaker, *A History of the Theories of Aether and Electricity*, Nelson, New York, *Vol. I* (1995) and *Vol. II* (1953).

[156] A.J. Fresnel, *Annls Chim. Phys.* **9** (1918) 57.

[157] H. L. Fizeau, *Compt. Rend.* **33** (1851) 349.

[158] H. L. Fizeau, *Ann. Chem. Phys.* **57** (1895) 385.

[159] A. A. Michelson and F. W. Merley, *Am. J. Sci.* **31** (1886) 377.

[160] P. Zeeman, *Proc. Roy. Acad. Amsterdam,* **17** (1914) 445; **18** (1915) 398.

[161] P. Zeeman, *Proc. Roy. Acad. Amsterdam,* **22** (1920) 462.

[162] P. Zeeman and Miss A. Snethlage, *Proc. Roy. Acad. Amsterdam,* **22** (1920) 512.

[163] P. Zeeman *et al.*, *Proc. Roy. Acad. Amsterdam,* **23** (1922) 1402.

[164] (a) R. V. Jones, *J. Phys.* A(GB) **4** (1971) L1.
(b) R. V. Jones, *Proc. R. Soc. Lond.* **A328** (1972) 337.

[165] R. V. Jones, *Proc. R. Soc. Lond.* **A345** (1975) 351.

[166] W. M. Macek *et al.*, *J. Appl. Phys.* **35** (1964) 2556.

[167] H. R. Bilger and A. T. Zavadny, *Phys. Rev.* **A5** (1972) 591.

[168] H.A. Lorentz, *The theory of electrons*, Teubner, Leipzig (1916).

[169] L. Rosenfeld, *Theory of electrons*, North–Holland, Amsterdam (1951).

[170] G. Sagnac, *C. r. hebd. Séanc. Acad. Sci., Paris,* **157** (1913) 708, 1410; *J. Phys. théor. appl.* (5) **4** (1914) 177.

[171] R. C. Jennison and P. A. Davies, *Nature* **248** (1974) 660.

[172] W. Kaufmann, *Nachr. Ges. Wiss. Göttingen, Math-Nat. Kl.* **143** (1901).

[173] M. Abraham, *Ann. d. Phys.* **10** (1903) 105.

[174] H. A. Lorentz, *Proc. Acad. Sci. Amst.* **6** (1904) 809.

[175] A. H. Bucherer, *Ann. d. Phys.* **28** (1909) 513; 585; **30** (1909) 974.

[176] G. Neumann, *Ann. d. Phys.* **45** (1914) 529.

[177] E. Hupka, *Ann. d. Phys.* **31** (1910) 169.

[178] C. E. Guye and C. Lavanchy, *Compt. Rend.* **161** (1915) 52.

[179] M. M. Rogers, A.W. McReynolds and F.T. Rogers, *Phys. Rev.* **57** (1940) 379.

[180] D. J. Grove and J. G. Fox, *Phys. Rev.* **90** (1953) 378.

[181] von V. Meyer, W. Reichart, H.H. Staub, H. Winkler, F. Zamboni and W. Zych, *Helv. Phys. Acta.* **36** (1963) 981.

[182] K. N. Geller and R. Koliarits, *Am. J. Phys.* **40** (1972) 1125.

[183] H.F. Dylla and John G. King, *Phys. Rev.* **A7** (1973) 1224.

[184] W. Bertozzi, *Am. J. Phys.* **32** (1964) 551.

[185] B. C. Brown *et al.*, *Phys. Rev. Lett.* **30** (1973) 763.

[186] Z. G. T. Guiragossián, G.B. Rothbart, and M.R. Yearian, *Phys. Rev. Lett.* **34** (1975) 335.

[187] F. C. Champion, *Proc. R. Soc.* **A136** (1932) 630.

[188] P. S. Faragó and L. Jánossy, *Nuovo Cimento* **5** (1957) 1411.

[189] S. Raboy and C. C. Trail, *Nuovo Cimento* **10** (1958) 797.

[190] G. Stephenson and C.W. Kilmister, *Special Relativity for Phyiscists* (Longmans, Green and Co. 1960).

[191] K. Glitscher, *Ann. d. Phys.* **52** (1917) 608.

[192] J. D. Cockcroft and G. T. S. Walton, *Proc. Roy. Soc.* **A137** (1932) 223.

[193] M. M. Smith, Jr. *Phys. Rev.* **56** (1939) 548.

[194] M. C. Hudson and W. H. Jocobson, Jr. *Phys. Rev.* **167** (1968) 1064.

[195] W. H. Wapstra et al., *Nacl. Data.* **A9** (1971) 364.

[196] R. V. Pound and G. A. Rebka, *Phys. Rev. Lett.* **4** (1960) 337.

[197] A.S. Goldhaber and M.M. Nieto, *Rev. Mod. Phys.* **43** (1971) 277.

[198] K. D. Froome and L. Essen, 1969, *The Velocity of Light and Radio Waves* (Academic, New York).

[199] E. F. Florman, *J. Res. Natl. Bur. Std.* **54** (1955) 335.

[200] M.A. Gintsburg, *Astron. Zh.* **40** (1963) 703 [*Sov. Astron. AJ* **7** (1964) 536].

[201] L. de Broglie, *La Méchanique Ondulatoire du Photon, Une Nouvelle Théorie de la Lumiére*, Vol. *1*, pp. 39–40, Hermann, Paris, 1940.

[202] I.Yu. Kobzarev and L.B. Okun', *Usp. Fiz. Nauk* **95** (1968) 131 [*Sov. Phys. Uspe.* **11** (1968) 338]

[203] D. H. Staelin and E. C. Reiferstein III, *Science* **162** (1968) 1481.

[204] G. Feinberg, *Science* **166** (1969) 879.

[205] J. C. Maxwell, *A Treatise on Electricity and Magnetism* (Oxford U. P.) **vol. 1**, 3rd ed. of 1891 reprinted by Dover, New York, 1954.

[206] S. J. Plimpton and W. E. Lawton, *Phys. Rev.* **50** (1936) 1066.

[207] D. F. Barlett, P. E. Goldhagen and E. A. Phillips, *Phys. Rev.* **D2** (1970) 483.

[208] E. R. Williams, J.E. Faller, and H.A. Hill, *Bull. Am. Phys. Soc.* **15** (1970) 586; *Phys. Rev. Lett.* **26** (1971) 721.

[209] G. D. Cochran and P. A. Franken, *Bull. Am. Phys. Soc.* **13** (1968) 1379.

[210] E. Schrödinger, *Proc. R. I. Acad.* **A49** (1943) 135; L. Bass and E. Schrödinger, *Proc. Roy. Soc.* **A232** (1955) 1.

[211] H. F. Finch and B. R. Leaton, *Geophys. J. Suppl.* **7** (1957) 314.

[212] E. Schrödinger, *Proc. R. I. Acad.* **49A** (1943) 43.

[213] A. S. Goldhaber and M. M. Nieto, *Phys. Rev. Lett.* **21** (1968) 567.

[214] L. Davis, Jr, A.S. Goldhaber and M.M. Nieto, *Phys. Rev. Lett.* **35** (1975) 1402.

[215] V. L. Patel and L.J. Cahill Jr., *Phys. Rev. Lett.* **12** (1964) 213.

[216] V. L. Patel, *Phys. Lett.* **14** (1965) 105.

[217] J. V. Hollweg, *Phys. Rev. Lett.* **32** (1974) 961.

[218] J. D. Scargle, *Astrophys. J.* **156** (1969) 401.

[219] A. Barnes and J. D. Scargle, *Phys. Rev. Lett.* **35** (1975) 1117.

[220] L.J. Lanzerotti, *Geophys. Res. Lett.* **1** (1974) 229.

[221] E. Williams and D. Park, *Phys. Rev. Lett.* **26** (1971) 1651.

[222] J. C. Byrne and R. R. Burman, *J. Phys.* **A**: *Gen. Phys.* **5** (1972) L109.

[223] J. C. Byrne and R. R. Burman, *J. Phys.* **A**: *Gen. Phys.* **6** (1973) L12.

[224] Y. Yamaguchi, *Prog. Theory. Phys. Suppl.* **11** (1959) 33.

[225] P. A. Franken and G. W. Ampulshi, *Phys. Rev. Lett.* **26** (1971) 115.

[226] A. S. Goldhaber and M. M. Nieto, *Phys. Rev. Lett.* **26** (1971) 1390.

[227] D. Park and E. R. Williams, *Phys. Rev. Lett.* **26** (1971) 1393.

[228] N. M. Kroll, *Phys. Rev. Lett.* **26** (1971) 1395.

[229] R. Van Dyck, P. Schwinberg, and H. Dehmelt, *Phys. Rev. Lett.* **38** (1977) 310.

[230] J. Wesley and A. Rich, *Phys. Rev.* **A4** (1971) 1341.

[231] P.S. Cooper *et al.*, *Phys. Rev. Lett.* **42** (1979) 1386.

[232] J. Bailey *et al.*, *Nuovo Cimento* **9** (1972) 369.

[233] J. Bailey *et al.*, *Phys. Lett.* **B68** (1977) 191.

[234] J. Schringer, *Phys. Rev.* **73** (1948) 416; ibid, **76** (1949) 790;
R. Karplus and N. Kroll, *Phys. Rev.* **77** (1950) 536;
C. Sommerfield, *Ann. Phys.* **5** (USA) (1958) 26;
A. Petermann, *Nucl. Phys.* **5** (1958) 677;
M. J. Levine, *Phys. Rev. Lett.* **26** (1971) 1351.

[235] D. Newman, G. W. Ford, A. Rich and E. Sweetman, *Phys. Rev. Lett.* **40** (1978) 1355.

[236] F. Combley *et al.*, *Phys. Rev. Lett.* **42** (1979) 1383.

INDEX

Aberration, 153
    in Edwards' theory, 94, 95, 97
    in Einstein's theory, 41, 42,
        153, 154, 179, 194
    test of, 153, 154

Aberration angle, 153, 154

Aberration constant, 153, 154

Acceleration, 22, 179, 181, 190,
        193, 198
    four-dimensional, 31, 35
    gravitational, 262
    three-dimensional, 22, 36, 49,
        52, 53, 135, 270, 271
    transformation of, 35, 36

Addition law of velocities
    Edwards', 88, 89, 94, 118
    Einstein's, 32–37, 41, 47, 53, 55,
        88, 89, 118, 181, 208, 234,
        236
    Mansouri–Sexl's (MS), 118
    Newtonian, 21, 34, 35, 37
    relation between Edwards' and
        Einstein's, 88, 89
    relation between Mansouri–Sexl's
        (MS) and Robertson's, 118
    Robertson's, 109, 110, 118

Alfvén wave (see wave)

Around–the–world atomic clock experiment (flight atomic clock experiment), 180–183

Ballistic hypothesis of light emission
(ballistic theory), 151, 152, 154,
        155

Brewster's angle [see also drag (effect)
experiments], 65, 208, 218

Cartesian frame (coordinate system),
        3–5, 7, 11, 12, 14–16

Causality (or cause), 9, 31, 37–39

Charge conservation law (see conservation law)

Charge–motion–independence, 225,
        233

Clock(s) [see also moving clock(s)]
    Edwards, 12, 17, 18, 79, 86
    Einstein, 12, 13, 17, 18, 87
    Mansouri–Sexl (MS), 15, 16, 121
    relation of Edwards and Einstein,
        11–14, 17, 18
    relation of Mansouri–Sexl (MS)
        and Robertson, 16
    Robertson, 14, 109, 111

Closed–path experiments, 109, 122,
        136–145, 147, 150, 170

Clock paradox, 175–179, 192

Clock synchronization (see also simultaneity), 7–16, 75, 76, 88, 89, 91, 129,
        168, 176

Comparison of Edwards' theory and
experiments, 90–98

Comparison of Mansouri–Sexl's theory and experiments, 117–123

Comparison of Robertson's theory and experiments, 109, 111–113

Conduction current density
(*see* current)

Conservation law
    charge, 57, 67, 70
    of angular momentum, 51
    of energy, 236, 238–240, 242
    of energy-momentum, 51–53
    of momentum, 51–53, 234–236, 238, 242, 262
    of (relativistic) mass, 240

Constitutive relations
    in moving medium, 59, 60, 207
    in stationary medium, 57–58

Contraction (of moving body)
    Fitzgerald–Lorentz, 40, 136, 140, 175
    in Edwards' theory, 98, 99
    in Einstein's theory, 40, 41, 99, 140
    Lorentz, factor, 24, 34, 42, 43, 189, 196, 198
    relation between Edwards' and Lorentz's, 98, 99

Convection current density
(*see* current)

Coordinate(s)
    Edwards, 11, 13
    Edwards–like, 115, 117, 124
    Einstein, 13
    Lorentz–like, 106, 124
    Mansouri–Sexl (MS), 16, 115
    Robertson, 14, 115

Coordinate time (interval)
    [*see* time (interval)]

Coulomb law, 248–252, 266

Covariant Form of Maxwell's Field Equations, 67–70

Covariant Form of Proca's Field Equations, 70

Current (density), electric, 57–60, 67–71, 174, 201–207, 230, 253, 256–258, 261, 262, 264, 267
    conduction, 60, 204
    convection, 60, 204
    displacement, 263

Dispersion
    of electromagnetic waves in moving medium, 59–62, 65, 201, 207, 208, 210–212, 214, 216, 217, 221
    of electromagnetic waves in plas-ma, 246
    of massive electromagnetic waves in vacuum (photon mass effect), 71, 72, 245–247, 259–261, 266
    theory, 163

Doppler effect (*or* Doppler shift),
    136, 146, 148, 149, 152, 163,
    166, 183, 189, 190, 194, 195
  in Edwards' theory, 94–98, 147
  in Einstein's theory, 41–45, 61,
    98, 113, 136, 145, 147, 157,
    179, 183–185, 187–189, 191–
    195, 198–200, 212, 223
  in Mansouri–Sexl's test theory,
    122, 123
  in Robertson's test theory, 112,
    113, 123, 147
  in stationary ether theory, 136,
    145–147
  relation between Edwards' and
    Einstein's, 97, 98
  relation between Mansouri–Sexl's
    and Robertson's, 121, 123
  tests of, 136, 145–149, 167, 180,
    183–195, 223
  transverse, 42, 43, 45, 113, 121,
    123, 136, 145–147, 179, 183–
    185, 187, 188, 193, 198–200

Drag coefficient (*or* drag effect)
  Einstein's, 65
  Fresnel's, 65, 208, 211
  Laub, 65
  Lorentz's, 210–212

Drag (effect) experiments (of Fizeau-
type), 201, 202, 207–223
  at Brewster's angle, 208, 218–222
  longitudinal, 208–214
  transverse, 208, 214–217

Edwards simultaneity
  (*see* simultaneity)

Einstein simultaneity

(*see* simultaneity)

Einstein's law of addition of velocities
  (*see* addition law of velocities)

Einstein's simultaneity factor
  (*see* simultaneity factor)

Electric current (density)
  (*see* current)

Electrodynamics (*or* electromagnetic
field theory)
  Maxwell, 67, 201, 204, 246, 248,
    254
  Maxwell–Minkowski
  (*or* of moving media),
    57–65, 201–203, 205, 206,
    208, 215, 222
  Proca's (massive), 67–72

Electromagnetic field(s), 53, 57, 203,
    246
  massive (Proca), 67–72, 245, 251,
    253, 261–263, 265
  Maxwell's, 57–61, 69, 201, 245,
    265
  transformation equations of
    [*see* transformation(s)]

Electromagnetic field equation(s)
  in moving media,
  (*or* Maxwell–Minkowski's),
    58–60, 201, 219
  Maxwell's, 22, 57, 67, 69–71, 245,
    258, 259
  Proca's (massive), 70, 71, 251,
    253, 259, 261–263, 265

Electromagnetic induction of moving bodies, 202–207

Electromagnetic wave
(*see* wave)

Energy
    binding, 174, 235, 241
    kinetic, 53, 54, 149, 191, 226, 230, 233, 234, 239–241, 273
    proper (rest), 54, 234, 240, 241, 242, 260
    radiation, 242
    relativistic (total), 54, 236, 260

Equivalence of mass and energy
(*see* mass–energy relation)

Ether, 22, 23, 136, 138, 139, 145, 146, 175, 201, 207, 208
    drift (wind), 136, 139–143, 145–147, 172, 174, 175, 208

Ether theory, 40, 136, 138, 139, 145–147, 185, 207, 225
    Fresnel's, 201, 208, 209

Event(s), 7, 21, 23, 24, 27, 30, 37, 38, 100, 121
    light–like, 30, 38
    space–like, 31, 38, 39
    time–like, 30, 38

Expansion factor, 24, 197

Extinction distances, 164

Extinction theorem, 163, 164

Fizeau's effect, 211

Fizeau's experiment, 201, 208, 209–211

Fizeau–type experiment
(*see* drag experiments)

Flight atomic clock (experiment)
(*see* Around–the–world atomic clock experiment)

Four–dimensional acceleration
(*see* acceleration)

Four–dimensional space (*or* space–time), 3, 27–31
    anisotropic, 82–88, 106–109, 123–126
    Minkowski's, 27–31, 82

Four–dimensional space interval
(*see* space interval)

Four–dimensional vector (*or* 4-vector)
(*see* vector)

Four–dimensional velocity
(*see* velocity)

Frame(s) 3–8, 10, 11, 14, 16–18, 23, 34, 51, 67, 75, 121, 136, 174, 203, 205, 207, 234
    Edwards, 11–14, 78–83, 85, 89, 90, 92, 94, 95, 99, 113, 137, 146
    Einstein, 10–14, 16–18, 23, 24, 29, 32, 36–43, 46, 47, 49, 51, 55, 57–59, 61–64, 69, 80–83, 85, 89–91, 113, 114, 115, 116, 118, 148, 156–182, 193–195, 208, 210, 235, 269
    instantaneous Einstein, 36, 179
    Mansouri–Sexl (MS), 15, 16, 114–118, 120, 128, 129, 148, 149

Newtonian, 10, 16, 21, 22, 36, 37, 45

"preferred" , 80, 104, 148, 154

relation between Edwards and Lorentz, 11–15, 17, 18, 81, 90, 91, 98

relation between Mansouri–Sexl (MS) and Robertson, 16, 114

Robertson, 14–16, 103–109, 111, 113, 114

Fresnel's drag coefficient (*see* drag coefficient)

Fresnel's ether dragging theory (*see* ether theory)

$g - 2$ factor, 270–272, 274

Group
full Lorentz, 27
Lorentz, 27
of orthogonal transformation, 5
Poincaré, 51
proper Lorentz, 27, 29

Group velocity, 62, 63, 72, 210, 215, 218, 245

Hydromagnetic wave (*see* wave)

Inertial system of reference [*see* frame(s)]

Interval
coordinate time, [*see* time (interval)]
proper time, [*see* coordinate (interval)]
space–time, (*see* space interval)

Kennedy–Thorndike Experiment, 145

Kronecker symbol, 4, 6

Law of inertia, 7, 8, 16, 21

Length contraction (*or* contraction of moving body) (*see* contraction)

Lifetime dilation of moving mesons, 195–200

Light–like (events) [*see* event(s)]

Line element (*see* metric)

Lorentz contraction (*see* contraction of moving body in Einstein's theory)

Lorentz contraction factor (*see* contraction)

Lorentz–Fitzgerald contraction (*see* contraction)

Lorentz gauge (condition), 68, 70

Lorentz group (*see* group)

Lorentz's theory of electron, 57, 194, 201, 202, 208, 209, 225

Lorentz transformation [*see also* transformation(s)]
general, 26, 46–48, 56, 58
infinitesimal, 36, 37, 48, 49
in four-dimensional form, 27, 28, 51

in three-dimensional form, 25, 26
special, 24, 25, 81
with rotation, 26, 47–49
without rotation, 24–26, 47–49,
    69

Magnetization vector, 58, 60, 202,
    205
    transformation of
    (*see* transformation)

Magnetosonic wave
    (*see* wave)

Mansouri–Sexl simultaneity
    (*see* simultaneity)

Mass
    inertial, 22, 23, 51
    relativistic (total), 52, 54, 225,
        226, 228, 231, 234, 240, 241
    rest, of photon, 70, 71, 245–247,
        251, 253, 255–262, 264–267
    rest (*or* proper), 34, 52, 54, 56,
        191, 196, 225, 233, 237, 240–
        242, 245, 269, 272

Mass defect, 240, 241

Mass–energy relation, 34, 53, 54, 67,
    239–242, 245
    tests of, 241, 242

Massive electromagnetic field(s)
    [*see* electromagnetic field(s)]

Massive electromagnetic theory
    (*see* electrodynamics)

Massive electromagnetic wave

(*see* wave)

Massive photon (*see also* electromagnetic field), 71, 258, 260

Mass–to–charge ratio, 225, 228,
    229

Mass–velocity relation, 34, 52, 53,
    67, 225–230, 233, 234, 238,
    240, 245
    tests of, 227–239, 243, 226

Maxwell–Minkowski electrodynamics
(*see* electrodynamics),

Maxwell's field equations (*see* electromagnetic field equations)

Metric (tensor)
    Euclidean, 4, 6, 28
    Minkowski, 27–30, 68, 82, 144
    in anisotropic space–time, 84, 85,
        87, 88, 108, 125, 126, 144

Michelson–Morley (type) experiment,
    40, 136–143, 154, 174, 201,
    208

Mössbauer effect experiment, 146,
147,
    190–194, 199, 200

Mössbauer rotor experiment, 123, 146,
    147, 150, 172, 192–194, 199,
    200
Moving clock(s)
    Around–the–world atomic clocks;
        180–183, 199
    hydrogen canal ray; 183–188
    moving absorber (*or/and* emit-
        ter); 192–195, 199, 200
    moving mesons; 195–198, 200

moving nucleus (*or* atom); 188–192, 198, 199

Moving light source experiments, 150–171
  astronomical observations of, 151–154
  by use of moving γ–ray sources, 163–171, 173
  by use of moving laboratory sources, 154–156
  by use of moving mirrors, 156–158
  by use of moving transparent substances, 158–163

Ohm's law, 58–60, 203

One-way experiment, 75, 123, 128, 145–150

Phase velocity, 61, 65, 72, 210, 213, 264

Photon rest mass (*see also* dispersion *or* Coulomb law) 245–267
  Magnetohydrodynamic Effects of, 258–264, 266, 267
  magnetostatic effect of, 253–258, 266
  Schrödinger's external field method; 247, 253–256, 266

Poincaré group (*see* group)

Polarization vector, 58, 60, 202, 204, 205
  transformation of (*see* transformation)

Principle of equivalence, 178, 193

Principle of relativity, 3, 11, 34, 79, 174
  Einstein, 3, 16, 17, 23, 34, 51, 80, 140, 174, 179
  Galilean, 3, 16, 21–23, 34
  tests of, 135, 136, 174

Proca vector field, 67–72

Proper energy *or* rest energy (*see* energy)

Proper frequency, 43, 44, 97, 113

Proper length (*or* rest length), 39, 40, 98

Proper lifetime, 195–198

Proper Lorentz group (*see* group)

Proper mass (*or* rest mass) (*see* mass)

Proper time (interval) [*see* time (interval)]

Proper wavelength, 43, 187

Reciprocity of relative velocities, 18, 18, 19, 89, 90, 110, 119

Red shift (*see also* transverse Doppler shift), 43, 163, 242
  gravitational, 181, 190, 193, 195, 199, 242

Reflection (law) for electromagnetic field
  experiment of, 223

on a stationary surface, 62, 63, 223
on a surface of moving medium, 63, 64, 222

Refraction (law) for electromagnetic field
in moving medium, 63–65, 215, 216, 218, 219
in stationary medium, 61–63
tests of, 214–222

Relationship among Lorentz transformation and generalized transformations, 126–128

Relativistic mechanics, 51–56
tests of, 225–243

Rest energy (*see* energy)

Rest length (*see* proper length)

Rest mass (*see* mass)

Rest wavelength
(*see* proper wave length)

Robertson simultaneity
(*see* simultaneity)

Römer Experiment, 91–94, 120–122, 124

Schrödinger's external field method
(*see* photon rest mass)

Simultaneity (clock synchronization), 3, 7–18, 34, 36, 37, 75, 76, 81, 82, 88–91, 94, 97–99, 118–120, 122, 127–129, 135, 136, 149, 164, 171, 176, 177, 209
Edwards, 11, 12, 15–17, 80, 97,

98, 176, 204
Einstein, 3, 10–12, 14–16, 37, 38, 41, 80, 82, 94, 98, 99, 111, 127, 176, 204, 209
Mansouri–Sexl (MS), 15, 16, 116, 118, 119
Newtonian, 9, 10, 21, 23
relativity of, 37–39
Robertson, 14–16, 97, 110, 119, 121, 127

Simultaneity factor, 129
Einstein's, 37, 179, 202

Slow transport of clocks, 120, 202, 127, 129, 135, 147

Space coordinate(s), 3, 4, 13, 16, 26, 27, 28, 32, 37, 39, 56, 70, 77, 90, 98, 111, 137

Space interval
four-dimensional, 27, 28, 81, 82, 84, 85, 87, 88, 106–108, 125, 126, 144
three-dimensional (Euclidean), 4, 28

Space-like (events) [*see* event(s)]

Space-time interval
(*see* space interval)

Space–time theory (*see also* test theory of special relativity)
flat, 3
Newton–Galileo (absolute), 17, 21, 37, 121, 122

Speed of light (*see* velocity of light)

Tachyon (faster-than-light), 31, 34, 35

Tensor (density), 6, 29, 30
    electromagnetic field-strength, 69, 70
    metric, [*see* metric (tensor)]
    Levi–Civita, 6

Test theory of special relativity, 103–131
    Edwards', 75–101, 113, 118, 120, 127, 136, 146
    Mansouri–Sexl's (MS), 76, 103, 113–130
    Robertson's, 76, 103–113, 121, 116

Thomas Precession, 45–49
    tests of, 269–274

Three-dimensional acceleration (*see* acceleration)

Three-dimensional (Euclidean) space, 3, 11, 27, 28

Three-dimensional vector (3-vector) (*see* vector)

Three-dimensional velocity (*see* velocity)

Time coordinate(s), 3, 7, 8, 10–17, 24, 28, 32, 37, 39, 40, 56, 70, 77, 90, 99, 121, 122, 137

Time dilation (retardation), 94, 111, 119, 122, 127, 136, 147, 183, 190, 197
    in Edwards' theory, 90, 91, 93, 97, 99
    in Einstein's theory, 40, 41, 43, 45, 91, 94, 176–181, 183–185, 189, 192–196, 198–200
    in Mansouri–Sexl's test theory, 119–121
    in Robertson's test theory, 111–113, 120, 121
    Larmor's, 175
    relation between Edwards' and Einstein's, 97, 98
    relation between Mansouri–Sexl's and Robertson's, 119, 120
    tests of, 122, 175–200

Time (interval)
    coordinate, 8, 17, 31, 33, 41, 43–45, 79, 82, 88, 90, 92, 93, 95–98, 111, 119, 120, 122, 123, 128, 135, 176, 177, 179, 181
    proper, 9, 31, 33, 40, 41, 43–45, 56, 90, 92, 93, 96, 97, 111, 119, 120, 122, 123, 117, 123, 176–178, 181

Time-like (events) [*see* event(s)]

Transformation(s)
    coordinate, 3, 4, 5, 11, 13, 16, 23
    Edwards, 19, 75, 78–82, 88, 90, 91, 99–104, 117, 118, 126, 202
    Edwards, in four-dimensional form, 85–87
    Edwards-like, 115, 117
    for 4-tensor (density), 29, 69
    for 3-tensor, 6
    for 4-vector, 28, 69
    for 3-vector, 5
    Galilean, 21, 22, 36, 37, 45, 52
    Infinitesimal Lorentz, 36, 37

Lorentz, 18, 21, 22–49, 52, 55–59, 63, 69, 75, 80–82, 86, 99–103, 105, 126, 179, 191

Lorentz-like, 106, 108, 109

Mansouri–Sexl (MS), 75, 103, 114 –119, 128, 126, 148, 150

matrix for, 4, 5, 27, 82–86, 108, 109, 124, 125

of acceleration, 35, 36

of electromagnetic quantities (electromagnetic fields, charge density, current density, polarization vector, and magnetization vector), 57–60, 191–203, 205

of energy, 55, 56, 225

of farce, 56

of frequency (*see* Doppler effect)

of mass, 55, 225

of momentum, 55, 56, 235, 237

of three-dimensional Cartesian frames, 4, 14

of velocity (*see* Addition law of velocities)

orthogonal, 4, 5, 6, 28

relation between Edwards and Lorentz, 76, 81, 82, 116–118, 121, 126–128

relation between Mansouri–Sexl (MS) and Edwards-like, 115–118, 126–128

relation between Mansouri–Sexl (MS) and Lorentz-like, 115, 117, 126–128

relation between Mansouri–Sexl (MS) and Robertson, 76, 115–118, 121, 126–128

relation between Robertson and Edwards-like, 126–128

relation between Robertson and

Lorentz-like, 105, 106, 108, 111, 115, 126, 127

Robertson, 75, 103–111, 114, 117, 126, 145, 150

Transformation matrix, 4, 5, 27, 69, 82–88, 107–109, 124–126

Transverse Doppler shift (*see* Doppler effect)

Two-way path experiments (*see* closed-path experiments)

Unipolar Induction (unipolar generator), 202–204

$U(1)$ gauge invariant, 70, 225, 245

$U(1)$ gauge transformation, 70

Vector
    four-dimensional contravariant, 28–31, 33, 35, 42, 51, 68–71, 82–87, 107–109, 124–126
    four-dimensional covariant, 28–30, 42, 56, 67, 68, 69–71, 87, 88
    magnetization, 58, 60, 202, 205
    polarization, 58, 60, 202, 204, 205
    three-dimensional, 5, 6, 25, 26, 47, 68

Vector potential
    four-dimensional, 68–70
    three-dimensional, 68, 71

Velocity
    Edwards, 18, 19, 79–81, 88, 89, 94, 96–99
    Einstein, 18, 19, 80–82, 89, 94, 98, 99, 104, 112, 118, 209

four-dimensional, 31, 33

Mansouri–Sexl (MS), 114, 117–
119, 121–123, 128

relation between Edwards and
Einstein, 18, 19, 79, 81, 89,
90, 94, 97, 99
relation between Mansouri–Sexl
and Robertson, 114, 118,
121–123
Robertson, 109, 110, 112, 114,
118, 121–126

Velocity and simultaneity, 17–19

Velocity of light
anisotropy of one-way, 11, 75,
121–123, 128, 131, 150
anisotropy of two-way, 14, 15,
105, 127–129, 131, 150
anisotropy of two-way and
one-way (*see also* Mansouri–
Sexl simultaneity)
constancy of, 3, 19, 23, 34, 37,
75, 78, 79, 104, 135–137, 143,
144, 154, 163, 164, 170–172,
185
constancy of one-way (*see also*
Einstein simultaneity), 10,
24, 34, 75, 82, 104, 135
constancy of two-way (*see also*
Edwards simultaneity), 11,
75, 78, 82, 94, 96, 137
frequency-independence of, 135,
246
independence of, on inertial frame,
67, 75, 103, 135
independence of, on motion of
light source, 10, 14, 23, 135
one-way, 3, 8–11, 13–16, 24, 34,
37, 67, 76, 78, 81, 94, 96, 99,
104, 105, 113, 116, 120–127,
129, 130, 135–137, 143, 144,
147, 150, 151, 168–171
relation between clock synchronization and one-way, 8–11, 14,
15
relationship of the one-way and
two-way, 9, 11, 14, 15, 76–78,
93, 94, 96, 105, 112–114, 116,
117, 120, 128, 129, 137
source–velocity–independence of,
23, 135

tests of isotropy (constancy) of
the two-way, 3, 15, 75, 128,
135–150

tests of source–velocity–independence of, (*see* Moving light
source experiments)
two-way, 3, 9, 10, 11, 13–16, 75,
76, 78, 79, 104, 105, 109, 111,
116, 117, 123, 127, 129, 130,
135–137, 143–145, 150, 169–
171

Wave
Alfvén, 258–270, 266, 267
electromagnetic (*or* light), 41, 43,
59, 62, 67, 95, 96, 135, 163,
179, 183–185, 210, 246
electromagnetic, in moving medium, 60–65
hydromagnetic, 135, 258–260
magnetosonic, 258, 259, 267

massive electromagnetic, 72, 245,
246

Wave equation
for Maxwell's electromagnetic
potential, 68

for Proca vector field, 70, 71, 253

Wilson–Wilson experiment, 57, 202,
        206, 207

World line, 30, 31

Zeeman's experiment, 208, 209,
        211–214